# Lecture Notes in Computer Sc

Edited by G. Goos, J. Hartmanis, and J. v

T0250661

**Springer**
*Berlin*
*Heidelberg*
*New York*
*Barcelona*
*Hong Kong*
*London*
*Milan*
*Paris*
*Tokyo*

Klaus Jansen   Stefano Leonardi
Vijay Vazirani (Eds.)

# Approximation Algorithms for Combinatorial Optimization

5th International Workshop, APPROX 2002
Rome, Italy, September 17-21, 2002
Proceedings

Springer

Series Editors

Gerhard Goos, Karlsruhe University, Germany
Juris Hartmanis, Cornell University, NY, USA
Jan van Leeuwen, Utrecht University, The Netherlands

Volume Editors

Klaus Jansen
Universität Kiel
Institut für Informatik und praktische Mathematik
Olshausenstr. 40, 24098 Kiel, Germany
E-mail: kj@informatik.uni-kiel.de

Stefano Leonardi
Universita di Roma La Sapienza
Dipartimento di Informatika e Sistemistica
Via Salaria 113, 00198 Roma, Italy
E-mail: leon@dis.uniroma1.it

Vijay Vazirani
Georgia Institute of Technology
College of Computing
801 Atlantic Drive, Atlanta, Georgia 30332-0280, USA
E-Mail: vazirani@cc.gatech.edu

Cataloging-in-Publication Data applied for

Die Deutsche Bibliothek - CIP-Einheitsaufnahme

Approximation algorithms for combinatorial optimization : 5th international
workshop ; proceedings / APPROX 2002, Rome, Italy, September 17 - 21, 2002.
Klaus Jansen ... (ed.). - Berlin ; Heidelberg ; New York ; Barcelona ; Hong
Kong ; London ; Milan ; Paris ; Tokyo : Springer, 2002
    (Lecture notes in computer science ; Vol. 2462)
    ISBN 3-540-44186-7

CR Subject Classification (1998): F.2, G.2, G.1

ISSN 0302-9743
ISBN 3-540-44186-7 Springer-Verlag Berlin Heidelberg New York

Springer-Verlag Berlin Heidelberg New York,
a member of BertelsmannSpringer Science+Business Media GmbH

http://www.springer.de

© Springer-Verlag Berlin Heidelberg 2002
Printed in Germany

Typesetting: Camera-ready by author, data conversion by Steingräber Satztechnik GmbH, Heidelberg
Printed on acid-free paper      SPIN: 10871225      06/3142      5 4 3 2 1 0

# Foreword

The Workshop on *Approximation Algorithms for Combinatorial Optimization Problems* **APPROX 2002** focused on algorithmic and complexity aspects arising in the development of efficient approximate solutions to computationally difficult problems. It aimed, in particular, to foster cooperation among algorithmic and complexity researchers in the field. The workshop, held at the University of Rome La Sapienza, Rome, Italy, was part of the ALGO 2002 event, in conjunction with ESA 2002 and WABI 2002. We would like to thank the local organizers at the University of Rome La Sapienza for this opportunity. APPROX is an annual meeting, with previous workshops in Aalborg, Berkeley, Saarbrücken, and Berkeley. Previous proceedings appeared as LNCS 1444, 1671, 1913, and 2129.

Topics of interest for APPROX 2002 were: design and analysis of approximation algorithms, inapproximability results, on-line problems, randomization techniques, average-case analysis, approximation classes, scheduling problems, routing and flow problems, coloring and partitioning, cuts and connectivity, packing and covering, geometric problems, network design, applications to game theory, and other applications. The number of submitted papers to APPROX 2002 was 54 from which 20 papers were selected. This volume contains the selected papers together with abstracts of invited lectures by Yuval Rabani (Technion) and R. Ravi (CMU). All papers published in the workshop proceedings were selected by the program committee on the basis of referee reports. Each paper was reviewed by at least three referees who judged the papers for originality, quality, and consistency with the topics of the conference.

We would like to thank all the authors who responded to the call for papers and our invited speakers Y. Rabani and R. Ravi. Furthermore, we thank the members of the program committee:

- Giorgio Ausiello, Rome
- Josep Diaz, Barcelona
- Ashish Goel, USC
- Ravi Kannan, Yale
- Sanjeev Khanna, UPenn
- Elias Koutsoupias, UCLA
- Ion Mandoiu, UCSD
- Kurt Mehlhorn, Saarbrücken
- Yuval Rabani, Technion
- Eva Tardos, Cornell
- Vijay Vazirani, Georgia Tech
- Gerhard Woeginger, Twente University
- Alexander Zelikovsky, Georgia State

and the subreferees Dimitris Achlioptas, Luca Becchetti, Piotr Berman, Gruia Calinescu, Eranda Cela, Chandra Chekuri, Julia Chuzhoy, W. Fernandez de la

Vega, Mathias Hauptmann, Volkan Isler, Marek Karpinski, Stefano Leonardi, Alberto Marchetti Spaccamela, Yakov Nekritch, Rudi Pendavingh, Jordi Petit i Silvestre, Kirk Pruhs, Amin Saberi, Jiri Sgall, and Luca Trevisan.

We gratefully acknowledge sponsorship from the University of Rome La Sapienza, Rome, the EU Thematic Network APPOL *Approximation and On-line Algorithms*, and the DFG Graduiertenkolleg *Effiziente Algorithmen and Mehrskalenmethoden* at the University of Kiel. We also thank Parvaneh Karimi-Massouleh, Marian Margraf, and Brigitte Preuss of the research group Theory of Parallelism at the University of Kiel, and Alfred Hofmann and Judith Freuden-berger of Springer-Verlag for supporting our project.

July 2002                    Klaus Jansen and Stefano Leonardi, Workshop Chairs
                             Vijay V. Vazirani, APPROX 2002 Program Chair

# Table of Contents

# Search and Classification
# of High Dimensional Data

Yuval Rabani

Computer Science Department, Technion – IIT, Haifa 32000, Israel,
rabani@cs.technion.ac.il

**Abstract.** Modeling data sets as points in a high dimensional vector space is a trendy theme in modern information retrieval and data mining. Among the numerous drawbacks of this approach is the fact that many of the required processing tasks are computationally hard in high dimension. We survey several algorithmic ideas that have applications to the design and analysis of polynomial time approximation schemes for nearest neighbor search and clustering of high dimensional data. The main lesson from this line of research is that if one is willing to settle for approximate solutions, then high dimensional geometry is easy. Examples are included in the reference list below.

# References

1. N. Alon, S. Dar, M. Parnas, and D. Ron. Testing of clustering. In *Proc. of the 41th Ann. IEEE Symp. on Foundations of Computer Science*, 2000, pages 240–250.
2. M. Bădoiu, S. Har-Peled, and P. Indyk. Approximate clustering via core-sets. In *Proc. of the 34th Ann. ACM Symp. on Theory of Computing*, 2002.
3. P. Drineas, A. Frieze, R. Kannan, S. Vempala, and V. Vinay. Clustering in large graphs and matrices. In *Proc. of the 10th Ann. ACM-SIAM Symp. on Discrete Algorithms*, 1999, pages 291–299.
4. W. Fernandez de la Vega, M. Karpinski, C. Kenyon, and Y. Rabani. Polynomial time approximation schemes for metric min-sum clustering. *Electronic Colloquium on Computational Complexity* report number TR02-025. Available at `ftp://ftp.eccc.uni-trier.de/pub/eccc/reports/2002/TR02-025/index.html`
5. S. Har-Peled and K.R. Varadarajan. Projective clustering in high dimensions using core-sets. In *Proc. of the 18th Ann. ACM Symp. on Computational Geometry*, 2002, pages 312–318.
6. P. Indyk and R. Motwani. Approximate nearest neighbors: Towards removing the curse of dimensionality. In *Proc. of the 30th Ann. ACM Symp. on Theory of Computing*, 1998, pages 604–613.
7. J. Kleinberg. Two algorithms for nearest-neighbor search in high dimensions. In *Proc. of the 29th Ann. ACM Symp. on Theory of Computing*, 1997, pages 599–608.
8. E. Kushilevitz, R. Ostrovsky, and Y. Rabani. Efficient search for approximate nearest neighbor in high dimensional spaces. *SIAM J. Comput.*, 30(2):457–474, 2000. Preliminary version appeared in STOC '98.

K. Jansen et al. (Eds.): APPROX 2002, LNCS 2462, pp. 1–2, 2002.

9. N. Mishra, D. Oblinger, and L. Pitt. Sublinear time approximate clustering. In *Proc. of the 12th Ann. ACM-SIAM Symp. on Discrete Algorithms*, January 2001, pages 439–447.

10. R. Ostrovsky and Y. Rabani. Polynomial time approximation schemes for geometric clustering problems. *J. of the ACM*, 49(2):139–156, March 2002. Preliminary version appeared in FOCS '00.

11. L.J. Schulman. Clustering for edge-cost minimization. In *Proc. of the 32nd Ann. ACM Symp. on Theory of Computing*, 2000, pages 547–555.

# Bicriteria Spanning Tree Problems

R. Ravi

GSIA, Carnegie Mellon University, Pittsburgh PA 15213, USA,
ravi@cmu.edu

Bicriteria spanning tree problems involve finding a spanning tree of a graph that simultaneously optimizes two objectives. We will survey approximability results for bicriteria spanning trees in undirected graphs for all pairs of minimization criteria drawn from the set (Total edge cost, Diameter, Maximum node degree). All of these results are bicriteria approximations in the sense that both objective values of the solution are approximately minimized. Hence, they are characterized by a pair of performance ratios, one per objective. We will highlight three different approaches that have been used in the design of these algorithms.

1. A cluster-growing approach for iteratively building a solution, coupled with a solution-based decomposition argument for bounding the performance ratios [7,9,10].
2. A black-box parametric search technique for converting an approximation algorithm with performance guarantee for a unicriterion problem to one with slightly weaker guarantees for both criteria, in the case when the two criteria are of the same type as the unicriterion problem (E.g., Using an MST algorithm for finding a spanning tree with approximately minimum edge cost under two different costs on the edges) [5].
3. An approach based on the Lagrangean relaxation of an integer-programming formulation of the problem [4,8].

# References

1. P. M. Camerini, G. Galbiati, F. Maffioli, On the complexity of finding multi-constrained spanning trees: Discrete Applied Math. **5** (1983) 39–50
2. M. Fürer, B. Raghavachari: Approximating the minimum degree Steiner tree to within one of optimal. Journal of Algorithms, **17** (1994) 409–423
3. S. Khuller, B. Raghavachari, N. E. Young: Balancing Minimum Spanning Trees and Shortest-Path Trees. Algorithmica **14(4)** (1995) 305–321
4. J. Könemann, R. Ravi: A matter of degree: improved approximation algorithms for degree-bounded minimum spanning trees. Proceedings of the 32nd Symp. on the Theory of Comput. (2000) 537–546
5. M. V. Marathe, R. Ravi, R. Sundaram, S. S. Ravi, D. J. Rosenkrantz, H. B. Hunt III: Bicriteria Network Design Problems. J. Algorithms **28(1)** (1998) 142–171
6. C. H. Papadimitriou, M. Yannakakis: On the Approximability of Trade-offs and Optimal Access of Web Sources. Proc. of the 41st Symp. on the Foundations of Comp. Sci. (2000) 86–92
7. R. Ravi: Rapid Rumor Ramification: Approximating the minimum broadcast time (Extended Abstract). Proc. of the 35th Symp. on the Foundations of Comp. Sci. (1994) 202–213

K. Jansen et al. (Eds.): APPROX 2002, LNCS 2462, pp. 3–4, 2002.
© Springer-Verlag Berlin Heidelberg 2002

8. R. Ravi, M. Goemans: The constrained minimum spanning tree problem. Proc. of the Scandinavian Workshop on Algorithmic Theory, LNCS **1097** (1996) 66–75
9. R. Ravi, M. V. Marathe, S. S. Ravi, D. J. Rosenkrantz, H. B. Hunt III: Many Birds with One Stone: Multi-Objective Approximation Algorithms (Extended Abstract). Proceedings of the 25th Symp. on the Theory of Comput. (1993) 438–447
10. R. Ravi, M. V. Marathe, S. S. Ravi, D. J. Rosenkrantz, H. B. Hunt III: Approximation Algorithms for Degree-Constrained Minimum-Cost Network-Design Problems. Algorithmica **31(1)** (2001) 58–78

# Improved Approximation Algorithms
# for Multilevel Facility Location Problems*

Alexander Ageev

Sobolev Institute of Mathematics, pr. Koptyuga 4, Novosibirsk, 630090, Russia,
ageev@math.nsc.ru

**Abstract.** We show that the metric multilevel facility location problem
is polynomial-time reducible within a factor of 3 to the metric uncapacitated facility location problem. This reduction together with recent
approximation algorithms for the latter problem, due to Jain, Mahdian
& Saberi, leads to a 4.83-approximation algorithm for the metric multilevel facility location problem and to a 9-approximation algorithm for
a capacitated version of it (where facilities have soft capacities). In the
class of combinatorial algorithms these performance ratios improve on
the previous ones due to Bumb and Kern (6 and 12 respectively).

## 1   Introduction

In the multilevel facility location problem we are given a set of demand points
and a set of facilities of $k$ types: one type of terminals and $(k-1)$ types of
transit stations. For every facility an opening cost is given. Each unit of demand
must be shipped from a terminal through transit stations of type $k-1,\ldots,1$ to
the demand points. The objective is to open a subset of facilities of each type
and to connect each demand point to a path along open facilities so that the
total cost of opening and shipping is minimized. We consider the metric case
of the problem where shipping costs are nonnegative, symmetric and satisfy the
triangle inequality. This case is a natural generalization of the metric (1-level)
uncapacitated facility location problem, which is known to be Max SNP-hard
[10,19]. The latter problem has been studied extensively from the perspective of
approximation algorithms [17,4,6,7,5,10,11,13,20,15,2,12,18,16]. The best (non-
combinatorial) approximation algorithm, due to Sviridenko [18], achieves a factor
of 1.582. Very recently, Mahdian, Ye & Zhang [16] claimed a combinatorial 1.52-
approximation algorithm.

Shmoys, Tardos & Aardal [17] and Aardal, Chudak & Shmoys [1] developed
first approximation algorithms for the metric multilevel facility location problem
with performance guarantees of 3.16 and 3, respectively. The drawback of these
algorithms is that they need solving linear programs and therefore can hardly be
viewed as practically efficient. Meyerson, Munagala & Plotkin [14] were the first

---

* Research was partially supported by the Russian Foundation for Basic Research,
  project codes 01-01-00786, 02-01-01153, by INTAS, project code 00-217, and by the
  Programme "Universities of Russia", project code UR.04.01.012.

who designed a combinatorial algorithm for the problem that finds a solution within $O(\log |D|)$ of the optimal value, where $D$ is the set of demand points. This result was improved on by Bumb & Kern [3] who developed a dual ascent algorithm for the problem with a performance guarantee of 6. From this result they also derived a 12-approximation algorithm for the following capacitated generalization of the problem (with facilities having soft capacities): at each site $i \in F$, instead of a single uncapacitated facility, it is allowed to open any number of copies of a facility with a restricted capacity $u_i \in \mathbb{N}$, which is an upper bound on the number of demand points served by one copy.

Very recently, Edwards [9] observed that the metric multilevel facility location problem admits a simple, albeit exponential in the number of levels $k$, factor 3 reduction to the 1-level case. In this paper we present a similar, but a bit more refined construction which proves that the problem is indeed polynomial-time reducible within a factor of 3 to the metric uncapacitated facility location problem. Our result together with approximation algorithms due to Jain, Mahdian & Saberi [12] leads to a combinatorial 4.83-approximation algorithm for the metric multilevel facility location problem (a better bound of 4.56 follows from the results claimed in [16]) and to a 9-approximation algorithm for the generalization of the problem where facilities have soft capacities.

## 2    Factor 3 Reduction to the 1-Level Case

### 2.1    Formal Setting of the Problem

Consider a complete $(k+1)$-partite graph graph $G = (D \cup F_1 \cup \ldots \cup F_k; E)$ whose vertex set is the union of $k+1$ disjoint sets $D, F_1, \ldots, F_k$ and the edge set $E$ consists of all edges between these sets. We call the vertices in $D$ the demand points and the vertices in $F = F_1 \cup \ldots \cup F_k$ the facilities (of level $1, \ldots, k$ respectively). We are given edge costs $c \in \mathbb{R}_+^E$ and opening costs $f \in \mathbb{R}_+^F$ ( i. e., opening a facility $i \in F$ incurs a cost $f_i \geq 0$). We assume that $c$ is induced by a metric on $V = D \cup F_1 \cup \ldots \cup F_k$.

Denote by $P$ the set of paths of length $k-1$ joining some vertex in $F_1$ to some vertex in $F_k$. For $j \in D$ and $p = (v_1, \ldots, v_k) \in P$, let $jp$ denote the path $(j, v_1, \ldots, v_k)$. Denote by $c(p)$ and $c(pj)$ the length of $p$ and $jp$ respectively (with respect to $c$). For a subgraph $H$ of $G$ (in particular, $H$ can be a path $p \in P$ or the union of such paths), denote by $f(H)$ the sum of opening costs of the facilities lying in $H$. We are now ready to give a formal formulation of the multilevel facility location problem:

Find an assignment $\omega : D \to P$ minimizing

$$\sum_{j \in D} c(\omega(j)j) + f\left(\bigcup_{j \in D} \omega(j)\right).$$

When $k = 1$ (all facilities are of level 1), the problem is nothing but the well-known uncapacitated facility location problem.

## 2.2   Reduction

The input is an instance $\mathcal{M}$ of the metric multilevel facility location problem with $k \geq 2$. For each $i \in F_1$ and $t = 1, \ldots, |D|$, the reduction algorithm finds a path $p_{it} \in P$ starting from $i$ and having the minimum value of $tc(p) + f(p)$ among all such paths, that is

$$tc(p_{it}) + f(p_{it}) = \min\{tc(p) + f(p) : p \in P \text{ such that } i \in p\}.$$

Then it constructs an instance $\mathcal{I}$ of the facility location problem as follows. In $\mathcal{I}$ the set of demand points is $D' = D$, the set of facilities $F' = F_1 \times \{1, \ldots, |D|\}$, the opening costs are $f'(it) = f(p_{it})$, and the shipping costs $c'_{it,j} = c_{i,j} + c(p_{it})$. Note that the shipping costs $c'$ satisfy the triangle inequality and thus $\mathcal{I}$ is an instance of the metric facility location problem. Now let $\varphi : D' \rightarrow F'$ be a feasible solution of $\mathcal{I}$ retrieved by some algorithm for solving the facility location problem. Recall that $\varphi(j)$ is a pair $it$ where $i \in F_1$, $1 \leq t \leq |D|$. Denote by $\varphi_1(j)$ the first member of this pair (corresponding to some facility in $F_1$). Define a feasible solution $\omega : D \rightarrow P$ of $\mathcal{M}$ by setting $\omega(j) = p_{\varphi(j)}$, $j \in D$. Output $\omega$.

*Remark 1.* The described reduction clearly runs in polynomial time. The most time-consuming phase of the algorithm is to find the paths $p_{it}$, which with the aid of the standard reduction to the shortest path problem can be done by $|D|$ calls of the classical algorithm of Dijkstra for finding a shortest distances tree [8].

## 2.3   Analysis

Our aim is to prove the following

**Lemma 1.** *If the cost of $\varphi$ is within of $\alpha$ of the optimal value of $\mathcal{I}$, then the cost of $\omega$ is within a factor of $3\alpha$ of the optimal value of $\mathcal{M}$.*

This lemma implies the main result of the paper:

**Theorem 1.** *An $\alpha$-approximation algorithm for the metric facility location problem yields a $3\alpha$-approximation algorithm for the metric multilevel facility location problem.*

Lemma 1 clearly follows from Lemmas 2 and 4 below.

**Lemma 2.** *The cost of $\omega$ is at most the cost of $\varphi$.*

*Proof.* The cost of $\varphi$ is equal to

$$f'(\bigcup_{j \in D}\{\varphi(j)\}) + \sum_{j \in D} c'_{\varphi(j),j} = \sum\{f(p_{it}) : it \in \bigcup_{j \in D}\{\varphi(j)\}\}$$
$$+ \sum_{j \in D}\left(c_{\varphi_1(j),j} + c(p_{\varphi(j)})\right).$$

The cost of $\omega$ is

$$f\left(\bigcup_{j\in D}\omega(j)\right) + \sum_{j\in D}c(\omega(j)j) = f\left(\bigcup_{j\in D}p_{\varphi(j)}\right) + \sum_{j\in D}c(\omega(j)j)$$

$$= f\left(\bigcup_{j\in D}p_{\varphi(j)}\right) + \sum_{j\in D}\left(c_{\varphi_1(j),j} + c(p_{\varphi(j)})\right).$$

Since

$$f\left(\bigcup_{j\in D}p_{\varphi(j)}\right) = f\left(\bigcup\{p_{it} : it \in \bigcup_{j\in D}\{\varphi(j)\}\}\right)$$

$$\leq \sum\{f(p_{it}) : it \in \bigcup_{j\in D}\{\varphi(j)\}\},$$

the cost of $\omega$ is at most the cost of $\varphi$.        □

*Remark 2.* Above we have implicitly shown that

$$f\left(\bigcup_{j\in D}\omega(j)\right) \leq f'\left(\bigcup_{j\in D}\{\varphi(j)\}\right),$$

$$\sum_{j\in D}c(\omega(j)j) = \sum_{j\in D}c'_{\varphi(j),j}.$$

**Lemma 3.** *Let $\mathcal{M}$ be an instance of the multilevel facility location problem. For every feasible solution $\omega'$ of $\mathcal{M}$, there exists a solution $\omega$ with at most the same cost and such that the following condition holds:*

**(F)** *for any two paths $\omega(j') = (i'_1, \ldots, i'_k)$ and $\omega(j'') = (i''_1, \ldots, i''_k)$ in $P$, if $i'_l = i''_l$ for some $l$, then $i'_r = i''_r$ for all $r \geq l$.*

*Proof.* Let $\omega'$ be a feasible solution of $\mathcal{M}$ and $F(\omega')$ be the set of open facilities in it. Consider the graph $H$ which is the subgraph of $G$ induced by $D \cup F(\omega')$ plus a new vertex *root* and edges of zero cost connecting *root* with all vertices in $F(\omega') \cap F_k$. Assume now that we forgot $\omega'$ but still remember the set $F(\omega')$. Then to find a feasible solution $\omega$ of cost at most the cost of $\omega'$ and whose set of open facilities coincides with $F(\omega')$ it just suffices to find a tree of shortest paths from *root* to all other vertices of $H$ [8]. This tree uniquely determines the assignment $\omega$ clearly satisfying condition **(F)**.        □

**Lemma 4.** *$\mathcal{I}$ has a feasible solution of cost at most thrice the optimal value of $\mathcal{M}$.*

*Proof.* Let $\omega^* : D \to P$ be an optimal solution of $\mathcal{M}$ satisfying condition **(F)**. Let $\{i'_1, \ldots, i'_r\}$ be the set of facilities open in $\omega^*$ and lying on level $k$ (e. g., in $F_k$). For every $l = 1, \ldots, r$, let $D_l$ denote the set of demand points assigned (by $\omega^*$) to a path finishing in $i'_l$ and let $j_l$ denote a demand point in $D_l$ having minimum value of $c(\omega^*(j)j)$ among all demand points $j \in D_l$.

Define another feasible solution $\tau$ of $\mathcal{M}$ as follows: for every $l = 1, \ldots, r$ and $j \in D_l$, set $\tau(j) = \omega^*(j_l)$.

We claim that the cost of $\tau$ is at most thrice the cost of $\omega^*$. Indeed, recall that $\omega^*$ satisfies condition $(\mathbf{F})$ and hence, by the triangle inequality, for every $l$ and $j \in D_l$,

$$\begin{aligned}
c(\tau(j)j) &\le c(\omega^*(j_l)j_l) + c_{jj_l} \\
&\le c(\omega^*(j_l)j_l) + c(\omega^*(j_l)j_l) + c(\omega^*(j)j) \\
&\le 3c(\omega^*(j)j).
\end{aligned} \tag{1}$$

Furthermore, by the construction of $\tau$, the set of open facilities in $\tau$ is a subset of open facilities in $\omega^*$. Together with (1) this proves the claim.

Observe that by condition $(\mathbf{F})$ the paths $\tau(j_1), \ldots, \tau(j_r)$ are pairwise disjoint. For every $l = 1, \ldots, r$, denote by $i_l$ the endpoint of path $\tau(j_l)$ lying in $F_1$.

Define a feasible solution $\psi : D \to F_1 \times \{1, \ldots, |D|\}$ of $\mathcal{I}$ as follows: for every $l = 1, \ldots, r$ and every $j \in D_l$, set $\psi(j) = i_l t_l$ where $t_l = |D_l|$.

We now claim that the cost of $\psi$ (in $\mathcal{I}$) is at most the cost of $\tau$. Indeed the cost of $\psi$ is equal to

$$\sum_{j \in D} c'_{\psi(j)j} + f'(\bigcup_{j \in D} \{\psi(j)\}) = \sum_{l=1}^{r} \sum_{j \in D_l} c'_{i_l t_l, j} + \sum_{l=1}^{r} f'(i_l t_l)$$

$$= \sum_{l=1}^{r} \sum_{j \in D_l} (c_{i_l j} + c(p_{i_l t_l})) + \sum_{l=1}^{r} f(p_{i_l t_l})$$

$$= \sum_{l=1}^{r} \sum_{j \in D_l} c_{i_l j} + \sum_{l=1}^{r} (f(p_{i_l t_l}) + t_l c(p_{i_l t_l})).$$

On the other hand, since the paths $\tau(j_1), \ldots, \tau(j_k)$ are pairwise disjoint, the cost of $\tau$ is equal to

$$\sum_{l=1}^{r} \sum_{j \in D_l} [c_{i_l j} + c(\tau(j))] + f(\bigcup_{l=1}^{r} \tau(j_l)) = \sum_{l=1}^{r} \sum_{j \in D_l} c_{i_l j} + \sum_{l=1}^{r} (f(\tau(j_l)) + t_l c(\tau(j_l))).$$

By the definition of paths $p_{it}$, for every $l = 1, \ldots, r$,

$$f(p_{i_l t_l}) + t_l c(p_{i_l t_l}) \le f(\tau(j_l)) + t_l c(\tau(j_l)).$$

Thus the cost of $\psi$ is at most the cost of $\tau$. $\square$

## 3  Problem with Soft Capacities

Bumb & Kern [3] presented a 12-approximation algorithm for the following generalization of the metric facility location problem. At each site $i \in F$, instead of one uncapacitated facility, we are allowed to open any number of copies of a facility with a capacity $u_i \in \mathbb{N}$, which is an upper bound on the number of paths served by one copy. This problem is a natural generalization of the capacitated facility location problem with soft capacities [11,12].

In [3], it is implicitly established that an $\alpha$-approximation algorithm for the multilevel facility location problem can be converted into a $2\alpha$-approximation algorithm for the multilevel facility location problem with soft capacities. By Theorem 1 it follows that an $\alpha$-approximation algorithm for the facility location problem can be transformed into a $6\alpha$-approximation algorithm for the multilevel facility location problem with soft capacities. This gives a 9.66-approximation algorithm for the multilevel facility location problem with soft capacities.

In this section we describe a 9-approximation algorithm for the capacitated problem. The algorithm is based on our reduction, a 2-approximation algorithm for the uncapacitated facility location problem in [12] and on a relevant adaptation of the Lagrangian multipliers technique used in [3] (which is an extension of that in [11]).

We start with an integer programming formulation of the multilevel facility location problem with soft capacities. Let us introduce for every $i \in F$, an integral variable $y_i$ denoting the number of open copies of facility $i$, and for every $j \in D$ and $p \in P$, a $(0,1)$ variable $x_{pj}$ denoting whether demand point $j$ is served along path $p$. Set

$$c(x) = \sum_{p \in P} \sum_{j \in D} c(pj) x_{pj}$$

and

$$f(y) = \sum_{i \in F} f_i y_i.$$

The multilevel facility location problem with soft capacities is now equivalent to the following integer program:

$$\text{minimize} \quad c(x) + f(y) \tag{2}$$

subject to:

$$\sum_{p \in P} x_{pj} = 1 \text{ for each } j \in D, \tag{3}$$

$$\sum_{j \in D} \sum_{p \ni i} x_{pj} \le u_i y_i \text{ for each } i \in F, \tag{4}$$

$$x_{pj} \in \{0,1\} \text{ for each } j \in D, p \in P, \tag{5}$$

$$y_i \ge 0 \text{ are integers for each } i \in F. \tag{6}$$

The (previously treated) uncapacitated problem arises when $u_i \equiv \infty$; in this case $y_i$ may be assumed to be at most 1 and constraints (4) can be replaced by

$$x_{pj} \le y_i \text{ for each } j \in D, p \ni P, i \in p. \tag{7}$$

Just as in [3], we associate with (2)–(6) the following uncapacitated problem:

$$C(\lambda) = \min \quad c(x) + f(y) + \sum_{i \in F} \lambda_i \left( \sum_{j \in D} \sum_{p \ni i} x_{pj} - u_i y_i \right)$$

$$= \min \quad \tilde{c}(x) + \tilde{f}(y) \tag{8}$$

$$\text{subject to} \quad (3),(5),(6), (7),$$

where $\tilde{f}_i = f_i - \lambda_i u_i$, for each $i \in F$ and $\tilde{c}$ is defined as follows: for every $i_1, i_2 \in F \setminus F_k$, $\tilde{c}_{i_1 i_2} = c_{i_1 i_2} + \frac{1}{2}(\lambda_{i_1} + \lambda_{i_2})$, for every $i_1 \in F_{k-1}$ and $i_2 \in F_k$, $\tilde{c}_{i_1 i_2} = c_{i_1 i_2} + \frac{1}{2}\lambda_{i_1} + \lambda_{i_2}$ and for every $i \in F_1$ and $j \in D$, $\tilde{c}_{ij} = c_{ij} + \frac{1}{2}\lambda_i$. We set

$$\lambda_i = \frac{2f_i}{3u_i} \tag{9}$$

for each $i \in F$. It is clear that $\tilde{c}$ can be extended to a metric on $D \cup F_1 \cup \ldots \cup F_k$ and so (8), (3), (5), (6), (7) is an instance of the metric multilevel facility location problem. Note that for every $\lambda \geq 0$, $C(\lambda) \leq OPT$ where $OPT$ is the optimal value of the original problem. To solve (8), (3), (5), (6), (7) we use our reduction in combination with the LMP (Lagrangian Multiplier Preserving) 2-approximation algorithm from [12, section 5] to find an approximate solution to the instance $\mathcal{I}$ of the (1-level) facility location problem (see subsection 2.2 of this paper). Let $(x, y)$ denote the solution retrieved by the resulting algorithm (the equivalent of $\omega$ in the previous section). By Remark 2 and the LMP property, we have

$$\tilde{c}(x) + 2\tilde{f}(y) \leq 2 \cdot OPT(\mathcal{I}),$$

where $OPT(\mathcal{I})$ is the optimal value of the instance $\mathcal{I}$. By Lemma 4, $OPT(\mathcal{I}) \leq 3 \cdot C(\lambda)$. Hence

$$\tilde{c}(x) + 2\tilde{f}(y) \leq 6 \cdot C(\lambda). \tag{10}$$

Then by (6), (7), and (9) we have

$$\tilde{c}(x) + 2\tilde{f}(y) = c(x) + \frac{2}{3} \sum_{i \in F} \frac{f_i}{u_i} \sum_{p \ni i} \sum_{j \in D} x_{pj} + \frac{2}{3} \sum_{i \in F} f_i y_i$$

$$= c(x) + \frac{2}{3} \sum_{i \in F} f_i \left( \frac{1}{u_i} \sum_{p \ni i} \sum_{j \in D} x_{pj} + y_i \right)$$

$$\geq c(x) + \frac{2}{3} \sum_{i \in F} f_i \bar{y}_i,$$

where

$$\bar{y}_i = \left\lceil \frac{1}{u_i} \sum_{p \ni i} \sum_{j \in D} x_{pj} \right\rceil.$$

It is clear that $(x, \bar{y})$ is a feasible solution of the capacitated problem and by (10), it satisfies

$$c(x) + \frac{2}{3} f(\bar{y}) \leq 6 \cdot C(\lambda) \leq 6 \cdot OPT.$$

It follows that

$$c(x) + f(\bar{y}) \leq \frac{3}{2}c(x) + f(\bar{y}) \leq 9 \cdot OPT.$$

# References

1. K. I. Aardal, F. Chudak and D. B. Shmoys, "A 3-approximation algorithm for the $k$-level uncapacitated facility location problem", Information Processing Letters **72** (1999), 161–167.
2. V. Arya, N. Garg, R. Khandekar, A. Meyerson, K. Munagala, and V. Pandit, "Local search heuristics for k-median and facility location problems", in: Proceedings of the 33rd ACM Symposium on Theory of Computing, ACM Press, pp. 21–29, 2001.
3. A. F. Bumb and W. Kern, "A simple dual ascent algorithm for the multilevel facility location problem", in: Proceedings of the 4th International Workshop on Approximation Algorithms for Combinatorial Optimization Problems (APPROX'2001), Lecture Notes in Computer Science, Vol. 2129, Springer, Berlin, pp. 55–62, 2001.
4. M. Charikar and S. Guha, "Improved combinatorial algorithms for facility location and k-median problems", in: Proceedings of the 40th Annual IEEE Symposium on Foundations of Computer Science, IEEE Computer Society, pp. 378–388, 1999.
5. M. Charikar, S. Khuller, D. Mount, and G. Narasimhan, "Facility location with outliers", in: Proceedings of the 12th Annual ACM-SIAM Symposium on Discrete Algorithms, Washington DC, pp. 642–651, 2001.
6. F. A. Chudak, "Improved approximation algorithms for uncapacited facility location", in: Proceedings of the 6th Integer Programming and Combinatorial Optimization Conference, Lecture Notes in Computer Science, Vol. 1412, Springer, Berlin, pp. 180–194, 1998.
7. F. A. Chudak and D. B Shmoys, "Improved approximation algorithms for the uncapacitated facility location problem", unpublished manuscript (1998).
8. E. W. Dijkstra, "A note on two problems in connection with graphs", Numerische Mathematik **1** (1959), 269–271.
9. N. J. Edwards, "Approximation algorithms for the multi-level facility location problem", Ph. D. Thesis (2001).
   http://www.orie.cornell.edu/~nedwards/thesis.oneside.ps
10. S. Guha and S. Khuller, "Greedy strikes back: Improved facility location algorithms", J. Algorithms **31** (1999), 228–248.
11. K. Jain and V. V. Vazirani, "Approximation Algorithms for Metric Facility Location and k-Median Problems Using the Primal-Dual Schema and Lagrangian Relaxation", J. ACM **48** (2001), 274–296.
12. K. Jain, M. Mahdian, and A. Saberi, "A new greedy approach for facility location problems", in: Proceedings of the 34th ACM Symposium on Theory of Computing (STOC'02), Montreal, Quebec, Canada, May 19-21, 2002.
13. M. R. Korupolu, C. G. Plaxton, and R. Rajaraman, "Analysis of a local search heuristic for facility location problems", in: Proceedings of the 9th Annual ACM-SIAM Symposium on Discrete Algorithms (SODA'98), ACM Press, pp. 1–10, 1998.
14. A. Meyerson, K. Munagala, and S. Plotkin, "Cost distance: Two metric network design", in: *Proceedings of the 41st Annual Symposium on Foundations of Computer Science*, IEEE Computer Society, pp. 624–630, 2000.

15. M. Mahdian, E. Markakis, A. Saberi, and V. Vazirani, "A greedy facility location algorithm analyzed using dual fitting", in: Proceedings of the 4th International Workshop on Approximation Algorithms for Combinatorial Optimization Problems (APPROX'2001), Berkeley, CA, USA, August 18-20, Lecture Notes in Computer Science, Vol. 2129, 127–137.

16. M. Mahdian, Y. Ye, and J. Zhang, "A 1.52-approximation algorithm for the uncapacitated facility location problem", manuscript (2001).
    http://www.math.mit.edu/~mahdian/floc152.ps

17. D. Shmoys, E. Tardos, and K. I. Aardal, "Approximation algorithms for facility location problems", in: Proceedings of the 29th Annual ACM Symposium on the Theory of Computing (STOC '97), ACM Press, 1997, pp. 265–274.

18. M. Sviridenko, "An improved approximation algorithm for the metric uncapacitated facility location problem", in: Proceedings of the 9th Integer Programming and Combinatorial Optimization Conference, Lecture Notes in Computer Science, Vol. 2337, Springer, Berlin, pp. 240–257, 2002.

19. M. Sviridenko, personal communication.

20. M. Thorup, "Quick k-median, k-center, and facility location for sparse graphs", in: Proceedings of the 28th International Colloquium on Automata, Languages and Programming (ICALP'2001), Lecture Notes in Computer Science, Vol. 2076, 249–260.

# On Constrained Hypergraph
# Coloring and Scheduling
## (Extended Abstract)

Nitin Ahuja* and Anand Srivastav

Mathematisches Seminar, Christian-Albrechts-Universität zu Kiel,
Ludewig-Meyn-Str. 4, 24098 Kiel, Germany,
{nia,asr}@numerik.uni-kiel.de

**Abstract.** We consider the design of approximation algorithms for multicolor generalization of the well known hypergraph 2-coloring problem
(property B). Consider a hypergraph $H$ with $n$ vertices, $s$ edges, maximum edge degree $\mathfrak{D}(\leq s)$ and maximum vertex degree $d(\leq s)$. We study
the problem of coloring the vertices of $H$ with minimum number of colors
such that no hyperedge $i$ contains more than $b_i$ vertices of any color. The
main result of this paper is a deterministic polynomial time algorithm
for constructing approximate, $\lceil (1 + \epsilon)OPT \rceil$-colorings ($\epsilon \in (0,1)$) satisfying all constraints provided that $b_i$'s are logarithmically large in $d$ and
two other parameters. This approximation ratio is independent of $s$. Our
lower bound on the $b_i$'s is better than the previous best bound. Due to
the similarity of structure these methods can also be applied to resource
constrained scheduling. We observe, using the non-approximability result for graph coloring of Feige and Killian[4], that unless $NP \subseteq ZPP$
we cannot find a solution with approximation ratio $s^{\frac{1}{2}-\delta}$ in polynomial
time, for any fixed small $\delta > 0$.

## 1  Introduction

Let $A \in \{0,1\}^{s \times n}$ and $b = (b_1, \ldots, b_s)^t \in \mathbb{N}^s$. We want to find vectors $x^{(k)} = (x_{1,k}, x_{2,k}, \ldots, x_{n,k})^t$ which minimize $T = \max\{z \mid x_{j,z} > 0 , j = 1, 2, \ldots, n\}$ such that

(i) $Ax^{(k)} \leq b \quad \forall k = 1, 2, \ldots, T$ ,
(ii) $\sum_{k=1}^{T} x_{j,k} = 1 \quad \forall j = 1, 2, \ldots, n$    and
(iii) $x_{j,k} \in \{0,1\} \quad \forall j, k$.

Let $T_{opt}$ be the value of the minimum solution of this integer program and let
$T^*$ be the value of the minimum solution of the *relaxed* problem where we allow
real variables $x_{j,k} \in [0,1]$, $\forall j, k$.

Let $H = (V, \mathcal{E})$ be a hypergraph with $|V| = n$ and $|\mathcal{E}| = s$. The integer
program described above models the following problem: using minimum number

---

* supported by a DFG scholarship through Graduiertenkolleg-357: Effiziente Algorithmen und Mehrskalenmethoden

K. Jansen et al. (Eds.): APPROX 2002, LNCS 2462, pp. 14–25, 2002.

of colors, color the vertices of $H$ such that in each hyperedge $i$ there are no more than $b_i$ vertices of any color. This coloring problem is a multicolor generalization of the usual (property B) hypergraph coloring problem where the aim is to avoid monochromatic edges. We will call it the constrained hypergraph coloring(CHC) problem.

Furthermore, the resource constrained scheduling (RCS) problem and a special case of multi-dimensional bin packing (MDBP) problem fit in this model. In a RCS problem we are given $n$ jobs and $s$ resources with resource constraint vector $b$. Each job requires at least one resource and the aim is to find a schedule for these jobs such that the resource constraints are never violated and the makespan is minimum. An instance of the MDBP with integral vectors consists of vectors $y_1, \ldots, y_n \in \{0, 1\}^s$ and an unlimited number of bins. The objective is to pack the vectors in bins using a minimum number of bins such that the sum of the $i$ th components of vectors in any bin is at most $b_i$ for all $i \in [s] = \{1, 2, \ldots, s\}$.

Special cases of CHC are well known and have been intensively studied. For a simple graph and $b_i = 1 \ \forall i$, CHC is the graph coloring problem. For a hypergraph $H$ with hyperedges $F_1, \ldots, F_s$, and $b_i = |F_i| - 1 \ \forall i$, CHC is closely related to the property B problem (*i.e* there exists a 2-coloring of $H$ in which no hyperedge is monochromatic): whenever $H$ has property B, CHC is equivalent to the problem of finding a non-monochromatic 2-coloring of $H$. Note that property B requires logarithmic lower bounds on the $b_i$'s. In fact if $H$ is $r$-uniform, then by the Lovász Local Lemma it has property B if its maximum edge degree $\mathfrak{D} \leq 2^{r-3}$, and this implies $b_i = r - 1 \geq \log \mathfrak{D} + 2 \ \forall i$. A coloring problem related to the CHC problem, which also generalizes the property B problem to multicolors has been studied by Lu [10]. The aim is to color the vertices of $H$ with $k$ given colors such that no color appears more than $b$ times in any edge. Let us call this the multicolor property B problem. Assuming $H$ to be $r$-uniform, the result of [10] says that if $H$ is $k$-colorable and $r/k = (\log(dr))^{1+\delta}$ for $\delta > 0$ then $b = \Theta\left((\log(dr))^{1+\delta}\right) \ (\geq r/k)$. The algorithm of Lu does not provide an approximation of the *optimum* $T_{opt}$ of the CHC problem.

For $b_i = 1 \ \forall i$, the previous approximation algorithms for $T_{opt}$ in this context are the following. In [5] Garey et al. use the First-Fit-Decreasing heuristic to give a polynomial time $(s + \frac{1}{3})$ approximation algorithm for MDBP problem. Subsequently De la Vega and Lueker [16] improved this result and gave a linear time algorithm which, for every $\epsilon > 0$, gives a $(1 + \epsilon)s$ approximate solution. We show that for $b_i = 1 \ \forall i$, constant approximation ratios are unachievable: by the non-approximability of graph coloring [4] for any fixed $\delta > 0$ it is not possible to find a solution of CHC problem with approximation ratio better than $s^{\frac{1}{2}-\delta}$ in polynomial time, unless $NP \subseteq ZPP$. For arbitrary $b_i$'s, Srivastav and Stangier [14,15] gave a polynomial time approximation algorithm for resource constrained scheduling problem with non-zero start times (problem class $P|res, \ldots 1, r_j, p_j = 1|T$). For every $\epsilon > 0$ their algorithm delivers a schedule of size at most $\lceil (1 + \epsilon)T_{opt} \rceil$ provided that for all $i$ $b_i \geq 3\epsilon^{-2}(1 + \epsilon) \log(8sT^*)$ and the number of processors is at least $3\epsilon^{-2}(1 + \epsilon) \log(8T^*)$. For a fixed $\epsilon$, the approximation ratio is a constant factor *independent* of $s$.

Our main result is the following. Given a hypergraph $H$ with $n$ vertices and $s$ hyperedges we show that for every $\epsilon \in (0,1)$, $H$ can be colored in polynomial time using at most $\lceil (1+\epsilon)T_{opt} \rceil$ colors provided that $b_i = \Omega\left(\epsilon^{-2}(1+\epsilon)\log \mathcal{D}\right)$ for all edges $i \in [s]$. The parameter $\mathcal{D} \leq d\eta\tau = O((sT^*)^2)$ where $d$ is the maximum vertex degree of the hypergraph and $\eta$, $\tau$ (defined in Section 4) depend on the structure of inequality constraints in the integer program. This improves the previous best lower bounds on $b_i$'s given by Srivastav and Stangier [15], specially so if $\mathcal{D} \ll (sT^*)^2$. For $\eta\tau = O(\text{poly}(r))$ we are able to construct a $\lceil (1+\epsilon)T_{opt} \rceil$-coloring provided that $b_i = \Omega\left(\epsilon^{-2}(1+\epsilon)\log(dr)\right)$ $\forall i$. We first give a randomized algorithm to construct such colorings using the techniques based on the algorithmic version of the Local Lemma (Beck [2], Alon [1], Molloy & Reed [11]). Our algorithm uses two ingredients: first we generate a feasible fractional coloring using at most $\lceil (1+\epsilon)T_{opt} \rceil$ colors. Then without violating any constraints we round the fractional solution to an integer one by generalizing the randomized/derandomized version of LLL given by Lu [10] and Leighton et al. [8] to minimax integer programs with a *fixed* right hand side (the $b_i$'s in our case).

## 2    Preliminaries

The Lovász Local Lemma has been successfully applied to show the existence of approximate solutions to integer programming (IP) problems. The lemma in its simplest form says

**Lemma 1.** *(Symmetric LLL) Let $\xi_1, \ldots, \xi_s$ be events in a probability space. Suppose that each event $\xi_i$ is mutually independent of all but at most $\mathcal{D}$ other events $\xi_j$ and that $Pr[\xi_i] \leq p$ for all $i \in [s]$. If $ep(\mathcal{D}+1) \leq 1$ then $Pr[\bigcap_{i=1}^{s} \bar{\xi_i}] > 0$.*

The extent of dependency (parameter $\mathcal{D}$) among events plays a vital role in almost all applications. Higher dependency among the events implies rarity of the structure we desire in some probability space and is an obstacle in many approximation schemes. Let $X_1, X_2, \ldots, X_n \in [0,1]$ be independent random variables with $X = \sum_{i=1}^{n} X_i$ and $\mathbb{E}[X] = \mu$. For $l = 1, \ldots, n$ define

$$S_l(X_1, \ldots, X_n) = \sum_{I \subseteq [n], |I| = l} \prod_{i \in I} X_i \tag{1}$$

and $S_0(X_1, \ldots, X_n) = 1$.

We state the large deviation bounds used in our analysis.

**Lemma 2.** *(i) (Chernoff-Hoeffding) For any $\delta > 0$, $Pr[X \geq \mu(1+\delta)] \leq G(\mu, \delta)$ where*

$$G(\mu, \delta) = \left(\frac{e^{\delta}}{(1+\delta)^{(1+\delta)}}\right)^{\mu}.$$

(ii) $\forall \mu > 0$ and $\forall p \in (0,1)$, there exists $\delta = H(\mu, p) > 0$ such that $G(\mu, \delta) \leq p$ and

$$
H(\mu, p) = \begin{cases} \Theta\left(\sqrt{\frac{\log p^{-1}}{\mu}}\right) & \text{if } \mu \geq \frac{\log p^{-1}}{2} \ ; \\ \Theta\left(\frac{\log p^{-1}}{\mu \log(\log(p^{-1})/\mu)}\right) & \text{otherwise} \ . \end{cases}
$$

**Lemma 3.** *(Schmidt, Siegel and Srinivasan, 1995)*

(i) *For any $\delta > 0$, any non-empty event $Z$ and any non-negative integer $l \leq \mu(1 + \delta)$*

$$
Pr[X \geq \mu(1 + \delta) \mid Z] \leq \frac{\mathbb{E}[S_l(X_1, \ldots, X_n) \mid Z]}{\binom{\mu(1+\delta)}{l}} \ .
$$

(ii) *If $l = \lceil \mu\delta \rceil$, then for any $\delta > 0$*

$$
Pr[X \geq \mu(1 + \delta)] \leq \frac{\mathbb{E}[S_l(X_1, \ldots, X_n)]}{\binom{\mu(1+\delta)}{l}} \leq G(\mu, \delta) \ .
$$

## 3  Generating Fractional Solutions

If we use randomized rounding to round the fractional solution of the relaxed problem it can lead to an infeasible solution of the original problem. Thus, to avoid infeasibility we define another LP problem with more restricted constraints than the original ones but with the same objective of minimizing the number of colors used. Let $\alpha \geq 1$ be a constant. The new constraints are:

- $Ax^{(k)} \leq b\alpha^{-1} \quad \forall k \in [T]$
- $\sum_{k=1}^{T} x_{j,k} = 1 \quad \forall j \in [n]$
- $x_{j,k} \in [0,1] \quad \forall j, k$.

Subsequently we will refer to this problem as the *reduced* problem.

It is intuitively clear that after reducing the $b_i$'s we will require more colors to properly color the hypergraph. In fact if we have a feasible solution of the relaxed problem with value $\tilde{T}$ then a feasible solution of the reduced problem with value $\lceil \alpha\tilde{T} \rceil$ can be constructed from it as follows: for all $k \in [\tilde{T}]$ let $\tilde{x}^{(k)} = (\tilde{x}_{1,k}, \ldots, \tilde{x}_{n,k})$ be a feasible solution of the relaxed problem with value $\tilde{T}$. To obtain the corresponding feasible solution of the reduced problem, for all $j \in [n]$ set

$$
\hat{x}_{j,l} = \begin{cases} \tilde{x}_{j,l}\alpha^{-1} & \text{if } l = 1, \ldots, \tilde{T}; \\ \sum_{k=1}^{\tilde{T}} \frac{\tilde{x}_{j,k}(\alpha-1)\alpha^{-1}}{\lceil(\alpha-1)\tilde{T}\rceil} & \text{if } l = \tilde{T}+1, \tilde{T}+2, \ldots, \lceil \alpha\tilde{T} \rceil. \end{cases} \tag{2}
$$

**Lemma 4.** *The vectors $\hat{x}^{(k)} = (\hat{x}_{1,k}, \ldots, \hat{x}_{n,k})$, $k = 1, \ldots, \lceil \alpha\tilde{T} \rceil$ form a feasible solution of the reduced problem.*

*Proof.* Observe that $\widehat{x}_{j,k} \in [0,1]$ $\forall j,k$ because we multiply the variables $\tilde{x}_{j,t} \in [0,1]$ by some positive number which is at most one. For each $j \in [n]$

$$\sum_{l=1}^{\lceil \alpha \tilde{T} \rceil} \widehat{x}_{j,l} = \sum_{l=1}^{\tilde{T}} \tilde{x}_{j,l}\alpha^{-1} + \sum_{l=\tilde{T}+1}^{\lceil \alpha \tilde{T} \rceil} \sum_{k=1}^{\tilde{T}} \frac{\tilde{x}_{j,k}(\alpha-1)\alpha^{-1}}{\lceil (\alpha-1)\tilde{T} \rceil}$$

$$= \alpha^{-1} + \sum_{k=1}^{\tilde{T}} \tilde{x}_{j,k}(\alpha-1)\alpha^{-1} = 1 .$$

Furthermore, other constraints are also satisfied because for $l = 1, \ldots, \tilde{T}$

$$\sum_{j=1}^{n} a_{i,j}\widehat{x}_{j,l} = \sum_{j=1}^{n} a_{i,j}\tilde{x}_{j,l}\alpha^{-1} \le b_i\alpha^{-1}$$

and for $l = \tilde{T}+1, \ldots, \lceil \alpha \tilde{T} \rceil$

$$\sum_{j=1}^{n} a_{i,j}\widehat{x}_{j,l} = \sum_{j=1}^{n} a_{i,j} \sum_{k=1}^{\tilde{T}} \frac{\tilde{x}_{j,k}(\alpha-1)\alpha^{-1}}{\lceil (\alpha-1)\tilde{T} \rceil}$$

$$\le \sum_{k=1}^{\tilde{T}} \frac{b_i(\alpha-1)\alpha^{-1}}{\lceil (\alpha-1)\tilde{T} \rceil}$$

$$\le b_i\alpha^{-1} .$$

$\square$

The optimal solution $(\widehat{x}_{j,k})$ of the reduced problem and the corresponding value $(\widehat{T})$ can be found by using binary search on the number of colors between 1 and $n$ and solving at most $\log n$ linear programs in polynomial time (see [7] and [9]). Lemma 4 reveals that $\widehat{T} \le \lceil \alpha T^* \rceil \le \lceil \alpha T_{opt} \rceil$. We will assume, w.l.o.g, that this optimum feasible solution is basic. The next step is to round this fractional solution to a feasible solution of our original integer program.

## 4   Rounding to Get Good Colorings

In this section we give a randomized algorithm to construct approximate, sub-optimal colorings. It is a generalization of the algorithmic version of Lovász local lemma [2] given by Lu [10] and Leighton et al [8] to minimax integer programs with *fixed* right hand side.

The randomized rounding procedure is quite apparent and simple. For each vertex $v_j$ independently round exactly one $\widehat{x}_{j,k}$ to one according to the probabilities $\widehat{x}_{j,1}, \ldots, \widehat{x}_{j,\widehat{T}}$. Define binary random variables $y_{j,k}$ for $j \in [n]$ and $k \in [\widehat{T}]$ such that $y_{j,k} = 1$ if vertex $v_j$ gets color $k$ and is zero otherwise. Let $(Ay^{(k)})_i$ denote the dot product of row $i$ of matrix $A$ with $y^{(k)} = (y_{1,k}, \ldots, y_{n,k})^t$, then $E[(Ay^{(k)})_i] \le b_i\alpha^{-1}$ by linearity of expectation.

Define $s\widehat{T}$ events $\xi_{i,k} \equiv$ "$\left(Ay^{(k)}\right)_i \geq b_i \alpha^{-1}(1 + \delta_i)$" where $\delta_i > 0 \; \forall i$. $\xi_{i,k}$ is the event that in edge $i$ of the hypergraph the number of vertices with color $k$ is at least $b_i \alpha^{-1}(1 + \delta_i)$. For a vertex $v_j$, $j \in [n]$, let $\rho_j$ be the number of variables to be rounded among $\widehat{x}_{j,1}, \ldots, \widehat{x}_{j,\widehat{T}}$, i.e, one among $\rho_j$ colors can be assigned to $v_j$. Let $\eta = \max_{j \in [n]} \rho_j$. For $i \in [s]$, $k \in [\widehat{T}]$ we define

$$\omega_{i,k} = |\{v_j \; : \; a_{i,j} \neq 0 \wedge \widehat{x}_{j,k} \in (0,1)\}| \, .$$

Let $\tau = \max_{i \in [s], \, k \in [\widehat{T}]} \omega_{i,k}$. So $\tau$ is the maximum number of variables to be rounded in any one of the inequality constraints $Ax^{(k)} \leq b_i \alpha^{-1}$. Intuitively, dependency among these events is due to the facts that $(i)$ each vertex $v$ with color $k$ will contribute to events $\xi_{i,k}$ corresponding to the edges containing it, and $(ii)$ $v$ will not contribute to events corresponding to other colors. But a closer look at the inequality constraints reveals more about the dependency structure of these events. Two events $\xi_{i,k}$ and $\xi_{i',k'}$ are mutually dependent if for a $j \in [n]$, $\widehat{x}_{j,k}, \widehat{x}_{j,k'} \in (0,1)$ and $a_{i,j}, a_{i',j} \neq 0$. Thus, maximum dependency among these events $\mathcal{D}$ is at most $d\eta\tau$, where $d$ is the maximum vertex degree of the hypergraph. We are now ready for

**Theorem 1.** *Given a CHC problem and the optimal solution $(\widehat{x}_{j,k})$ of the corresponding reduced problem,*

*(1) there exist vectors $\breve{y}^{(k)} = (\breve{y}_{1,k}, \ldots, \breve{y}_{n,k}) \in \{0,1\}^n$ such that $\left(A\breve{y}^{(k)}\right)_i < b_i \alpha^{-1}(1 + \delta_i)$ for all $i \in [s]$, $k \in [\widehat{T}]$, and*

*(2) the inequality constraints of the original integer program are satisfied if $b_i = \Omega\left(\frac{(1+\epsilon) \log \mathcal{D}}{\epsilon^2}\right) \; \forall i \in [s]$.*

*Proof.* (1)    By part (ii) of Lemma 2, if $b_i \geq \alpha \log(e\gamma(\mathcal{D}+1))/2$ then for

$$\delta_i = \Theta\left(\sqrt{\frac{\alpha \log(e\gamma(\mathcal{D}+1))}{b_i}}\right) \tag{3}$$

direct calculation shows

$$Pr[\xi_{i,k}] \leq G(b_i \alpha^{-1}, \delta_i) \leq p = \frac{1}{(e\gamma(\mathcal{D}+1))} \, ,$$

where $\gamma \geq 1$ is also a constant. Since $ep(\mathcal{D}+1) = \gamma^{-1} < 1$, the Lovász local lemma ensures the existence of vectors $\breve{y}^{(k)} = (\breve{y}_{1,k}, \ldots, \breve{y}_{n,k}) \in \{0,1\}^n$ such that for all $i \in [s]$, $k \in [\widehat{T}]$, no event $\xi_{i,k}$ occurs.
(2)    Since we want to satisfy the original inequality constraints, therefore, we want $b_i \alpha^{-1}(1 + \delta_i) \leq b_i \; \forall i$, which yields

$$(\alpha - 1)^2 = \Omega\left(\frac{\alpha \log(e\gamma(\mathcal{D}+1))}{b_i}\right) \, . \tag{4}$$

Thus, for any $\epsilon \in (0,1)$ and $\alpha = 1 + \epsilon$ we get

$$b_i = \Omega \left( \frac{(1+\epsilon) \log(e\gamma(\mathcal{D}+1))}{\epsilon^2} \right). \tag{5}$$

$\square$

Our objective is to use the algorithmic version of LLL to obtain good approximate colorings. For this purpose we will need the weaker condition $p\mathcal{D}^4 \leq 1$ rather than the usual $ep(\mathcal{D}+1) \leq 1$. Thus, we need $p = (e\gamma(\mathcal{D}+1))^{-4}$. Notice that this changes $\delta_i$ (or the lower bound on $b_i$'s) by just a constant factor. Since rounding does not affect the number of colors used, we end up using $\widehat{T} \leq \lceil (1+\epsilon)T_{opt} \rceil$ colors. This is a definite improvement over the lower bounds $b_i = \Omega((1+\epsilon)\epsilon^{-2} \log(sT^*))$ of [15] because $\mathcal{D} \leq d\eta\tau$ where $d \leq s$, $\eta \leq \widehat{T}$ and $\tau \leq s\widehat{T}$(here the assumption of Section 3, that the feasible solution is basic, is used). Next, we will construct an approximate coloring assuming the lower bound (5) on $b_i$'s efficiently, first in a randomized, then in a deterministic (derandomized) way. Note that Theorem 1 gives a probabilistic way of generating a good solution, but there the success probability is too small.

## 4.1   Randomized Construction

For each $i \in [s]$ and $k \in [\widehat{T}]$ define $n$ 0/1, independent random variables $z_{i,j,k} = a_{i,j}y_{j,k}$. The event $\xi_{i,k}$ can be re-written as "$\sum_{j=1}^{n} z_{i,j,k} \geq b_i\alpha^{-1}(1+\delta_i)$". Let $G(V,E)$ be the dependency graph with vertex set $V = [s\widehat{T}]$, each vertex corresponds to an event $\xi_{i,k}$ and two vertices are adjacent iff one affects the other. Let $G^{a_1,a_2}$ represent a graph with vertex set $V$ where two vertices are adjacent iff they are at a distance of $a_1$ or $a_2$ from each other in $G$. $\widehat{V} \subseteq V$ is called an $(a_1,a_2)$-tree if vertices in $\widehat{V}$ form a connected component in $G^{a_1,a_2}$. In the following discussion whenever we talk of graphs and trees it pertains to the dependency graph.

**Theorem 2.** *Given a constrained hypergraph coloring problem, for any $\epsilon \in (0,1)$ a coloring with at most $\lceil (1+\epsilon)T_{opt} \rceil$ colors can be found in randomized polynomial time provided that $b_i = \Omega \left( \frac{(1+\epsilon) \log \mathcal{D}}{\epsilon^2} \right)$ for all $i \in [s]$.*

*Proof.* Call a vertex $v_{i,k} \in V$ bad if

$$\sum_{j=1}^{n} z_{i,j,k} \geq \frac{b_i}{\alpha} \left( 1 + H \left( \frac{b_i}{\alpha}, \frac{1}{6\mathcal{D}^4} \right) \right). \tag{6}$$

Thus a vertex $v_{i,k} \in V$ is bad with probability at most $\frac{1}{6\mathcal{D}^4}$ where $\mathcal{D} \leq d\eta\tau$ is the maximum dependency. Let $\mathcal{T}$ be a $(1,2)$-tree and consider a vertex $v_{i,k} \in V$ and let

$$I_{i,k,\mathcal{T}} = \{j \in [n] \mid \exists v_{\hat{i},\hat{k}} \in \mathcal{T}\backslash\{v_{i,k}\}, \text{ s.t. } a_{i,j}, a_{\hat{i},j} \neq 0 \text{ and } \widehat{x}_{j,k}, \widehat{x}_{j,\hat{k}} \notin \{0,1\}\}.$$

$I_{i,k,\mathcal{T}}$ is the set of indices of those vertices of the hypergraph which when re-colored, effect the event $\xi_{i,k}$. Let $\widetilde{I}_{i,k,\mathcal{T}} = [n]\backslash I_{i,k,\mathcal{T}}$. We call a vertex $v_{i,k} \in V$ bad for $\mathcal{T}$ if

$$\sum_{j \in \widetilde{I}_{i,k,\mathcal{T}}} z_{i,j,k} \geq \mathbb{E}[\sum_{j \in \widetilde{I}_{i,k,\mathcal{T}}} z_{i,j,k}] + \frac{b_i}{\alpha} H\left(\frac{b_i}{\alpha}, \frac{1}{6\mathcal{D}^4}\right). \qquad (7)$$

This also happens with probability at most $\frac{1}{6\mathcal{D}^4}$. Notice that the probability bounds on these events hold because we assume that the $b_i$'s satisfy the required lower bounds. We say that a $(1,2)$-tree $\mathcal{T}$ is bad if every vertex in $\mathcal{T}$ is bad or bad for $\mathcal{T}$.

To round the fractional solution properly we require at most three phases. Phase 1 requires the following lemma.

**Lemma 5.** *With probability at least $1 - \frac{1}{s\widehat{T}}$ all bad $(1,2)$-trees have size at most $2\mathcal{D}\log(s\widehat{T})/\log \mathcal{D}$.*

*Proof (Lemma 5).* Consider a $(1,2)$-tree of size $\mathcal{D}r$ where $r = 2\log(s\widehat{T})/\log \mathcal{D}$. By removing unnecessary vertices (and the corresponding edges) at a distance one from each vertex in the tree we can obtain a $(2,3)$-tree from this $(1,2)$-tree. But we can remove at most $\mathcal{D}$ neighbours from each vertex. Thus, we obtain a $(2,3)$-tree of size at least $r$ from this $(1,2)$-tree. Since the vertices are not adjacent in a $(2,3)$-tree, a $(1,2)$-tree of size $\mathcal{D}r$ is bad with probability at most $\frac{1}{(3\mathcal{D}^4)^r}$. The number of $(1,2)$-trees of size $\mathcal{D}r$ is at most $\frac{s\widehat{T}}{(\mathcal{D}^2-1)\mathcal{D}r+1}\binom{\mathcal{D}^3 r}{r} < s\widehat{T}(3\mathcal{D}^3)^r$. So the probability of obtaining a bad $(1,2)$-tree of size $\mathcal{D}r$, after rounding, is at most $s\widehat{T}(3\mathcal{D}^3)^r \frac{1}{(3\mathcal{D}^4)^r} \leq \frac{s\widehat{T}}{\mathcal{D}^r} = \frac{1}{s\widehat{T}}$. □

Thus randomized rounding produces no large bad $(1,2)$-trees. We continue the proof of Theorem 2.

In Phase 2 we take a maximal bad $(1,2)$-tree $\mathcal{T}$ and recolor all vertices of the hypergraph contained in $\mathcal{T}$. But this may harm the neighbours of $\mathcal{T}$ which are not bad. Let $N(\mathcal{T})$ be the set of neighbours of $\mathcal{T}$. The following simple but crucial observation is the key to Phase 2: No vertex $v_{i,k}$ of $G$ is adjacent to two vertices belonging to different maximal bad $(1,2)$-trees and no vertex in $N(\mathcal{T})$ can harm any other bad $(1,2)$-tree except $\mathcal{T}$. This means that we can deal with each bad $(1,2)$-tree independently.

Consider $v_{i,k} \in N(\mathcal{T})$, since $v_{i,k}$ is not bad and is not bad for $\mathcal{T}$, therefore

$$\sum_{j \in \widetilde{I}_{i,k,\mathcal{T}}} z_{i,j,k} < \mathbb{E}[\sum_{j \in \widetilde{I}_{i,k,\mathcal{T}}} z_{i,j,k}] + \frac{b_i}{\alpha} H\left(\frac{b_i}{\alpha}, \frac{1}{6\mathcal{D}^4}\right).$$

After recoloring, let a vertex $v_{i,k} \in \mathcal{T} \cup N(\mathcal{T})$ be bad if

$$\sum_{j=1}^{n} z_{i,j,k} \geq \frac{b_i}{\alpha}\left(1 + 2H\left(\frac{b_i}{\alpha}, \frac{1}{6\mathcal{D}^4}\right)\right). \qquad (8)$$

If $v_{i,k} \in \mathcal{T}$ then this happens with probability at most $\frac{1}{6\mathcal{D}^4}$ and if $v_{i,k} \in N(\mathcal{T})$ then

$$\sum_{j \in I_{i,k,\mathcal{T}}} z_{i,j,k} \geq \mathbb{E}[\sum_{j \in I_{i,k,\mathcal{T}}} z_{i,j,k}] + \frac{b_i}{\alpha} H\left(\frac{b_i}{\alpha}, \frac{1}{6\mathcal{D}^4}\right) \tag{9}$$

because of (8) and this also happens with probability at most $\frac{1}{6\mathcal{D}^4}$. Now there are $\frac{1}{\log \mathcal{D}}(2\mathcal{D}\log(s\widehat{T}) + 2\mathcal{D}^2 \log(s\widehat{T}))$ vertices in $\mathcal{T} \cup N(\mathcal{T})$ and if the dependency $\mathcal{D} \geq \sqrt{\log s\widehat{T} / \log\log s\widehat{T}}$ then the probability of having a bad vertex is at most $2\mathcal{D}(\mathcal{D}+1)\frac{\log s\widehat{T}}{\log \mathcal{D}}\frac{1}{6\mathcal{D}^4} < 1$. Now, we can use the method of conditional probabilities (Section 4.2) to find a good recoloring in deterministic polynomial time. Otherwise, since $\mathcal{D} < \sqrt{\log s\widehat{T} / \log\log s\widehat{T}}$ we can apply Phase 1 to all bad $(1,2)$-trees to obtain smaller bad $(1,2)$-trees of size $O(\sqrt{\log s\widehat{T} \log\log s\widehat{T}} / \log \mathcal{D})$.

In Phase 3 the LLL guarantees that we can still re-round the remaining $\widehat{x}_{j,k}$'s to get a good coloring. Each bad vertex corresponds to a row of the inequality constraints and the number of variables to be rounded in each such constraint is at most $\tau$. Now, since $\tau \leq \mathcal{D} < \sqrt{\log s\widehat{T} / \log\log s\widehat{T}}$ and the size of each bad $(1,2)$-tree is $O(\sqrt{\log s\widehat{T} \log\log s\widehat{T}} / \log \mathcal{D})$, the number of variables to be rounded in each bad $(1,2)$-tree is $O(\log s\widehat{T} / \log \mathcal{D})$. This is small enough for us and we can now try all possible roundings in polynomial time to find a suitable $\lceil (1+\epsilon)T_{opt} \rceil$-coloring. $\qquad\square$

This algorithm can be derandomized using the pessimistic estimator construction of [10]. The key to derandomization is to round the fractional solution deterministically, in polynomial time, such that no $(1,2)$-tree of size $2\mathcal{D}\log(s\widehat{T}) / \log \mathcal{D}$ becomes bad. Next we describe the derandomized version of Lemma 5.

## 4.2   Derandomization

Let $\mathcal{Q}$ be the set of all $(1,2)$-trees of size $2\mathcal{D}\log(s\widehat{T}) / \log \mathcal{D}$. The idea is to round the variables in such a way that no $(1,2)$-tree $\mathcal{T} \in \mathcal{Q}$ becomes bad. We denote $\widehat{x}_j$ to be the vector $(\widehat{x}_{j,1}, \widehat{x}_{j,2}, \ldots, \widehat{x}_{j,\widehat{T}})$ and $y_j$ to be $(y_{j,1}, y_{j,2}, \ldots, y_{j,\widehat{T}})$ for any $j \in [n]$. Recall that rounding the variables is nothing but assigning colors to the vertices of the hypergraph. For all $j \in [n]$ we choose $y_j$ so as to minimize the probability of some $(1,2)$-tree $\mathcal{T} \in \mathcal{Q}$ turning bad if we randomly round $\widehat{x}_{j+1}, \ldots, \widehat{x}_n$ conditional on the already fixed $y_1, \ldots, y_{j-1}$. Let $\mathcal{T}_{2,3}$ be an arbitrary maximal $(2,3)$-tree in $\mathcal{T}$ and suppose that we have already fixed $y_1, \ldots, y_{j-1}$ then this probability i.e

$$Pr_{j-1} = Pr_{y_j,\ldots,y_n}[(\exists \mathcal{T} \in \mathcal{Q}) \wedge (\mathcal{T} \text{ is bad}) \mid y_1, \ldots, y_{j-1}]$$
$$\leq \sum_{\mathcal{T} \in \mathcal{Q}} \prod_{v_{i,k} \in \mathcal{T}_{2,3}} [(v_{i,k} \text{ is bad}) \vee (v_{i,k} \text{ is bad for } \mathcal{T}) \mid y_1, \ldots, y_{j-1}].$$

Let $e_{i,k,\mathcal{T}} = \mathbb{E}[\sum_{j \in \tilde{I}_{i,k,\mathcal{T}}} z_{i,j,k}] + b_i \alpha^{-1} H(b_i \alpha^{-1}, \frac{1}{6\mathcal{D}^4}),\, l = \lceil b_i \alpha^{-1} H(b_i \alpha^{-1}, \frac{1}{6\mathcal{D}^4}) \rceil$
and $g = b_i \alpha^{-1}(1 + H(b_i \alpha^{-1}, \frac{1}{6\mathcal{D}^4})$. Let $S_l^{(i,k)}(I)$ denote $S_l$ on input $\mathbb{E}[z_{i,j,k}]$ where $j \in I$ and $I \subseteq [n]$. Consider the following candidate $PE_{j-1}$ for the pessimistic estimator when $y_1, \ldots, y_{j-1}$ have been fixed:

$$PE_{j-1} = \sum_{\mathcal{T} \in \mathcal{Q}} \prod_{v_{i,k} \in \mathcal{T}_{2,3}} \left( \frac{S_l^{(i,k)}([n])}{\binom{g}{l}} + \frac{S_l^{(i,k)}(\tilde{I}_{i,k,\mathcal{T}})}{\binom{e_{i,k,\mathcal{T}}}{l}} \right).$$

By Lemma 3 it is clear that $Pr_{j-1} \leq PE_{j-1}$ so it is a pessimistic estimator. Observe that

$$PE_{j-1} = \sum_{\mathcal{T} \in \mathcal{Q}} \prod_{v_{i,k} \in \mathcal{T}_{2,3}} \mathbb{E}_{y_j} \left[ \frac{S_l^{(i,k)}([n])}{\binom{g}{l}} + \frac{S_l^{(i,k)}(\tilde{I}_{i,k,\mathcal{T}})}{\binom{e_{i,k,\mathcal{T}}}{l}} \right] \qquad (10)$$

$$= \sum_{\mathcal{T} \in \mathcal{Q}} \mathbb{E}_{y_j} \left[ \prod_{v_{i,k} \in \mathcal{T}_{2,3}} \left( \frac{S_l^{(i,k)}([n])}{\binom{g}{l}} + \frac{S_l^{(i,k)}(\tilde{I}_{i,k,\mathcal{T}})}{\binom{e_{i,k,\mathcal{T}}}{l}} \right) \right]$$

$$= \mathbb{E}_{y_j}[PE_j].$$

Thus, at each step we choose $y_j$ to minimize $PE_j$ and obtain

$$Pr_n \leq PE_n \leq PE_{n-1} \leq \cdots \leq PE_1 \leq PE_0 \leq \frac{1}{s\widehat{T}}.$$

Since $Pr_n$ can be either 0 or 1, we successfully find a rounding such that no $(1,2)$-tree of size $2\mathcal{D}\log(s\widehat{T})/\log \mathcal{D}$ is bad. The upper bound on $PE_0$ follows from Lemmas 3 and 5.

The number of $(1,2)$-trees in $\mathcal{Q}$ is $O(poly(s\widehat{T}))$ and they can be enumerated in polynomial time (see [8]). Further $S_l$ can be efficiently computed in polynomial time using dynamic programming approach. Thus $PE_j$s can be computed in polynomial time, giving us the deterministic version of Theorem 2 :

**Theorem 3.** *Given a constrained hypergraph coloring problem, for any $\epsilon \in (0,1)$ a coloring with at most $\lceil (1+\epsilon)T_{opt} \rceil$ colors can be found in polynomial time provided that $b_i = \Omega\left( \frac{(1+\epsilon)\log \mathcal{D}}{\epsilon^2} \right)$ for all $i \in [s]$ and $\mathcal{D} \leq d\eta\tau = O((sT^*)^2)$.*

## 5   Inapproximability

For $b_i = 1 \,\forall i$ the algorithm of de la Vega and Lueker [16] gives a $(1 + \epsilon)s$ approximation of $T_{opt}$. We show in the following that for $b_i = O(1) \,\forall i$, the approximation ratio cannot be independent of $s$. Let $G = (V, E)$ be a simple graph. The problem of properly coloring $G$ with minimum colors can be viewed as a CHC problem because $G$ can be viewed as a 2-uniform hypergraph with $|V|$ vertices and $|E|$ edges and a proper coloring can be obtained by putting $b_e = 1$ for all $e \in E$. Similarly it can also be viewed as a RCS problem with $|V|$ jobs and $|E|$ resources where each resource is required by exactly two jobs and exactly one unit of each resource is available. Feige and Kilian [4] showed that if $NP \nsubseteq ZPP$ then it is impossible to approximate the chromatic number of a $n$ vertex graph within a factor of $n^{1-\epsilon}$, for any fixed $\epsilon > 0$, in time polynomial in $n$. Therefore, the same applies for the CHC/RCS problem. Since $s \leq n^2$ in simple graphs, the following holds:

**Theorem 4.** *The CHC (resp. RCS) problem with $n$ vertices (resp. jobs) and $s$ edges (resp. resources) has no polynomial time approximation algorithm with approximation ratio at most $s^{\frac{1}{2}-\epsilon}$, for any fixed $\epsilon > 0$, unless $NP \subseteq ZPP$.*

# 6   Open Questions

On one hand we have a $(1 + \epsilon)s$ approximation of de la Vega and Lueker [16] when $b_i = 1$ $\forall i$ and on the other hand we have our $(1 + \epsilon)$ approximation for $b_i = \Omega((1 + \epsilon)\epsilon^{-2} \log \mathcal{D})$ (Theorem 3). But, we don't know the approximation quality for all values of $b_i$. So two interesting questions arise

1. How exactly does the approximation ratio behave when $b_i \in (1, \log \mathcal{D}]$ $\forall i$ and
2. Is it possible to get better approximation bounds for the number of colors without loosing much on $b_i$'s?

# References

1. N. Alon, *A parallel algorithmic version of the local lemma*, Random Structures and Algorithms, 2(1991), 367 - 378.
2. J. Beck, *An algorithmic approach to the Lovász Local Lemma*, Random Structures and Algorithms, 2(1991), 343 - 365.
3. P. Erdös and L. Lovász, *Problems and results on 3-chromatic hypergraphs and some related questions in infinite and finite sets*, A. Hajnal et al, eds, Colloq. Math. Soc. J. Bolyai 11, North Holland, Amsterdam, 1975, 609 - 627.
4. U. Feige and J. Kilian, *Zero knowledge and the chromatic number*, Journal of Computer and System Sciences, 57(1998), 187 - 199.
5. M.R. Garey, R.L. Graham, D.S. Johnson and A.C.-C. Yao, *Resource constrained scheduling as generalized bin packing*, JCT Ser. A, 21(1976), 257 - 298.
6. M.R. Garey and D.S. Johnson, *Computers and Intractability*, W.H. Freeman and Company, New York, 1979.
7. M. Grötschel, L. Lovász and A. Schrijver, *Geometric algorithms and combinatorial optimization*, Springer-Verlag, 1988.
8. T. Leighton, Chi-Jen Lu, S. Rao and A. Srinivasan, *New algorithmic aspects of the local lemma with applications to routing and partitioning*, to appear in SIAM Journal on Computing.

9. J. K. Lenstra, D.B. Shmoys and E. Tardos, *Approximation algorithms for scheduling unrelated parallel machines*, Math. Programming, 46(1990), 259 - 271.

10. Chi-Jen Lu, *Deterministic hypergraph coloring and its applications*, Proceedings of the 2nd International Workshop on Randomization and Approximation Techniques in Computer Science, 1998, 35 - 46.

11. M. Molloy and B. Reed, *Further algorithmic aspects of the Lovász Local Lemma*, Proc. 30th Annual ACM Symposium on Theory of Computing, 1998, 524 - 529.

12. P. Raghavan and C.D Thompson, *Randomized rounding : a technique for provably good algorithms and algorithmic proofs*, Combinatorica, 7(4)(1987), 365 - 374.

13. A. Srinivasan, *An extension of the Lovász Local Lemma and its applications to integer programming*, ACM-SIAM Symposium on Discrete Algorithms, 1996, 6 - 15.

14. A. Srivastav and P. Stangier, *Algorithmic Chernoff-Hoeffding inequalities in integer programming*, Random Structures and Algorithms, 8(1)(1996), 27 - 58.

15. A. Srivastav and P. Stangier, *Tight approximations for resource constrained scheduling and bin packing*, Discrete Applied Math, 79(1997), 223 - 245.

16. W.F. de la Vega and C.S Lueker, *Bin packing can be solved within* $(1 + \epsilon)$ *in linear time*, Combinatorica, 1(1981), 349 - 355.

# On the Power of Priority Algorithms for Facility Location and Set Cover

Spyros Angelopoulos and Allan Borodin

Department of Computer Science, University of Toronto,
Toronto, Ontario, Canada M5S 3G4,
{spyros,bor}@cs.toronto.edu

**Abstract.** We apply and extend the priority algorithm framework introduced by Borodin, Nielsen and Rackoff to define "greedy-like" algorithms for (uncapacitated) facility location and set cover. These problems have been the focus of extensive research from the point of view of approximation algorithms, and for both problems greedy algorithms have been proposed and analyzed. The priority algorithm definitions are general enough so as to capture a broad class of algorithms that can be characterized as "greedy-like" while still possible to derive non-trivial lower bounds on the approximability of the problems. Our results are orthogonal to complexity considerations, and hence apply to algorithms that are not necessarily polynomial-time.

## 1 Introduction

We follow the framework of Borodin, Nielsen and Rackoff [1] so as to to characterize "greedy and greedy-like algorithms" for the uncapacitated facility location problem and the set cover problem. These well studied and related NP-hard problems are central problems in the study of approximation algorithms. (See the Appendix for discussion of some of the relevant results.) The best known polynomial time computable approximation ratio (essentially $\ln n$, where $n$ is the size of the underlying universe) for the (weighted) set cover problem is achieved by a natural greedy algorithm [8] [9] [3] and quite good approximation ratios (namely, 1.61 [7] and 1.861 [10]) have been derived by greedy algorithms for the metric uncapacitated facility location problem. For these two optimization problems, we apply the "priority algorithm" framework in [1] and derive lower bounds on the approximation ratio for algorithms in this class. Informally, priority algorithms are characterized by two properties:

1. The algorithm chooses an ordering of "the inputs" and each input is considered in this order.
2. As each input is considered, the algorithm must make an "irrevocable decision" concerning this input.

To make these concepts precise, one has to apply the framework to particular problems (see section 2.2). The goal is to make the definitions sufficiently general

K. Jansen et al. (Eds.): APPROX 2002, LNCS 2462, pp. 26–39, 2002.

so as to capture known algorithms that could be classified as "greedy-like" while still being sufficiently restrictive that interesting lower bounds can be derived. As in the competitive analysis of online algorithms, priority algorithm lower bounds are orthogonal to complexity bounded approximation ratios. That is, although greedy algorithms tend to be time efficient, the definition of priority algorithms permits arbitrarily complex (in terms of computing time) computation while deriving lower bounds by exploiting the structure of the algorithm.

Two classes of priority algorithms can be defined:

- Algorithms in the class FIXED PRIORITY decide the ordering before any input is considered and this ordering does not change during the execution of the algorithm.
- Algorithms in the more general class ADAPTIVE PRIORITY are allowed to specify a new ordering after each input is processed. The new ordering can thus depend on inputs already considered. For adaptive priority algorithms, a further distinction can be made. The ordering and irrevocable decision in a *memoryless* adaptive priority algorithm depends only on the "current configuration"; that is, only on inputs already considered *and utilized* in the solution being constructed. For instance, in the context of the facility location problem, the ordering in any iteration of a memoryless adaptive priority algorithm depends only on the set of facilities that have been opened and not on any facilities that have been considered but not opened. The known greedy algorithms for facility location are indeed memoryless.

As in [1], "greedy algorithms" are (fixed or adaptive) priority algorithms which satisfy an additional property: the irrevocable decision is such that the objective function is *locally optimized*. More specifically, the objective function must be optimized as if the the input currently being considered is the last input.

Within this framework, we prove the following lower bounds:

1. The set cover problem over a universe of $n$ elements.
   (a) No adaptive priority algorithm can achieve a better approximation ratio than the precise bound $(\ln n - \ln \ln n + \Theta(1))$ obtained by Slavík [16] for *the* greedy algorithm (see the Appendix). This lower bound applies to the uniform set cover problem in which all set costs are the same. Note that Feige [4] showed that under the (reasonable) complexity assumption that NP is not contained in $DTIME(n^{O(\log \log n)})$, it is not possible to obtain a polynomial-time approximation of $(1 - \epsilon) \ln n$ for any $\epsilon > 0$.
   (b) For any $\epsilon > 0$, no fixed priority algorithm can achieve an approximation ratio of $(1 - \epsilon)n$.
   Since the set cover problem can be viewed as a special case of the facility location problem (with distances in $\{0, \infty\}$), the set cover lower bounds hold for facility location when arbitrary distances are allowed.
2. The metric uncapacitated facility location problem. The following lower bounds will apply to the unweighted case (i.e., each city has weight 1) and where all distances are in $\{1, 3\}$. The best known corresponding greedy upper bounds apply to the weighted case and for an arbitrary metric distance function.

(a) No adaptive priority algorithm can achieve an approximation ratio better than 4/3 for the uniform (i.e., all opening costs are identical) metric facility location problem.

(b) For the non uniform case, no memoryless or greedy adaptive priority algorithm can achieve an approximation ratio better than 1.463 for the metric facility location problem, matching the complexity-based bound (under the same assumption used by Feige [4]) of Guha and Khuller [5].

(c) For all $\epsilon > 0$, no fixed priority greedy algorithm can achieve an approximation ratio better than $3 - \epsilon$ for the uniform metric facility location problem, matching the 3-approximation upper bound of Mettu and Plaxton [12] which applies to the non uniform case.

## 2 Preliminaries

### 2.1 Problem Statements

In the *(uncapacitated, unweighted) facility location* problem, the input consists of a set $\mathcal{F}$ of facilities and a set $\mathcal{C}$ of cities with $\mathcal{F} \cap \mathcal{C} = \emptyset$ [1]. Each facility $i \in \mathcal{F}$ is associated with an *opening cost* $f_i$ which reflects the cost that must be paid to utilize the facility. Furthermore, for every facility $i \in \mathcal{F}$ and city $j \in \mathcal{C}$, the non-negative *distance* or *connection cost* $c_{ij}$ is the cost that must be paid to connect city $j$ to facility $i$. The objective is to open a subset of the facilities in $\mathcal{F}$ and connect each city in $\mathcal{C}$ to an open facility so that the total cost incurred, namely the sum of the opening costs and the connection costs, is minimized. In the *metric* version of the problem, the connection costs satisfy the triangle inequality.

In the *set cover* problem, we are given a universe $U$ of $n$ elements, and a collection $\mathcal{S}$ of subsets of $U$. Each set $S \in \mathcal{S}$ is associated with a cost $c(S)$. We seek a minimum-cost subcollection of $\mathcal{S}$ that covers all elements of $U$; that is, a collection of sets $\mathcal{S}' \subseteq \mathcal{S}$ of minimum total cost such that for every $e \in U$ there exists a set $S \in \mathcal{S}'$ with $e \in S$.

### 2.2 Definitions of Priority Algorithms

What precisely constitutes the input for the problems we study? For the uncapacitated facility location problem the cost of a solution is fully determined by the set of facilities the algorithm decides to open, since each city will be connected to the nearest open facility. Hence, we will assume that the input to the problem consists only of facilities where each facility is identified by a unique id, its distance to every city and its opening cost. When considering a facility (of highest priority), the algorithm must make an irrevocable decision as to whether or not to open this facility. The decision is irrevocable, in the sense that once a facility is opened, its opening cost will count towards the total cost that the

---

[1]  We assume that $\mathcal{F} \cap \mathcal{C} = \emptyset$ noting that if $x \in \mathcal{F} \cap \mathcal{C}$, then we can replace $x$ by $x_f \in \mathcal{F}$ and $x_c \in \mathcal{C}$ with zero connection cost between $x_f$ and $x_c$.

algorithm incurs and if a facility is not opened when considered, it cannot be opened at a later stage.

For the class of scheduling problems considered in [1], there is no issue as to what is an "input". But for the facility location problem (and the set cover problem), there is at least one other very natural way to view the inputs. Namely, as in Meyerson [13], we can think of the cities as being the inputs, with each city being identified by its id, and its distance to each facility. (We could treat the opening costs of all facilities as global information.) The irrevocable decision would then be to assign each input city to some open facility, possibly opening a new facility if desired[2]. Indeed, this model is very natural in the sense that one would expect the number of cities to be much larger than the number of facilities. We have chosen our model as it abstracts the known $O(1)$ approximation greedy algorithms[3] and since the objective function is completely determined by the choice of which facilities to open.

The input for the set cover problem consists of sets where each set is identified by its cost and by the elements it covers. We impose the constraint that the priority algorithm must irrevocably select any set it considers if the set covers a new element, that is, an element not covered by the sets already selected. This requirement is motivated by the observation that if the set that covers the new element is the only such set and the algorithm does not select it, the resulting solution will not be feasible. The well-known greedy algorithm for set cover [8] [9] observes the above requirement. It is worth mentioning that in this context every priority set cover algorithm belongs in the class of GREEDY algorithms[4].

### 2.3 Orderings and the Role of the Adversary

To show a lower bound on the approximation ratio we must evaluate the performance of every priority algorithm for an appropriately constructed nemesis input set. The construction of such a nemesis set can be seen as a game between an adversary and the algorithm and is conceptually similar to the construction of an adversarial input in competitive analysis. However, in the setting of priority algorithms, it is the algorithm and not the adversary that chooses the ordering of the inputs. But we do not want the algorithm to be able to chose an optimal ordering for each input set (e.g. chose the optimal set of facilities as those of highest priority). One possibility is to define the "allowable orderings" to be those that are induced by functions mapping the set of all inputs into the non negative reals. The nature of the adversary described below provides a more inclusive definition for what orderings are allowed.

---

[2] Meyerson assumes that every city is also a potential facility but this assumption is not important for our purpose in abstracting the class of algorithms under consideration.

[3] Meyerson's $O(1)$ bound applies to "randomized priority algorithms" where both the ordering of the inputs and the irrevocable decisions are determined by using randomization.

[4] In contrast, every solution for an instance of facility location in which at least one facility is open, is feasible. Thus, not every priority facility location algorithm is necessarily greedy.

We consider the basic framework where the priority algorithm has no additional global information (such as the total number of inputs [5] ,the sum of all opening costs, etc.) beyond the inputs themselves. In this setting, the game between the adversary and a FIXED PRIORITY algorithm can be described as follows. The adversary presents a large set of potential inputs $S$. The algorithm determines a total ordering on this set. The adversary then selects a subset $S'$ of $S$ as the actual input set; that is, the adversary discards part of the potential input set. We enforce the condition that the removal of inputs does not affect the priority of the remaining inputs. As an example of this, in our framework of a priority algorithm for facility location, a facility is not defined in terms of other facilities; therefore, when the adversary removes a facility, the priority of every other input is unaffected.

For the class ADAPTIVE PRIORITY, even though the algorithm can determine a new ordering after considering each input, the adversary can adaptively remove inputs from $S$ (in each iteration) so as to derive the actual input set $S'$.

# 3   Adaptive Priority Algorithms

In this section we show lower bounds on the approximability of set cover and facility location by adaptive priority algorithms. All our lower-bound constructions for metric facility location use connection costs in $\{1, 3\}$, suggesting the following definitions. Let $C_f$ be the set of cities at distance 1 from facility $f$. We say that $f$ covers $C_f$. The complement of $f$ with respect to $C$, where $C_f \subseteq C$, is defined as the facility that covers all and only the cities in $C \setminus C_f$. For simplicity, when $C$ is the set $\mathcal{C}$ of all cities, we say that $f$ and $\overline{f}$ are complementary.

## 3.1   Adaptive Priority Facility Location
with Uniform Opening Costs

We first consider the uniform case where the facilities have identical opening costs. (Throughout this paper we ignore floors and ceilings and assume that $n$ is appropriately divisible.) We note that Guha and Khuller [5] show a 1.278 approximation lower bound for the uniform case under the complexity assumption that NP is not contained in $DTIME(n^{O(\log \log n)})$.

**Theorem 1.** *No priority algorithm for the uniform metric facility location problem can achieve an approximation ratio better than $\frac{4}{3}$.*

*Proof.* Define an instance of the uniform metric facility location problem as follows. The set of cities $\mathcal{C}$ consists of $n$ cities. Each facility is identified with the set of cities it covers. More precisely, for every $C \subset \mathcal{C}$ with $|C| = \frac{n}{2}$, there exists a facility $f$ that covers $C$ (and only cities in $C$). For every city in $\mathcal{C} \setminus C$, the cost of connecting the city to $f$ is equal to 3. Note that for every facility $f$ in the instance, the complement $\overline{f}$ of $f$ is also in the instance.

---

[5] For the problems considered in this paper, it is not hard to show that we can extend our lower bounds to the case that the algorithm does know the number of inputs.

Every time the algorithm considers a facility $f$ it must decide whether to open it. If it opens $f$, the adversary removes all facilities which do not cover exactly $\tilde{n}/2$ uncovered cities, where $\tilde{n}$ is the number of currently uncovered cities. The adversary also removes $\overline{f}$ unless $\overline{f}$ is the last remaining facility in the input set. If the algorithm does not open $f$, the adversary removes all remaining facilities except for $\overline{f}$. Let all facility costs be equal to $\frac{n}{4}$. The optimal algorithm will open a pair of complementary facilities at an opening cost of $2 \cdot \frac{n}{4}$, and connect all cities at a cost of $n$, for a total cost of $\frac{3}{2}n$. Suppose that the algorithm opens $k$ facilities, with $k \geq 1$. It is easy to see that the union of these $k$ facilities covers $n \sum_{i=1}^{k} 2^{-i} = n(1 - 2^{-k})$ cities. Hence the algorithm pays a facility-opening cost of $k \cdot \frac{n}{4}$ and a connection cost of $n(1 - 2^{-k}) + 3n \cdot 2^{-k}$, and thus a total cost of $(1 + \frac{k}{4} + 2^{-k+1})n$. This expression is minimized at either $k = 2$ or $k = 3$, giving a $\frac{4}{3}$ ratio between the algorithm's cost and the optimal cost.                                     $\square$

Using an argument along the same lines one can prove the following:

**Theorem 2.** *The approximation ratio of every priority algorithm for facility location in arbitrary spaces is $\Omega(\log n)$.*

*Proof.* We consider the same set of cities $C_f$ covered by a facility $f$ as in the proof of Theorem 1, except now any city not covered by $f$ (i.e. not at distance 1) has $\infty$ distance from $f$. In this case, any priority algorithm must open every facility it considers. Hence it will consider and open $\log n$ facilities while the optimal algorithm will again open two facilities.                                     $\square$

Note that the bound is tight, since Hochbaum [6] showed that a simple greedy algorithm, which can be classified as ADAPTIVE PRIORITY, is an $O(\log n)$ approximation for the problem.

## 3.2   Adaptive Priority Set Cover

In this section we prove that no (adaptive) priority algorithm for set cover can perform better than *the* greedy set-cover algorithm.

A remarkably tight analysis of the greedy set cover algorithm due to Slavík [16] shows that its approximation ratio is exactly $\ln n - \ln \ln n \pm \Theta(1)$, where $n = |U|$ is the size of the universe. For the proof of the lower (as well as the upper) bound, Slavík considered an instance of set cover with sets of uniform costs on a universe $U$ of $n \geq N(k, l)$ elements. Here, $N(k, l)$ is defined as the smallest size of $U$ such that there is a set cover instance over $U$ with the property that the cost of the greedy algorithm is exactly $k$, while the cost of the optimal algorithm is exactly $l$, for given $k, l$ with $k \geq l$. The collection of input sets for this specific instance consists of two subcollections, each covering all $n$ elements of $U$. The first subcollection, denoted by $\mathcal{S}_1$, contains $n_1, n_2, \ldots, n_r$ disjoint sets of sizes $s_1, s_2, \ldots, s_r$, respectively, so that $\sum_{i=1}^{r} n_i s_i = n$. The numbers $n_i$'s and sizes $s_i$'s of the sets are selected in [16] appropriately, but for the purposes of our discussion it suffices to mention the following: i) $n_i > 0$, for all $i$ with $1 \leq i \leq r$, and $r$ is a postitive integer; ii) $\sum_{i=1}^{r} n_i = k$; iii) $s_1 > s_2 > \ldots > s_r > 0$; and iv) $s_1 = \lceil \frac{n}{l} \rceil$. The second subcollection, denoted by $\mathcal{S}_2$, consists of $l$ sets of sizes in $\{\lceil \frac{n}{l} \rceil, \lceil \frac{n}{l} \rceil - 1\}$. Slavík argued that the greedy algorithm must select all sets in $\mathcal{S}_1$, at a total cost of $\sum_{i=1}^{r} n_i$. In addition, the greedy algorithm selects the sets in $\mathcal{S}_1$ in non-increasing order of sizes, breaking ties arbitrarily. On the other hand, the cost of the optimal algorithm is the total cost of the subcollection $\mathcal{S}_2$, namely, $l$. Slavík showed that for every $n$ sufficiently large, one can find $k, l$ such that $N(k, l) \leq n < N(k+1, l)$, and $\frac{k}{l} \geq \ln n - \ln \ln n + \Theta(1)$. We will show that, for every such $n$ (and then $k$ and $l$ chosen as in Slavík's analysis), there exists an instance of set cover with $|U| = n$ for which the cost of every priority algorithm is at least $k = \sum_{i=1}^{r} n_i$, while the cost of the optimal algorithm is at most $l$. Then Slavík's analysis carries over to yield the same lower bound on the approximation ratio for every priority algorithm.

Consider an instance of set cover which consists of a universe $U$ of size $n$ and all $\binom{n}{d}$ sets of size $d$, where $d = s_1 = \lceil \frac{n}{l} \rceil$. All sets have identical costs. Given an adaptive priority algorithm for the problem, we describe the actions of the adversary, which takes place in $r$ phases. As shown in Lemma 1, it is feasible to define such an adversary. Phase 1 begins with the first set selected by the algorithm; it terminates when the algorithm has selected exactly $n_1$ sets that cover at least one new element each. If the number $c_1$ of elements covered by the algorithm in phase 1 is less than $n_1 s_1$, the adversary chooses any $n_1 s_1 - c_1$ uncovered elements, which together with the $c_1$ covered elements form a set of elements denoted by $C_1$. Without loss of generality, we will consider all elements in $C_1$ as being covered. The adversary removes all sets from the input except for sets that contain at least $d - s_2$ elements of $C_1$, namely a set $S$ remains in the input only if $|S \cap C_1| \geq d - s_2$ (recall that $d = s_1 > s_2$). Phase 2 then begins. Likewise, phase $i$, with $1 \leq i \leq r-1$ terminates when the algorithm has selected exactly $n_i$ sets that cover at least one uncovered element each. If the number $c_i$ of elements covered in phase $i$ is less than $n_i s_i$, the adversary chooses any of the remaining $n_i s_i - c_i$ elements, which together with the $c_i$ covered elements are the contents of a set denoted by $C_i$. (We again consider all elements in $C_i$ as being covered.) The adversary then removes all input sets except for sets that contain at least $d - s_{i+1}$ elements of $\bigcup_{j=1}^{i} C_i$, namely, a set $S$ remains in the input only if $|S \cap \bigcup_{j=1}^{i} C_i| \geq d - s_{i+1}$. The adversary does not remove any sets at the end of (the last) phase $r$. All sets not explicitly removed by the adversary comprise the actual input that is presented to the algorithm.

We need to show that it is feasible to define an adversary as above. Denote by $R_i$ the collection of sets that remain in the input at the beginning of phase $i$, with $1 \leq i \leq r$. Note that every set in $R_i$ can cover at most $s_i$ new (so far uncovered) elements.

**Lemma 1.** *At the beginning of phase $i$, at least $n_i s_i$ elements are uncovered, and the sets in $R_i$ can cover these elements.*

*Proof.* By induction on $i$. Recall that $\sum_{j=1}^{r} n_j s_j = n$. Clearly, at the beginning of phase 1 there are $n \geq n_1 s_1$ uncovered elements, and $R_1$ consists of all $\binom{n}{d}$ sets, hence the $n_1 s_1$ uncovered elements can be covered by sets in $R_1$. Consider the beginning of phase $i$, with $1 < i \leq r$. In each phase $j$, with $1 \leq j < i$, the algorithm covered $|C_j| = n_j s_j$ elements (by the definition of the adversary and the induction hypothesis). Therefore, at the beginning of phase $i$, $n - \sum_{j=1}^{i-1} n_j s_j = \sum_{j=i}^{r} n_j s_j \geq n_i s_i$ elements are uncovered. Let $P$ be a set that contains exactly $n_i s_i$ uncovered elements. Let $P_1, P_2, \ldots, P_{n_i}$ be any disjoint partition of $P$ in sets of size $s_i$. We claim that for any $m$ with $1 \leq m \leq n_i$ there exists at least one set in $R_i$ that covers $P_m$. Define the set $S$ as the set that contains exactly $s_h - s_{h+1}$ elements of $C_h$, for every $h$ with $1 \leq h < i$ and also contains $P_m$. Since the $C_h$'s with $h < i$ and $P_m$ are all disjoint, the set $S$ has size

$$\sum_{h=1}^{i-1} (s_h - s_{h+1}) + s_i = (s_1 - s_i) + s_i = s_1 = d$$

hence $S$ belongs in $R_1$. It remains to show that $S$ is in $R_i$. To this end, note that for all $j < i$,

$$\left| S \cap \bigcup_{h=1}^{j} C_h \right| = \sum_{h=1}^{j} (s_h - s_{h+1}) = s_1 - s_{j+1},$$

therefore $S$ is in $R_{j+1}$, for all $j < i$ (and in particular, for $j = i - 1$). Thus the adversary will not remove $S$ before the end of phase $i$.                    □

Denote by $cost(ALG)$, $cost(OPT)$ the cost of the priority algorithm and the cost of the optimal algorithm, respectively. We then have[6]:

**Lemma 2.** $cost(ALG) \geq \sum_{i=1}^{r} n_i = k$.

The following combinatorial lemma will be useful in upper-bounding the optimal cost.

**Lemma 3.** *Suppose that there exists a collection of (not necessarily disjoint) sets $P_1, P_2, \ldots, P_l \subseteq U$, each of size $d$, with the following properties:[7]*

- *$P_1 \ldots P_l$ cover all elements of $U$, namely $\bigcup_{h=1}^{l} P_h = U$.*
- *There exist disjoint sets $G_1, G_2, \ldots, G_r \subseteq U$, with $|G_j| = n_j s_j$, such that $|P_h \cap \bigcup_{j=1}^{i} G_j| \geq d - s_{i+1}$, for all $i \leq r - 1$ and all $h$ with $1 \leq h \leq l$.*

*Then $cost(OPT) \leq l$.*

We can now show that the cost of the optimal algorithm is upper-bounded by $l$.

**Lemma 4.** $cost(OPT) \leq l$.

*Proof.* It suffices to show that the conditions of Lemma 7 are satisfied. Consider an instance of set cover, with a universe $U$ of size $n$ and the two subcollections of sets $\mathcal{S}_1$ and $\mathcal{S}_2$, as defined earlier in the section. Let $G_i$ be the set of elements contained in the $n_i$ sets of size $s_i$ in $\mathcal{S}_1$, for $i \leq r$. Clearly, all $G_i$'s are disjoint.

---

[6] Because of space limitations we omit some proofs in this version of the paper.

[7] We emphasize that the $P_i$'s do not necessarily belong to a particular input instance of set cover. However, after applying a suitable bijection of $U$, these sets will become part of the input created by the adversary.

We will assume that all sets in $\mathcal{S}_2$ are of size $d$. This is not a restrictive assumption, as will become evident later. Let $\mathcal{S}_2$ consist of sets $P_1, \ldots, P_l$. We claim that $|P_h \cap \bigcup_{j=1}^i G_j| \geq d - s_{i+1}$, for all $i \leq r - 1$ and $h \leq l$. To this end, we look at the choices of the greedy algorithm on the set cover instance that consists of sets in $\mathcal{S}_1 \cup \mathcal{S}_2$ over $U$. As argued in [16] the greedy algorithm selects only sets in $\mathcal{S}_1$ in non-increasing order of size, breaking ties arbitrarily. Therefore, the greedy algorithm covers all elements in $G_i$ prior to covering any element of $G_{i'}$, for all $i < i' \leq r$. Let the $i$-th phase of the greedy algorithm correspond to the selection of the $n_i$ sets (each of size $s_i$) that cover the elements in $G_i$. Assume, by way of contradiction, that for one of the sets in $\mathcal{S}_2$, say $P_h$, $|P_h \cap \bigcup_{j=1}^i G_j| < d - s_{i+1}$. This implies that at the beginning of phase $i + 1$ of the greedy algorithm, $P_h$ covers at least $d - (d - s_{i+1}) + 1 = s_{i+1} + 1$ new elements. Since every set in $\mathcal{S}_1$ not yet selected by the greedy algorithm at the beginning of phase $i + 1$ covers exactly $s_{i+1}$ new elements, the greedy algorithm must select a set from $\mathcal{S}_2$, a contradiction.

We now show how to waive the assumption that all sets in $\mathcal{S}_2$ have size $d$. We append one element from some set (of size $d$) that the greedy algorithm selects in its first phase to every set in $\mathcal{S}_2$ of size $d - 1$. It is easy to see that the previous argument goes through, and all sets in $\mathcal{S}_2$ have size precisely $d$.

Using the analysis in [16], it follows that for $n$ sufficiently large, $cost(OPT) = l$. □

From Lemma 2 and Lemma 4, the approximation ratio of every priority algorithm for set cover is at least $\frac{\sum_{i=1}^r n_i}{l} = \frac{k}{l}$. From Slavík's analysis, this is precisely the approximation ratio of the greedy set cover algorithm which shows:

**Theorem 3.** *No priority algorithm for set cover performs better than the greedy algorithm. The approximation ratio for priority set cover is thus* $\ln n - \ln \ln n + \Theta(1)$ *and this bound is tight.*

## 3.3   Adaptive Priority Metric Facility Location

The result of Theorem 3 has important implications for the approximability of metric facility location by priority algorithms. Guha and Khuller [5] established a polynomial time reduction from set cover to metric facility location. Using the reduction, Guha and Khuller argued that if there exists a polynomial-time algorithm for metric facility location with approximation ratio better than 1.463, then set cover can be approximated (by some polynomial-time algorithm) with approximation ratio $(1 - \epsilon) \ln n$ for $\epsilon > 0$, where $n$ is the size of the universe. By Feige's inapproximability result for set cover [4], this implies that $NP \subseteq DTIME(n^{O(\log \log n)})$. It is worth noting that the reduction introduces different facility opening costs, even if the set cover instance contains sets of identical cost. In fact, the Guha-Khuller reduction can be used to show that one can derive a priority algorithm for set cover from any memoryless or greedy priority algorithm for metric facility location. In view of this observation, we can interpret the argument of Guha and Khuller as follows: If there exists such a priority algorithm for metric facility location with approximation ratio better than 1.463, then set cover is $(1-\epsilon)\ln n$-approximable by a priority algorithm for some $\epsilon > 0$, contradicting Theorem 3. We thus established the following (a more detailed discussion concerning priority algorithms and the Guha-Khuller reduction will be provided in the full paper):

**Theorem 4.** *Metric facility location cannot be approximated within a factor smaller than 1.463 by any memoryless or greedy priority algorithm.*

## 4   Fixed Priority Algorithms

In this section we present a tight lower bound on the approximation ratio of algorithms in the class FIXED PRIORITY, GREEDY for facility location (for both metric and arbitrary spaces). We also derive a tight lower bound for FIXED PRIORITY set cover.

Consider the following instance $(\mathcal{F}, \mathcal{C})$ of the metric facility location problem. Let $C_1, C_2, \ldots, C_d$ be a partition of the $n$ cities into $d$ (disjoint) sets of size $\frac{n}{d}$ each, for some large constant $d$. For every pair $(k, l)$, with $1 \leq k, l \leq d$, and $k \neq l$ we identify the sets of facilities $F_{k,l}$ and $\tilde{F}_{k,l}$ as follows. Denote by $f_{ij}^{k,l}$ the facility that covers every city in $C_k$, except for city $i \in C_k$, and also covers city $j \in C_l$. In addition, denote by $\tilde{f}_{ij}^{k,l}$ the facility that covers every city in $C_l$ with the exception of city $j \in C_l$, and also covers city $i \in C_k$. $F_{k,l}$ and $\tilde{F}_{k,l}$ are defined as the sets $\{f_{ij}^{k,l} \mid i \in C_k, j \in C_l\}$ and $\{\tilde{f}_{ij}^{k,l} \mid i \in C_k, j \in C_l\}$, respectively. Note that by definition, the complement wrt $C_k \cup C_l$ of a facility in $F_{k,l}$ is in $\tilde{F}_{k,l}$ (and vice versa). The cost for connecting every city in $\mathcal{C}$ not covered by a facility in $F_{k,l}$ (respectively, $\tilde{F}_{k,l}$) to a facility in $F_{k,l}$ (respectively, $\tilde{F}_{k,l}$) is set to 3. The set of facilities $\mathcal{F}$ in the instance is defined as the union of all $F_{k,l}$ and $\tilde{F}_{k,l}$, for all pairs $k, l$, with $k \neq l$. Every facility is assigned an opening cost of $2 - \epsilon$. We emphasize that all connection costs are in $\{1, 3\}$ and that the facilities have uniform opening costs.

Let $\sigma$ be the ordered sequence of facilities in $\mathcal{F}$ produced by a fixed priority algorithm.

**Lemma 5.** *For every pair $k, l$, with $1 \leq k, l \leq d$ and $k \neq l$, there exists a set $S_{k,l}$ of $\frac{n}{d}$ pairs of facilities $(f_1, \overline{f}_1), \ldots, (f_{\frac{n}{d}}, \overline{f}_{\frac{n}{d}}) \in F_{k,l} \cup \tilde{F}_{k,l}$ such that: For every $m$ with $1 \leq m \leq \frac{n}{d}$, the facilities $f_m$ and $\overline{f}_m$ are complementary wrt $C_k \cup C_l$, $f_m$ precedes $\overline{f}_m$ in $\sigma$, and at least one of the following holds:*

1. *Each $f_m$ covers a city in $C_k$ which is not covered by any facility of the form $f_{m'} \in S_{k,l}$ with $m' \neq m$; or*
2. *Each $f_m$ covers a city in $C_l$ which is not covered by any facility of the form $f_{m'} \in S_{k,l}$ with $m' \neq m$.*

Note that for every fixed pair $k, l$, either case (1) or case (2) of Lemma 5 (or possibly both) apply. For the purposes of our proof we will assume, without loss of generality, that only one of the cases applies, for every fixed pair $k, l$. We use the notation $C_l \to C_k$ (resp. $C_k \to C_l$) to denote that case (1) (resp. case (2)) applies.

Define a digraph $G = (V, E)$ on $d$ vertices $v_1, \ldots, v_d$ as follows: the directed edge $(v_k, v_l)$, with $k \neq l$ is in $E$ iff $C_k \to C_l$. From Lemma 5, any two vertices in $G$ are adjacent to one common directed edge, and thus $G$ has exactly $\binom{d}{2}$ edges.

**Lemma 6.** *Let $G$ be defined as above. There exists a set $V' \subseteq V$ of size at most $\log d$ that dominates $V \setminus V'$. Namely, for every $v_l \in V \setminus V'$, there exists $v_k \in V'$ such that $(v_k, v_l) \in E$.*

Lemma 6 implies that there exists a collection $D$ of at most $\log d$ partitions $C_{k_1}, \ldots, C_{k_p}$ such that for every partition $C_l \notin D$, there exists $C_{k_i} \in D$ such that $C_{k_i} \to C_l$. We now describe how the adversary constructs the input that is presented to the algorithm. Recall from Lemma 5 that $S_{k_i, l}$ is the set of facilities that enforce $C_{k_i} \to C_l$. For fixed $l$, with $C_l \notin D$, denote by $I_l$ the set $S_{k_i, l}$ with the property that $i$ is the minimum among all $j$'s for which $C_{k_j} \to C_l$, and $C_{k_j} \in D$. Let $I = \bigcup \{I_l \mid C_l \notin D\}$. The input to the algorithm contains only facilities in $I$. In addition, for every $f \in I$ which belongs in $S_{k_i, l}$, the adversary removes $f$'s complement wrt $C_{k_i} \cup C_l$, except for the case $f$ and its complement are the last pair of complementary facilities wrt $C_{k_i} \cup C_l$. Note that from Lemma 5, $f$'s complement wrt $C_{k_i} \cup C_l$ follows $f$ in $\sigma$, so the removal of the facilities by the adversary is feasible. All facilities $I' \subset I$ which are not explicitly removed by the adversary comprise the actual input.

It remains to bound the cost of the priority algorithm and the optimal algorithm. From the construction of $I'$ and Lemma 5, it follows that the first facility that can cover a city $j \in C \setminus D$ covers only $j$ among the cities in $C \setminus D$. The greedy criterion dictates that when the algorithm considers the facility in question, it will open it: the algorithm will pay a total of $3 - \epsilon$ for opening the facility and connecting $j$ to the open facility, which improves upon the cost of $3$ that must be paid if the algorithm does not open the facility. Since there are $\frac{n}{d}(d - |D|) \geq \frac{n}{d}(d - \log d)$ cities in $C \setminus D$ we get

$$cost(ALG) = (2 - \epsilon) \cdot \frac{n}{d} \cdot (d - |D|) + n \geq (2 - \epsilon) \cdot \frac{n}{d} \cdot (d - \log d) + n.$$

The optimal algorithm, on the other hand, opens only pairs of facilities that are complementary with respect to partitions of cities. The open facilities cover all cities in the instance, and hence the total optimal cost is

$$cost(OPT) \leq 2 \cdot (2 - \epsilon) \cdot (d - |D|) + n \leq 2 \cdot (2 - \epsilon) \cdot d + n.$$

Observe that the ratio $\frac{cost(ALG)}{cost(OPT)}$ can be made arbitrarily close to 3, for large, albeit constant $d$. We thus showed the following:

**Theorem 5.** *The approximation ratio of every* FIXED PRIORITY, GREEDY *algorithm for metric facility location is at least $3 - \epsilon$, for arbitrarily small $\epsilon$.*

Using an argument along the same lines, one can derive the following.

**Theorem 6.** *The approximation ratio of every* FIXED PRIORITY, GREEDY *algorithm for facility location in arbitrary spaces (resp.* FIXED PRIORITY *algorithm for set cover) is at least* $(1 - \epsilon)n$, *where $n$ is the number of cities (resp. the size of the universe) in the input instance.*

## 5 Future Directions and Open Problems

Several interesting issues and open problems are left to investigate. A natural direction is to improve the lower bound for adaptive-priority metric facility location, if this is indeed possible. Can the memoryless assumption be removed in the 1.463 lower bound? In addition, although the $3 - \epsilon$ bound we showed is tight for the class FIXED PRIORITY GREEDY, we do not know whether a bound better than 1.463 can be obtained for FIXED PRIORITY (not necessarily greedy) metric facility location algorithms. Our lower-bound constructions use a $\{1, 3\}$ cost metric, and all instances are unweighted. Can we improve the results by considering arbitrary metric distances or weighted instances, where each city has a weight and the connection cost from a city is the product of the distance and the weight?

As already discussed, there is another natural way to model facility location priority algorithms, namely by letting the cities be the inputs. Meyerson [13] gives a randomized $O(1)$-approximation priority algorithm in this model where both the ordering and the irrevocable decisions use randomization. Does there exist a deterministic $O(1)$-approximation priority algorithm in this setting? More generally, we need to study the power of randomization in the context of priority algorithms.

There are a number of variants of the facility location problem as well as the related $k$-median problem that can be studied within our framework. For example, the capacitated facility location problem where each facility has a capacity bound on the number (weight) of the cities it can serve. Of course, our lower bounds apply to the capacitated version (i.e. by setting all capacities to exceed the sum of all city weights). But can we derive better results for this variant? As in the set cover problem, it seems appropriate to only consider greedy priority algorithms since not opening a facility may result in an infeasible solution. We are also considering the $k$-facility location problem in which a feasible solution allows at most $k$ opened facilities, generalizing the facility location and $k$-median problems. Finally, we are interested in an offline version of incremental clustering, as introduced in Charikar *et al* [2].

### Acknowledgments

This work began in discussions with Yuval Rabani and our first result, namely Theorem 1, was a result of that collaboration. We also thank Éva Tardos and David Williamson for their suggestions and references.

# References

1. A. Borodin, M. Nielsen, and C. Rackoff. (Incremental) priority algorithms. In *Proceedings of the 13th Annual ACM-SIAM Symposium on Discrete Algorithms*, pages 752–761, 2002.
2. M. Charikar, C. Chekuri, T. Feder, and M. Motwani. Incremental clustering and dynamic information retrieval. In *Proceedings of the 29th Annual ACM Symposium on Theory of Computing*, pages 626–634, 1997.
3. V. Chvátal. A greedy heuristic for the set covering problem. *Mathematics of Operations Research*, 4(3):233–235, 1979.
4. U. Feige. A threshold of ln n for approximating set cover. *Journal of the ACM*, 45(4):634–652, 1998.
5. S. Guha and S. Khuller. Greedy strikes back: Improved facility location algorithms. In *Proceedings of the 9th ACM-SIAM Symposium on Discrete Algorithms*, pages 649–657, 1998.
6. D. Hochbaum. Heuristics for the fixed cost median problem. *Mathematical Programming*, 22:148–162, 1982.
7. K. Jain, M. Mahdian, and A. Saberi. A new greedy approach for facility location problems. In *Proceedings of the 34th Annual ACM Symposium on Theory of Computation*, pages 731–740, 2002.
8. D.S. Johnson. Approximation algorithms for combinatorial problems. *Journal of Computer and System Sciences*, 9(3):256–278, 1974.
9. L. Lovász. On the ratio of optimal integral and fractional covers. *Discrete Mathematics*, 13:383–390, 1975.
10. M. Mahdian, E. Markakis, A. Saberi, and V. V. Vazirani. A greedy facility location algorithm analyzed using dual fitting. In *Proceedings of the 4th International Workshop on Approximation Algorithms for Combinatorial Optimization Problems (APPROX)*, pages 127–137, 2001.
11. M. Mahdian, J. Ye, and J. Zhang. A 1.52-approximation algorithm for the uncapacitated facility location problem. Available at `http://www.mit.edu/~mahdian/pub.html`.
12. R. R. Mettu and C. G. Plaxton. The online median problem. In *Proceedings of the 41st Annual IEEE Symposium on Foundations of Computer Science*, pages 339–348, 2000.
13. A. Meyerson. Online facility location. In *Proceedings of the 42nd Annual IEEE Symposium on Foundations of Computer Science*, pages 426–431, 2001.
14. D.B. Shmoys. Approximation algorithms for facility location problems. In K. Jansen and S. Khuller, editors, *Approximation Algorithms for Combinatorial Optimization*, volume 1913 of *Lecture Notes in Computer Science*. Springer, Berlin, 2000.
15. D.B. Shmoys, E. Tardos, and K. Aardal. Approximation algorithms for facility location problems (extended abstract). In *Proceedings of the 29th Annual ACM Symposium on Theory of Computing*, pages 265–274, 1997.
16. P. Slavík. A tight analysis of the greedy algorithm for set cover. *Journal of Algorithms*, 25:237–254, 1997.

# Appendix (Related Work)

Facility location problems have been the focus of extensive research from the operations research and computer science communities for several decades. However, it was only recently that Shmoys, Tardos and Aardal [15] gave the first polynomial time $O(1)$ approximation algorithm. The past several years have witnessed a series of improvements on the approximability of metric facility location. The approaches used include LP-rounding, the primal-dual method, local search, dual fitting, and combinations of the above. Due to space limitations we do not cite a complete history of recent developments; the interested reader is referred to the survey of Shmoys [14]. Currently, the best-known approximation ratio (1.52) for metric facility location is due to Mahdian, Ye and Zhang [11], and is achieved by a non-priority algorithm. For the special case of connection costs in $\{1, 3\}$, the LP-based (non-priority) algorithm of Guha and Khuller [5] provides a 1.463-approximation, matching the complexity based hardness result presented in the same paper.

We identify some algorithms that follow our paradigm of greedy (adaptive) priority algorithms. Hochbaum [6] presented an $O(\log n)$ approximation algorithm for facility location in arbitrary spaces, and showed that the analysis for that particular algorithm is tight. Mahdian, Markakis, Saberi and Vazirani [10] showed that a natural algorithm, which they call a small modification of Hochbaum's algorithm, is a 1.861 approximation algorithm for metric facility location; the analysis is performed by an elegant application of the dual-fitting technique. To our knowledge, the best approximation achieved by a priority algorithm is due to Jain, Mahdian and Saberi [7] and it is also analyzed by using the dual fitting technique. Mettu and Plaxton [12] present a greedy fixed priority 3-approximation algorithm for the metric facility location problem.

Set cover is one of the oldest and most well-studied NP-hard problems. Johnson [8] and Lovász [9] proposed a simple greedy algorithm which they showed provides a $H(n)$-approximation, where $H(n)$ is the $n$-th harmonic number, and $n = |U|$. Chvátal [3] extended their results to the weighted case. A tight analysis of the greedy algorithm due to Slavík [16] is of particular interest, as discussed in section 3.2. This specific greedy adaptive priority algorithm iteratively selects the most cost-effective set, i.e., the set that minimizes the average cost at which it covers new (i.e., currently uncovered) elements. In this paper, we refer to this algorithm as *the* greedy algorithm for set cover, to distinguish it from other priority algorithms that one can define.

It is interesting to note that certain primal-dual algorithms can be seen as priority algorithms, or, more precisely, certain priority algorithms have an alternative primal-dual statement. For instance, the greedy algorithm for set cover and the greedy-like algorithms for facility location due to Mahdian *et al* [10] and Jain, Mahdian and Saberi [7] can be stated as primal-dual algorithms. Of course, not every primal-dual algorithm is a priority algorithm; for example, primal-dual algorithms that apply a reverse-delete (or, generally speaking, reverse-reconstruction) step do not appear to be priority algorithms.

# Two Approximation Algorithms
# for 3-Cycle Covers

Markus Bläser and Bodo Manthey*

Institut für Theoretische Informatik, Universität zu Lübeck,
Wallstraße 40, 23560 Lübeck, Germany,
`blaeser/manthey@tcs.mu-luebeck.de`

**Abstract.** A cycle cover of a directed graph is a collection of node disjoint cycles such that every node is part of exactly one cycle. A $k$-cycle cover is a cycle cover in which every cycle has length at least $k$. While deciding whether a directed graph has a 2-cycle cover is solvable in polynomial time, deciding whether it has a 3-cycle cover is already NP-complete. Given a directed graph with nonnegative edge weights, a maximum weight 2-cycle cover can be computed in polynomial time, too. We call the corresponding optimization problem of finding a maximum weight 3-cycle cover Max-3-DCC.

In this paper we present two polynomial time approximation algorithms for Max-3-DCC. The heavier of the 3-cycle covers computed by these algorithms has at least a fraction of $\frac{3}{5} - \epsilon$, for any $\epsilon > 0$, of the weight of a maximum weight 3-cycle cover.

As a lower bound, we prove that Max-3-DCC is APX-complete, even if the weights fulfil the triangle inequality.

## 1 Introduction

A cycle cover of a directed or undirected graph $G$ is a spanning subgraph consisting of node-disjoint cycles. (In the case of undirected graphs, cycle covers are also called 2-factors.) Cycle covers have been intensively studied for many decades, see e.g. Lovász and Plummer [12] and Graham et al. [8] and the abundance of references given there.

A $k$-cycle cover is a cycle cover in which each cycle has at least length $k$. Such cycle covers are also called $(k - 1)$-restricted. In this paper, we are concerned with approximating maximum weight 3-cycle covers in complete directed graphs with nonnegative edge weights. To be specific, we call this problem Max-3-DCC. As our main contribution, we devise approximation algorithms for Max-3-DCC. On the other hand, we show that Max-3-DCC is APX-complete.

### 1.1 Previous Results

The problem of deciding whether an unweighted directed graph has a 2-cycle cover can be solved in polynomial time by computing a maximum bipartite

---

* Birth name: Bodo Siebert. Supported by DFG research grant Re 672/3.

matching. The corresponding optimization problem Max-2-DCC is polynomial time computable, too. On the other hand, deciding whether an unweighted directed graph has a 3-cycle cover is already NP-complete. This follows from the work of Valiant [16] (see also Garey and Johnson [7, GT 13]). Thus, considering optimization problems, Max-3-DCC is the next interesting problem. If the given graph has only edge weights zero and one, then there is a $\frac{2}{3}$-approximation algorithm for finding a maximum weight 3-cycle cover. This algorithm can be obtained from the algorithm presented by Bläser and Siebert [3] for the minimum weight 3-cycle cover problem with weights one and two by replacing weight zero with weight two.

What is known for undirected graphs? The problem of finding a 3-cycle cover in undirected graphs can be solved in polynomial time by Tutte's reduction [15] to the classical perfect matching problem in undirected graphs. The classical perfect matching problem can be solved in polynomial time (see Edmonds [5]). The corresponding maximization problem can be solved in polynomial time, even if we allow arbitrary nonnegative weights. Hartvigsen [9] has designed a powerful polynomial time algorithm for deciding whether an undirected graph has a 4-cycle cover. He has also presented a polynomial time algorithm that finds 5-cycle covers in bipartite graphs [10]. Both algorithms also work for the maximization version, if we only have the two possible edge weights zero and one. On the other hand, Cornuéjols and Pulleyblank [4] have reported that Papadimitriou showed the NP-completeness of finding a $k$-cycle cover in undirected graphs for $k \geq 6$.

One possibility to obtain approximation algorithms for Max-3-DCC is to modify algorithms for the maximum asymmetric TSP. Often it is sufficient just to modify the analysis. Taking the algorithm of Lewenstein and Sviridenko [11], which is the currently best approximation algorithm for this problem, one gets a polynomial time $\frac{7}{12}$-approximation algorithm for Max-3-DCC. For undirected graphs, the algorithm of Serdyukov [14] gives a polynomial time $\frac{7}{10}$-approximation algorithm for finding a 4-cycle cover of maximum weight.

### 1.2   Our Results

We present two approximation algorithms for Max-3-DCC. The heavier one of the cycle covers produced by these algorithms has at least a fraction of $\frac{3}{5} - \epsilon$ of the weight of an optimal 3-cycle cover. Thus, combining the two algorithms yields a $\left(\frac{3}{5} - \epsilon\right)$-approximation algorithm for Max-3-DCC (for any $\epsilon > 0$) whose running time is proportional to that of computing a maximum weight bipartite matching (and is particularly independent of $\epsilon$). This improves the previously best algorithm for this problem, which achieves a factor of $\frac{7}{12}$. As a lower bound, we prove that Max-3-DCC is APX-complete.

## 2   Approximation Algorithms for Max-3-DCC

We present two approximation algorithms for Max-3-DCC. The heavier one of the 3-cycle covers computed by these algorithms will have at least a fraction of $\frac{3}{5} - \epsilon$ of the weight of a maximum weight 3-cycle cover.

To avoid lengthy repetitions, we define some names that we use throughout this section. The input graph is called $G$, its node set is denoted by $V$, and the cardinality of $V$ is $n$. Both algorithms start with computing a 2-cycle cover on the input graph $G$. We call this cycle cover $\mathcal{C}$. Technically, we treat a cycle cover as the set of its edges. The cycles in $\mathcal{C}$ are $C_1, \ldots, C_\ell$. The total weight of $\mathcal{C}$ is denoted by $W$. Since a 3-cycle cover is also a 2-cycle cover, $W$ is an upper bound for the weight of an optimum 3-cycle cover. Let $I_2 \subseteq \{1, \ldots, \ell\}$ be the set of all $i$ such that $C_i$ is a 2-cycle (a 2-cycle is a cycle of length two). For each $i \in I_2$, we choose $b_i, c_i \in [0, 1]$ such that $b_i \cdot W$ and $c_i \cdot W$ are the weight of the heavier and lighter edge, respectively, of the 2-cycle $C_i$. Moreover, $b := \sum_{i \in I_2} b_i$ and $c := \sum_{i \in I_2} c_i$.

## 2.1   Algorithm 1

Algorithm 1 is a simple factor $\frac{1}{2}$-approximation algorithm. It starts with computing a 2-cycle cover. Then it discards the lightest edge of each 2-cycle and patches the obtained edges together to form one big cycle. If there is only one 2-cycle, then also one longer cycle will be broken. The edge of the 2-cycle and the path obtained from this longer cycle are then patched together to form a cycle of length at least five. The worst case for Algorithm 1 is $b = c = \frac{1}{2}$. However, if $\mathcal{C}$ contains cycles of length three or more that have a significant portion of the total weight of $\mathcal{C}$ or $b$ is much larger than $c$, then Algorithm 1 yields a better approximation ratio. More precisely, the amount of weight contained in $\mathcal{C}'$ is at least $\left(1 - c - \frac{1}{n}\right) \cdot W$. The loss of $c \cdot W$ comes from discarding the lightest edge in each 2-cycle and the loss of $\frac{1}{n} \cdot W$ is incurred when we have to break one cycle of length at least three. Since all edge weights are nonnegative, we do not loose any weight when patching the edges and paths together. This proves the following lemma.

**Lemma 1.** *Algorithm 1 computes a* $(1 - c - \frac{1}{n})$*-approximation to a maximum weight 3-cycle cover.*                                                                                      □

---

**Input:**   a complete directed graph $G$ with weight function $w$
**Output:** a 3-cycle cover $\mathcal{T}$

1. Compute a maximum weight 2-cycle cover $\mathcal{C}$ of $G$.
2. Discard the lightest edge of each 2-cycle in $\mathcal{C}$ to obtain a collection $\mathcal{C}'$ of node-disjoint edges and of cycles of length at least three. If there is only one 2-cycle in $\mathcal{C}$, take the lightest edge contained in any of the cycles of length at least three and discard it, too.
3. Construct a 3-cycle cover $\mathcal{T}$ by patching the paths in $\mathcal{C}'$ arbitrarily together.

---

**Fig. 1.** Algorithm 1

## 2.2   Algorithm 2

In Algorithm 2, we pay special attention to the 2-cycles. Like Algorithm 1, Algorithm 2 starts with computing a maximum weight cycle cover $\mathcal{C}$ and then transforms it into a 3-cycle cover. To this aim, we define a new weight function $w'$ by setting the weight of the edges of every 2-cycle to zero. Then we compute a maximum weight matching $\mathcal{M}$ with respect to the new weight function. That means, we replace the two edges between each pair of nodes by an undirected edge with weight equal to the maximum of the weight of the two replaced edges. Then we compute a matching of maximum weight on that graph. Finally, we translate everything back into the directed world by replacing each undirected edge by the directed one for which the maximum was attained (breaking ties arbitrarily). Then we color the edges of $\mathcal{M}$ and $\mathcal{C}$ with two colors in such way that each color class forms a collection of node disjoint paths and cycles of length at least three. This is the reason why we give the edges of the 2-cycles weight zero under $w'$. Otherwise, we could get triple edges and would consequently need three colors instead of two. Then we take the color class with larger weight and patch the paths arbitrarily together to obtain a 3-cycle cover. If this color class contains only one path and this path has length one, then we have to break one of the cycles.

The next lemma shows that the coloring described always exists. To apply this lemma, we temporarily remove all edges from $\mathcal{M}$ that are also edges in $\mathcal{C}$. Call the resulting set $\mathcal{M}' = \mathcal{M} \setminus \mathcal{C}$. The graph $(V, \mathcal{M}' \cup \mathcal{C})$ fulfills the premise of the lemma. Thus, we can color the edges in $\mathcal{M}' \cup \mathcal{C}$ with two colors. Thereafter, we deal with the edges in $\mathcal{M} \setminus \mathcal{M}'$ that are not part of a 2-cycle. These edges are already in one color class (because of their occurrence in $\mathcal{C}$) and can be safely placed into the other color class, too, without creating any 2-cycles. Edges in $\mathcal{M} \setminus \mathcal{M}'$ that are part of a 2-cycle in $\mathcal{C}$ are ignored, since we only consider the modified weight $w'(\mathcal{M})$.

---

**Input:**   a complete directed graph $G$ with weight function $w$
**Output:** a 3-cycle cover $\mathcal{T}$

1. Compute a maximum weight 2-cycle cover $\mathcal{C}$ of $G$.
2. Define a new weight function $w'$ on $G$ as follows: for each $i \in I_2$ assign both edges in $C_i$ the new weight zero. All other edges keep their old weight.
3. Compute a maximum weight matching $\mathcal{M}$ on $G$ with respect to $w'$.
4. Let $\mathcal{M}' = \mathcal{M} \setminus \mathcal{C}$.
5. Color the edges of $\mathcal{C} \cup \mathcal{M}'$ according to Lemma 2 with two colors.
6. Add each edge $e \in \mathcal{M} \setminus \mathcal{M}'$ that is not contained in a 2-cycle of $\mathcal{C}$ to the color class that does not already contain $e$.
7. Patch the paths in the color class with the larger weight arbitrarily together to obtain a 3-cycle cover $\mathcal{T}$. (If neccessary, break one longer cycle as in Algorithm 1.)

---

**Fig. 2.** Algorithm 2

**Lemma 2.** *Let $G = (V, E)$ be a directed loopless graph such that*

1. *every node in $V$ has indegree at most two,*
2. *every node in $V$ has outdegree at most two, and*
3. *every node in $V$ has total degree at most three.*

*Then the edges of $G$ can be colored with two colors such that each color class consists solely of node-disjoint paths and cycles of length at least three.*

*Proof.* To be specific, we call the colors red and blue. We construct an auxiliary undirected graph $H = (E, Z)$ whose nodes are the edges of $G$. There is an edge $\{e, f\}$ in $Z$ iff there are nodes $u$, $v$, and $x$ in $V$ such that $e = (u, x)$ and $f = (v, x)$ or $e = (x, u)$ and $f = (x, v)$, in other words, if $e$ is red, then $f$ has to be blue and vice versa. By assumption, every node in $H$ has degree at most two. Hence, $H$ consists solely of simple cycles and paths. By construction, all cycles in $H$ have even length. This is due to the fact that an edge in $Z$ corresponds to the event that either the heads or the tails of two edges from $E$ meet in one node in $G$. This already shows that we can color the edges of $G$ with the colors red and blue such that each of the two color classes consists of node-disjoint paths and cycles. We just color the edges of each path and cycle in an alternating way.

But each class could still contain 2-cycles which we now have to eliminate. Note that for each cycle and each path in $H$, there are two possible colorings and we can choose one of them arbitrarily. This is the key to eliminate the 2-cycles.

Figure 3 shows how a 2-cycle in $G$ can interact with the other edges in $G$. Due to the degree restrictions in $G$, either of the two nodes of a 2-cycle can only have at most one other edge. The case depicted on the left-hand side also includes the case where both edges not in the 2-cycle are reversed. The right-hand side also treats the case where one or two edges are missing. The solid edges are edges in the graph $G$. The dashed edges are the edges in $H$ connecting the edges of $G$, i.e, the nodes of $H$.

The case on the right-hand side in Fig. 3 is easy, here we have in $H$ one long path and one single node which is one of the edges of the 2-cycle. Since we can

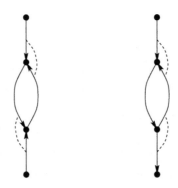

**Fig. 3.** The two ways how a 2-cycle can interact with the other edges. (Solid edges are edges from the graph $G$, dashed edges represent the resulting edges in $H$.)

choose its color arbitrarily, we color this single node in $H$ red if the other edge of the 2-cycle is colored blue and vice versa.

In the case on the left-hand side, we have two paths in $H$ whose end-nodes are the edges of the 2-cycle in $G$. To ensure that these two end-nodes get different colors, we "connect" these end-nodes in $H$, i.e., we add the additional edge $\{e, f\}$ to the edges of $H$, where $e$ and $f$ denote the two edges of the 2-cycle. This can of course create new cycles in $H$. But whenever we add such an edge $\{e, f\}$, then either two tails or two heads are connected. It follows that all the newly generated cycles have even length. Thus, we can color the edges of $G$ with the colors red and blue such that each of the two color classes consists of node-disjoint paths and cycles of length at least three.                                    $\square$

In order to estimate the approximation performance of Algorithm 2, we have to bound the weight of the matching $\mathcal{M}$. This is done in the following lemma.

**Lemma 3.** *Let $\mathcal{T}_{\mathrm{opt}}$ be a maximum weight 3-cycle cover on $G$ and let $w(\mathcal{T}_{\mathrm{opt}}) = L$. Let $I_2' \subseteq I_2$ be the set of all 2-cycles $C$ of $\mathcal{C}$ such that $C$ and $\mathcal{T}_{\mathrm{opt}}$ have a common edge. Furthermore, set $b' = \sum_{i \in I_2'} b_i$ and $c' = \sum_{i \in I_2'} c_i$. Then the weight of the matching $\mathcal{M}$ computed by Algorithm 2 with respect to $w'$ is at least*

$$w'(\mathcal{M}) \geq \tfrac{1}{2} \cdot L - \tfrac{1}{6} \cdot W + \tfrac{1}{6}(c' - 2b') \cdot W .$$

*Proof.* We divide the cycles of $\mathcal{T}_{\mathrm{opt}}$ into two sets $\mathcal{S}$ and $\overline{\mathcal{S}} = \mathcal{T}_{\mathrm{opt}} \setminus \mathcal{S}$. The set $\mathcal{S}$ contains all cycles that have an edge with a 2-cycle from $\mathcal{C}$ in common. With respect to $w'$, $\mathcal{T}_{\mathrm{opt}}$ has weight at least $L - b' \cdot W$, since a cycle of length at least three can run through only one edge of a 2-cycle. On the other hand, the total weight of the cycles in $\overline{\mathcal{S}}$ is at most $(1 - c' - b') \cdot W$. Otherwise we could add the cycles $C_i$ with $i \in I_2'$ to $\overline{\mathcal{S}}$ and would obtain a cycle cover of weight more than $W$, contradicting the optimality of $\mathcal{C}$. Let $D := w(\overline{\mathcal{S}})$. (Note that also $D = w'(\overline{\mathcal{S}})$ holds.) With respect to $w'$, $\mathcal{S}$ contains weight $w'(\mathcal{S}) \geq L - b' \cdot W - D$.

We now construct a matching $\mathcal{N}$ with $w'(\mathcal{N}) \geq \tfrac{1}{2} \cdot L - \tfrac{1}{6} \cdot W + \tfrac{1}{6}(c' - 2b') \cdot W$. This implies the assertion of the lemma. We can color the edges of $\overline{\mathcal{S}}$ with three colors such that each color class forms a (partial) matching. The worst-case for $\overline{\mathcal{S}}$ is a collection of 3-cycles. Let $\mathcal{N}_2$ be the color class with maximum weight, breaking ties arbitrarily. We have $w'(\mathcal{N}_2) \geq \tfrac{1}{3}D$. Since all cycles in $\mathcal{S}$ have one edge of weight zero under $w'$, we can color the edges with nonzero weight of $\mathcal{S}$ with two colors such that each color class forms a (partial) matching. Let $\mathcal{N}_1$ be the color class of larger weight. We have $w'(\mathcal{N}_1) \geq \tfrac{1}{2}(L - b' \cdot W - D)$. Then $\mathcal{N} = \mathcal{N}_1 \cup \mathcal{N}_2$ has weight at least

$$w'(\mathcal{N}) \geq \tfrac{1}{2} \cdot L - \tfrac{1}{2}b' \cdot W - \tfrac{1}{6}D$$
$$\geq \tfrac{1}{2} \cdot L - \tfrac{1}{6} \cdot W + \tfrac{1}{6}(c' - 2b') \cdot W ,$$

where the last inequality follows by plugging in $D \leq (1 - c' - b') \cdot W$.                $\square$

The next lemma bounds the approximation performance of Algorithm 2.

**Lemma 4.** *Let $b'$ and $c'$ be defined as in Lemma 3. Algorithm 2 computes a $(\tfrac{2}{3} + \tfrac{1}{12}(c' - 2b') - \tfrac{1}{n})$-approximation to a maximum weight 3-cycle cover of $G$.*

*Proof.* After step 6 of Algorithm 2, both color classes together contain all edges from $\mathcal{C}$ and those edges from $\mathcal{M}$ that are not part of any 2-cycle of $\mathcal{C}$. Thus, the total weight in both color classes is at least $w(\mathcal{C}) + w'(\mathcal{M})$. We have $w(\mathcal{C}) = W$. The weight contained in $\mathcal{M}$ with respect to $w'$ is $w'(\mathcal{M}) \geq \frac{1}{2} \cdot L - \frac{1}{6} \cdot W + \frac{1}{6}(c' - 2b') \cdot W$ by Lemma 3. In step 7, we perhaps loose weight at most $\frac{1}{n} \cdot W$. Thus the total weight of the heavier color class is at least

$$\frac{1}{2}\left(W + \frac{1}{2} \cdot L - \frac{1}{6} \cdot W + \frac{1}{6}(c' - 2b') \cdot W\right) - \frac{1}{n} \cdot W$$
$$\geq \frac{2}{3} \cdot L + \frac{1}{12}(c' - 2b') \cdot L - \frac{1}{n} \cdot L,$$

because $W \geq L$ and the sum of the coefficients of $W$ is positive. □

## 2.3   Combining the Algorithms

The final algorithm runs Algorithm 1 and 2 on $G$ and returns the heavier 3-cycle cover.

For the analysis, we skip the terms $\frac{1}{n} \cdot W$. We will pay for this at the end of the analysis by subtracting an arbitrarily small constant $\epsilon > 0$. The approximation factor of the combined algorithm can be bounded as follows:

$$\begin{aligned}
\text{minimize} \quad & \max\{1 - c, \tfrac{2}{3} + \tfrac{1}{12}(c' - 2b')\} \\
\text{subject to} \quad 0 \leq \quad & c' \quad \leq b', \\
& c' + b' \leq 1, \\
0 \leq \quad & c \quad \leq \tfrac{1}{2}(1 - b' + c').
\end{aligned}$$

Some simple calculations show that the above minimum is $\frac{3}{5}$. It is attained for $b' = \frac{3}{5}$ and $c = c' = \frac{2}{5}$.

**Theorem 1.** *For any $\epsilon > 0$, there is a factor $\left(\frac{3}{5} - \epsilon\right)$-approximation algorithm for* Max-3-DCC *running in polynomial time.* □

The running time of the above algorithm is dominated by the time needed to compute a maximum weight cycle cover. This time is proportional to the time needed to compute a maximum weight bipartite matching. Hence it is $O(n^3)$ or $O(n^{5/2} \log(nB))$ where $B$ is the largest weight in the given graph (see Ahuja et al. [1]).

## 3   Max-3-DCC is **APX**-complete

In this section we prove that Max-3-DCC is APX-complete. For this purpose we reduce E3-Max-Cut to Max-3-DCC. An instance of Max-Cut is an undirected graph $H = (X, Z)$. The goal is to find a subset $\tilde{X} \subseteq X$ of nodes such that the number of edges connecting $\tilde{X}$ and $X \setminus \tilde{X}$ is maximized. In the following, we denote the set of edges between $\tilde{X}$ and $X \setminus \tilde{X}$ by cut$(\tilde{X})$. The problem Max-Cut is known to be APX-complete, even if restricted to cubic graphs (see Alimonti and Kann [2]). We call Max-Cut restricted to cubic graphs E3-Max-Cut.

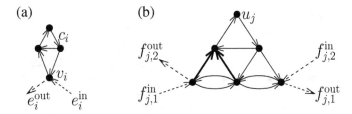

**Fig. 4.** (a) The node gadget $X_i$ representing the node $x_i$. All edges drawn have weight 2. (b) The edge gadget $Z_j$ representing edge $z_j \in Z$. The thick edges have weight $\frac{11}{6}$, all other edges drawn have weight 2. Edges left out in the figures have weight 1.

We present an L-reduction (see Papadimitriou and Yannakakis [13] for a definition) from E3-Max-Cut to Max-3-DCC to prove the APX-hardness of Max-3-DCC.

Let $H = (X, Z)$ be an instance for E3-Max-Cut, i.e., an undirected cubic graph. Let $X = \{x_1, x_2, \ldots, x_n\}$ be the set of nodes and $Z = \{z_1, z_2, \ldots, z_m\}$ be the set of edges. Since $H$ is a cubic graph, we have $n = \frac{2}{3} \cdot m$.

Let us now construct a directed edge weighted graph $G = (V, E)$ as an instance for Max-3-DCC. For each $x_i \in X$ create a node $v_i \in V$. This node $v_i$ is connected with a cycle $c_i$ of length 3 (see Fig. 4a). Furthermore, there is an edge $e_i^{\text{out}}$ starting at $v_i$ and an edge $e_i^{\text{in}}$ ending at $v_i$. Such a subgraph is called *node gadget* $X_i$ for $x_i$. For each edge $z_j \in Z$ create an *edge gadget* $Z_j$ as depicted in Fig. 4b.

The gadgets and the nodes are connected as follows. We order the nodes incident with an edge arbitrarily. Analogously, we order the edges incident with a node arbitrarily. Assume that $z_j$, $z_{j'}$, and $z_{j''}$ are the first, second, and third edge, respectively, incident with a node $x_i$. If $x_i$ is the first node of edge $z_j$ then $e_i^{\text{out}}$ and $f_{j,1}^{\text{in}}$ are identical, if $x_i$ is the second node of $z_j$ then $e_i^{\text{out}}$ and $f_{j,2}^{\text{in}}$ are identical. The corresponding outgoing edge of $Z_j$ is identical with one incoming edge of $Z_{j'}$ depending on whether $x_i$ is the first or the second node of $z_{j'}$. The gadgets $Z_{j'}$ and $Z_{j''}$ are connected in a similar manner. Finally, one of the edges $f_{j'',1}^{\text{out}}$ and $f_{j'',2}^{\text{out}}$ is identical with $e_i^{\text{in}}$. Figure 5 shows an example of a graph $H$ and the corresponding graph $G$.

We call a cycle cover $\mathcal{C}$ of $G$ *consistent with a subset* $\tilde{X} \subseteq X$, if the following properties hold:

1. All edge gadgets are traversed as depicted in Fig. 6.
2. If $x_i \in \tilde{X}$ then both $e_i^{\text{in}}$ and $e_i^{\text{out}}$ are in $\mathcal{C}$ and $c_i$ forms a cycle of length 3. If $x_i \notin \tilde{X}$ then $v_i$ and $c_i$ form a cycle of length 4.

We call a cycle cover $\mathcal{C}$ *consistent* if there exists a subset $\tilde{X}$ such that $\mathcal{C}$ is consistent with $\tilde{X}$.

The weight of a node in $V$ with respect to a cycle cover $\mathcal{C}$ is the sum of half the weight of its incoming edge plus half the weight of its outgoing edge. The weight $w(Z_j)$ of the edge gadget $Z_j$ is the sum of the weights of its nodes. Likewise, the weight $w(X_i)$ of the node gadget $X_i$ is the sum of the weights of

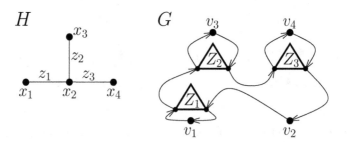

**Fig. 5.** A graph $H$ and the corresponding graph $G$. For the sake of readability $H$ is not cubic. The edge gadgets are symbolized by triangles, the cycles of the node gadgets are left out. The node $x_2$ is the second node of $z_1$ and the first node of both $z_2$ and $z_3$.

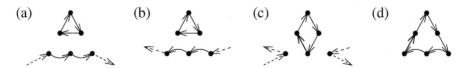

**Fig. 6.** Consistent traversals of the gadget $Z_j$ representing the edge $z_j = \{x_i, x_{i'}\}$. (a) only $x_i$ is in $\tilde{X}$, (b) only $x_{i'}$ is in $\tilde{X}$, (c) both $x_i$ and $x_{i'}$ are in $\tilde{X}$, (d) neither node of $z_j$ is in $\tilde{X}$.

its nodes. Assume that $z_j = \{x_i, x_{i'}\}$. Then we call

$$b(Z_j) := w(Z_j) + \tfrac{1}{3} \cdot (w(X_i) + w(X_{i'}))$$

the *burden of* $Z_j$. We have

$$w(\mathcal{C}) = \sum_{j=1}^{m} w(Z_j) + \sum_{i=1}^{n} w(X_i) = \sum_{j=1}^{m} b(Z_j) \,.$$

In a cycle cover consistent with $\tilde{X}$ we have $w(X_i) = 8$ for all $i$. For every edge gadget $Z_j$, we have either $w(Z_j) = 12$ and $b(Z_j) = \frac{104}{6}$ if $z_j \in \mathrm{cut}(\tilde{X})$ or $w(Z_j) = \frac{71}{6}$ and $b(Z_j) = \frac{103}{6}$ if $z_j \notin \mathrm{cut}(\tilde{X})$. Thus, we have the following lemma.

**Lemma 5.** *Let $\mathcal{C}$ be the cycle cover consistent with $\tilde{X}$. Then $w(\mathcal{C}) = \frac{103}{6} \cdot m + \frac{1}{6} \cdot |\mathrm{cut}(\tilde{X})|$.* □

Consider an arbitrary cycle cover $\mathcal{C}$ of $G$ with weight $w(\mathcal{C}) = \frac{103}{6} \cdot m + \frac{1}{6} \cdot \gamma$. We describe how to construct a subset $X'$ with the property that if $b(Z_j) > \frac{103}{6}$ then $z_j \in \mathrm{cut}(X')$. Thus, $|\mathrm{cut}(X')| \geq \gamma$.

In the following, assume that $z_j = \{x_i, x_{i'}\}$. If a node gadget $X_i$ is inconsistent then we have $w(X_i) \leq \frac{15}{2}$. If an edge gadget $Z_j$ is inconsistent then we have $w(Z_j) \leq \frac{23}{2}$. Hence, if the edge gadget $Z_j$ or one of the node gadgets $X_i$ or $X_{i'}$ is inconsistent, we have $b(Z_j) \leq \frac{103}{6}$.

Let us now consider all edge gadgets $Z_j$ with burden $b(Z_j) > \frac{103}{6}$. Due to the considerations above, all these edge gadgets are consistent. Furthermore, if $b(Z_j) > \frac{103}{6}$ then both $X_i$ and $X_{i'}$ are consistent. Thus, all edge gadgets with $b(Z_j) > \frac{103}{6}$ are consistent and fulfil $b(Z_j) = \frac{104}{6}$. If we can find a subset $\tilde{X}$ fulfilling $\mathrm{cut}(\tilde{X}) \supseteq \{z_j \mid b(Z_j) = \frac{104}{6}\}$, we are done.

An edge gadget $Z_j$ with $b(Z_j) = \frac{104}{6}$ is called a *witness for $x_i \in \tilde{X}$*, if it is traversed as shown in Fig. 6a or 6c. Otherwise, it is called a *witness for $x_i \notin \tilde{X}$*. Likewise, we call $Z_j$ a witness for $x_{i'} \in \tilde{X}$ if it is traversed as shown in Fig. 6b or 6c, and for $x_{i'} \notin \tilde{X}$, otherwise.

Assume that for some node $x_i$ we have two edge gadgets $Z_j$ and $Z_{j'}$ such that $Z_j$ is a witness for $x_i \in \tilde{X}$ and $Z_{j'}$ is a witness for $x_i \notin \tilde{X}$. Since both gadgets are consistent, $z_j$ and $z_{j'}$ are the first and the third edge, respectively, of $x_i$ or vice versa. But this implies, that the node gadget $X_i$ cannot be consistent. Hence, $b(Z_j) \leq \frac{103}{6}$ and $b(Z_{j'}) \leq \frac{103}{6}$, a contradiction.

The witnesses induce a partition $X_\in$, $X_\notin$, $X_?$ of $X$, such that all nodes in $X_\in$ have only witnesses for $x \in \tilde{X}$, all nodes in $X_\notin$ have only witnesses for $x \notin \tilde{X}$, and all nodes in $X_?$ do not have any witness. Due to the construction, there are at least $\gamma$ edges between $X_\in$ and $X_\notin$. Choose $\tilde{X}$ arbitrarily with $X_\in \subseteq \tilde{X} \subseteq X_\in \cup X_?$. Then $|\mathrm{cut}(\tilde{X})| \geq \gamma$. Hence, we have proved the following lemma.

**Lemma 6.** *Given an arbitrary cycle cover $\mathcal{C}$ with weight $w(\mathcal{C}) = \frac{103}{6} \cdot m + \frac{1}{6} \cdot \gamma$ we can construct a consistent cycle cover $\mathcal{C}'$ with weight $w(\mathcal{C}') \geq w(\mathcal{C})$ in polynomial time.*   □

Now we can prove the main theorem of this section.

**Theorem 2.** Max-3-DCC *is* APX-*complete.*

*Proof.* We show that the reduction presented above is an L-reduction.

We denote the size of the maximum cut of $H$ with $\mathrm{opt}(H)$ and the weight of the maximum cycle cover of $G$ with $\mathrm{opt}(G)$.

For any graph $H = (X, Z)$ there exists a subset $\tilde{X}$ of $X$ such that $|\mathrm{cut}(\tilde{X})| \geq \frac{m}{2}$. Thus, $\mathrm{opt}(H) \geq \frac{m}{2}$ and $\mathrm{opt}(G) \leq \frac{104}{6} \cdot m \leq \frac{104}{3} \cdot \mathrm{opt}(H)$.

On the other hand, given a cycle cover $\mathcal{C}$ with weight $w(\mathcal{C}) = \frac{103}{6} \cdot m + \frac{1}{6} \cdot \gamma$ we can construct a subset $\tilde{X}$ of $X$ with $|\mathrm{cut}(\tilde{X})| \geq \gamma$ in polynomial time, see Lemma 6. Thus, the following holds:

$$\left| \mathrm{opt}(H) - |\mathrm{cut}(\tilde{X})| \right| = 6 \cdot \left| \mathrm{opt}(G) - \left( \tfrac{103}{6} \cdot m + \tfrac{1}{6} \cdot |\mathrm{cut}(\tilde{X})| \right) \right|$$
$$\leq 6 \cdot \left| \mathrm{opt}(G) - w(\mathcal{C}) \right|.$$

This proves that the reduction presented is an L-reduction. Since E3-Max-Cut is APX-complete and Max-3-DCC is in APX, the theorem is proved.   □

## 4   Open Problems

A problem that remains open is the approximability of computing minimum weight 3-cycle covers. Without any restrictions, this problem is NPO-complete.

With the triangle inequality, we can perform two iterations of the algorithm of Frieze, Galbiati, and Maffioli [6] for the asymmetric TSP. This yields an approximation factor of two. It seems to be a challenging problem to improve this factor of two, since this could also yield new insights into the approximability of the asymmetric TSP.

Another open problem is the status of maximum weight undirected 4-cycle covers. The question is whether this problem can be solved in polynomial time or has at least a polynomial time approximation scheme, provided that $P \neq NP$.

# References

1. R. K. Ahuja, T. L. Magnanti, and J. B. Orlin. *Network Flows: Theory, Algorithms, and Applications*. Prentice Hall, 1993.
2. P. Alimonti and V. Kann. Some APX-completeness results for cubic graphs. *Theoret. Comput. Sci.*, 237(1-2):123–134, 2000.
3. M. Bläser and B. Siebert. Computing cycle covers without short cycles. In *Proc. 9th Ann. European Symp. on Algorithms (ESA)*, volume 2161 of *Lecture Notes in Comput. Sci.*, pages 368–379. Springer, 2001.
4. G. P. Cornuéjols and W. R. Pulleyblank. A matching problem with side conditions. *Discrete Math.*, 29:135–159, 1980.
5. J. Edmonds. Paths, trees, and flowers. *Canad. J. Math.*, 17:449–467, 1965.
6. A. M. Frieze, G. Galbiati, and F. Maffioli. On the worst-case performance of some algorithms for the traveling salesman problem. *Networks*, 12(1):23–39, 1982.
7. M. R. Garey and D. S. Johnson. *Computers and Intractability: A Guide to the Theory of NP-Completeness*. W. H. Freeman and Company, 1979.
8. R. L. Graham, M. Grötschel, and L. Lovász, editors. *Handbook of Combinatorics*, volume 1. Elsevier, 1995.
9. D. Hartvigsen. *An Extension of Matching Theory*. PhD thesis, Carnegie-Mellon University, 1984.
10. D. Hartvigsen. The square-free 2-factor problem in bipartite graphs. In *Proc. 7th Int. Conf. on Integer Programming and Combinatorial Optimization (IPCO)*, volume 1620 of *Lecture Notes in Comput. Sci.*, pages 234–241. Springer, 1999.
11. M. Lewenstein and M. Sviridenko. A 5/8 approximation algorithm for the asymmetric maximum TSP. Manuscript, 2002.
12. L. Lovász and M. D. Plummer. *Matching Theory*. Elsevier, 1986.
13. C. H. Papadimitriou and M. Yannakakis. Optimization, approximation, and complexity classes. *J. Comput. System Sci.*, 43(3):425–440, 1991.
14. A. I. Serdyukov. An algorithm with an estimate for the traveling salesman problem of the maximum. *Upravlyaemye Sistemy*, 25:80–86, 1984. (in Russian).
15. W. T. Tutte. A short proof of the factor theorem for finite graphs. *Canad. J. Math.*, 6:347–352, 1954.
16. L. G. Valiant. The complexity of computing the permanent. *Theoret. Comput. Sci.*, 8(2):189–201, 1979.

# Approximation Algorithms
# for the Unsplittable Flow Problem[*]

Amit Chakrabarti[1], Chandra Chekuri[2], Anuptam Gupta[2], and Amit Kumar[3]

[1] Computer Science Dept., Princeton University,
amitc@cs.princeton.edu
[2] Bell Labs, Lucent Tech.,
{chekuri,anupamg}@research.bell-labs.com
[3] Computer Science Dept., Cornell University,
amitk@cs.cornell.edu

**Abstract.** We present approximation algorithms for the *unsplittable flow problem* (UFP) on undirected graphs. As is standard in this line of research, we assume that the maximum demand is at most the minimum capacity. We focus on the *non-uniform capacity* case in which the edge capacities can vary arbitrarily over the graph. Our results are:

- For undirected graphs we obtain a $O(\Delta\alpha^{-1}\log^2 n)$ approximation ratio, where $n$ is the number of vertices, $\Delta$ the maximum degree, and $\alpha$ the expansion of the graph. Our ratio is capacity independent and improves upon the earlier $O(\Delta\alpha^{-1}(c_{\max}/c_{\min})\log n)$ bound [15] for large values of $c_{\max}/c_{\min}$. Furthermore, if we specialize to the case where all edges have the same capacity, our algorithm gives an $O(\Delta\alpha^{-1}\log n)$ approximation, which matches the performance of the best-known algorithm [15] for this special case.
- For certain strong constant-degree expanders considered by Frieze [10] we obtain an $O(\sqrt{\log n})$ approximation for the uniform capacity case, improving upon the current $O(\log n)$ approximation.
- For UFP on the line and the ring, we give the first constant-factor approximation algorithms. Previous results addressed only the uniform capacity case.
- All of the above results improve if the maximum demand is bounded away from the minimum capacity.

Our results are based on randomized rounding followed by greedy alteration and are inspired by the use of this idea in recent work [21, 9].

## 1 Introduction

In the *unsplittable flow problem* (UFP), we are given an $n$-vertex graph $G$ and a set of $k$ vertex pairs (terminals) $\mathcal{T} = \{(u_i, v_i) : i = 1, \ldots, k\}$; each pair $(u_i, v_i)$ in $\mathcal{T}$ has a *demand* $\rho_i$ and a *weight* (or profit) $w_i$, and furthermore, the graph is *capacitated* with edge capacities $\{c_e\}$. The goal is to find the maximum weight subset of pairs from $\mathcal{T}$, along with a path for each chosen pair, so that the

---

[*] Part of this work was done while the first and the last author visited Bell Labs.

K. Jansen et al. (Eds.): APPROX 2002, LNCS 2462, pp. 51–67, 2002.
© Springer-Verlag Berlin Heidelberg 2002

entire demand for each such pair can be routed on its path while respecting the capacity constraints.

We note at the outset that even very special cases of UFP are NP-hard. For instance, when $G$ is just a single edge, UFP specializes to the KNAPSACK problem. When each $c_e = 1$ and each $\rho_i = w_i = 1$, UFP specializes to the well-known *maximum edge-disjoint paths* problem (MEDP), where the goal is simply to find the largest number of pairs from $\mathcal{T}$ which can be simultaneously connected by edge-disjoint paths in $G$. MEDP is NP-hard even for planar graphs.

A substantial amount of research has focused on obtaining good approximation algorithms for MEDP and UFP due to their importance in network routing and design. For MEDP, the best known approximation ratio on general graphs is $O(\sqrt{m})$, where $m$ is the number of edges in the graph. In *directed* graphs it is NP-hard to approximate it to a ratio better than $\Omega(m^{1/2-\varepsilon})$ [11]. However, in *undirected* graphs, which will be the focus of this paper, MEDP is only known to be hard to approximate to within constant factors. Improved approximation ratios have been obtained for several classes of graphs including trees, mesh-like planar graphs, and graphs with high expansion (see, e.g., [12,15] for references).

Let $\rho_{\max} = \max_i \rho_i$ be the maximum demand among the pairs and $c_{\min} = \min_e c_e$ be the minimum capacity of an edge. We only consider instances with $\rho_{\max} \leq c_{\min}$; this is a standard assumption in the literature and is sometimes referred to as the *no-bottleneck* assumption. In its absence, we can embed an UFP instance on a given graph $G = (V, E)$ into any other graph $G' = (V, E')$ with $E \subseteq E'$, thus limiting our ability to study the role of the graph structure on the approximability. It is also known that UFP in directed graphs without the no-bottleneck assumption is provably harder than MEDP; in fact, it is hard to approximate to within $\Omega(n^{1-\varepsilon})$ [3]. Finally, this restriction is useful in many applications and still includes MEDP as a special case. Henceforth we will assume that $c_{\min} = 1$ and that $0 < \rho_i \leq 1$ for all $i$.

A special case of UFP is the *uniform capacity unsplittable flow* problem (UCUFP) in which all edges have the same capacity. UCUFP has received more attention and its approximability is usually related to the corresponding MEDP problem. In contrast, probably the only well understood case for UFP is when $c_{\min}/\rho_{\max}$ is $\Omega(\log n / \log \log n)$. In this case, an offline constant-factor approximation [18] (via randomized rounding) as well as an $O(\log n)$-competitive online algorithm [2] are known.

**Our Results:** In this paper we address UFP with non-uniform edge capacities on undirected graphs. For general graphs, we obtain algorithms with approximation ratios that depend on the *flow number* $F$ of the graph $G$. This parameter was first defined by Kolman and Scheideler [15] who also established some of its basic properties. In particular, they showed that $F$ is related to the expansion of the graph by $F = O(\Delta \alpha^{-1} \log n)$, where $\alpha$ is the edge expansion and $\Delta$ is the maximum degree. Thus, bounds stated in terms of $F$ are typically most interesting when $G$ is an expander, although, as noted in [15], there are other interesting cases such as meshes and hypercubes where a logarithmic bound on flow number holds even though the expansion is large.

Our results for general graphs (see Corollary 2) are:

- An $O(F \log n) = O(\Delta \alpha^{-1} \log^2 n)$ approximation for UFP. Our bound is capacity-independent, unlike the $O(\Delta \alpha^{-1}(c_{\max}/c_{\min}) \log n)$ bound of [15].
- An $O(F) = O(\Delta \alpha^{-1} \log n)$ approximation for UCUFP. This matches the bound of [15] and provides an alternative algorithm.
- When $\rho_{\max} \leq c_{\min}/B$, for integer $B$, the above approximation bounds improve to $O((\Delta \alpha^{-1} \log^2 n)^{1/B})$ for UFP, and $O((\Delta \alpha^{-1} \log n)^{1/B})$ for UCUFP,

Additionally, we obtain even better approximation ratios on special classes of graphs, by further exploiting some of the techniques used in proving the above. We obtain:

- An $O(\sqrt{\log n})$ approximation for UCUFP on "sufficiently strong" constant degree expanders, as defined by Frieze [10] (see Definition 1 and Theorem 3). This improves on the current $O(\log n)$ ratio [10,15] and is the first sublogarithmic approximation for constant degree expanders that we are aware of.
- An $O(1)$ approximation for UFP on line and ring networks (see Theorems 5 and 7). Only UCUFP results were known earlier.

**Our Techniques:** Most approaches to approximating MEDP and UCUFP on expanders rely on proving the existence of near-optimal solutions to the multicommodity flow relaxation of the problem that use *short* (polylogarithmic in length) flow paths. Kolman and Scheideler [15] generalize this to UFP through their notion of flow number. However, their upper bound on the length of the flow paths depends on $c_{\max}/c_{\min}$ which can be quite large. We take a different approach by showing that there exist flow paths that use only a *few* (polylogarithmic number of) edges of *low* capacity, even though the overall length of the flow path might be large. High capacity edges (of capacity $\Omega(\log n)$) behave well under randomized rounding and this leaves us to worry only about the behavior of the low capacity edges under randomized rounding.

Our second idea, which proves useful for the case of the line and the ring as well, is to perform the randomized rounding step with more care. The naïve rounding involves scaling down the fractional solution before randomized rounding, where the scaling factor is chosen to be large enough to argue that none of the constraints are violated. Typically the events corresponding to the violation of these constraints are not independent and the union bound is too weak to estimate the failure probability of the randomized rounding. To overcome these problems, either the Lovász Local Lemma [19,15] or a correlation type inequality like the FKG inequality [19] is used. These approaches are technically involved and add substantial complexity to the algorithm. The approach we will use, called the method of *alteration* [1] is applicable to monotone problems. Applications to approximation algorithms have recently been shown independently by Srinivasan [21] (who applied it to general packing *and* covering problems) and Calinescu et al. [9] (who applied it to a specific packing problem). In this approach, the first step is the same as above: scaling followed by randomized rounding. However, instead of desiring that all constraints are satisfied (feasible), we look at the random solution and alter it if it is not feasible: we simply change 1's in the random solution to 0's to ensure feasibility. This greedy (problem-dependent) alteration is designed to obtain a feasible solution, and hence the burden shifts to analyzing the expected loss in the alteration step. This turns out to be simple and effective for various problems and we believe that this idea will find more applications in the future.

**Relation to Previous Work:** We first discuss work connecting UFP and related problems with expansion of the underlying graph. Culminating a long line of work, Frieze [10] recently showed that for regular expanders with sufficiently strong expansion and sufficiently large (but constant) degree, there exists a constant $c$ such that *any* $cn/\log n$ pairs can be connected via edge-disjoint paths provided no vertex appears in more than a constant (depending on and less than the degree) number of pairs. This result is optimal to within constant factors, and has been also has been extended to expander digraphs [7]. An immediate consequence of this is an $O(\log n)$ approximation for MEDP on such expanders. Kleinberg and Rubinfeld [13] in 1996 had used an earlier result of Broder, Frieze, and Upfal [8] to show that a deterministic *online* algorithm, the so-called bounded greedy algorithm (BGA), gave an $O(\log n \log \log n)$ approximation guarantee. (In fact Frieze's result mentioned above implies an $O(\log n)$ bound for BGA.) In the same paper, Kleinberg and Rubinfeld also showed the existence of a near-optimal fractional solution to any multicommodity flow instance on an expander that used only *short* paths of length $O(\log^3 n)$. This latter result formed the basis of an $O(\log^3 n)$ approximation for UCUFP on expanders by Srinivasan [19]. The above results do not explicitly specify the dependence of the approximation ratio on $\Delta$ and $\alpha$. Kolman and Scheideler [15] suggest a dependence of the form $\Omega(\Delta^2 \alpha^{-2})$.

The results on short flow paths (for UCUFP) have been recently improved by Kolman and Scheideler, who show the existence of near-optimal solutions to multicommodity flow instances that use paths of length $O(\Delta R)$ where $R$ is the *routing number* of the graph [14], or solutions that use paths of length $O(F)$, where $F$ is the *flow number* of the graph [15]. It is known that $R = O(\Delta \alpha^{-1} \log n)$ and, as we have mentioned earlier, $F = O(\Delta \alpha^{-1} \log n)$. In addition to improving the approximation ratio for UCUFP, these new results offer other advantages. The dependence on $\alpha$ and $\Delta$ is improved and is made explicit. The bound on the flow path lengths is stronger, and the proof is direct and simple and is based on the work of Leighton and Rao [16]. It is also shown that the flow number is more appropriate than the routing number in capturing the flow path lengths (see [15] for more details).

We now turn to the problem of UFP on the line network. In this case, the MEDP problem corresponds to the maximum independent set problem on interval graphs, which has a polynomial time algorithm. However, the UCUFP

problem on the line is NP-hard (it generalizes KNAPSACK), and is equivalent
to the task assignment problem on a single machine with fixed time windows.
Generalizations of the task assignment problem to multiple machines and time
windows have been studied in the recent past [4,17], and most of these problems
have $O(1)$-approximation algorithms as well as $O(1)$ integrality gaps. This is
not necessarily the case with UFP, and in particular, if the demands are not
constrained to be less than the minimum capacity, the integrality gap of the
natural LP could be $\Omega(\log \rho_{max}) = \Omega(n)$. Two of the techniques that have been
used to develop $O(1)$ approximations for the task assignment problem, i.e., the
local-ratio method [4] and LP rounding [17], do not seem to extend to the case
of UFP. However, the recent techniques of Calinescu et al. [9] using a combination of randomized rounding, alteration and dynamic programming, can be
extended, and we build upon these ideas to give an $O(1)$ approximation for UFP
when $\rho_{max} \le c_{min}$. Without this restriction, we show that the LP integrality gap
can be $\Omega(\log \rho_{max})$ and provide a matching approximation ratio. We extend the
results on the line to the ring via a simple reduction.

## 2    Preliminaries

UFP has a natural IP formulation based on multicommodity flow. Let $\mathcal{P}_i$ denote
the set of all paths in $G$ from $u_i$ to $v_i$. The IP is:

$$\max \quad \sum_{i=1}^{k} w_i x_i, \quad \text{s.t.}$$
$$\sum_{\pi \in \mathcal{P}_i} f_\pi = x_i \qquad i = 1, \ldots, k$$
$$\sum_{i=1}^{k} \sum_{\pi \in \mathcal{P}_i : \pi \ni e} \rho_i f_\pi \le c_e \qquad e \in E(G)$$
$$x_i \in \{0, 1\} \qquad i = 1, \ldots, k$$
$$f_\pi \in \{0, 1\} \qquad \pi \in \cup_{i=1}^{k} \mathcal{P}_i.$$

The LP relaxation, which we shall call LPMAIN, is obtained by allowing $x_i$
and $f_\pi$ to lie in the real interval $[0, 1]$. Let $(x_1, \ldots, x_k, f_{\pi_1}, f_{\pi_2}, \ldots)$ be a fractional
solution to LPMAIN. We shall refer to $\sum_{i=1}^{k} w_i x_i$ as the *profit* or the *value* of the
solution. We say that the solution *uses* a flow path $\pi$ if $f_\pi > 0$. An optimal solution to LPMAIN can be obtained in polynomial time[1], and efficient combinatorial
methods that obtain a $(1 + \varepsilon)$-approximation are also known. We can also find
efficient $(1 + \varepsilon)$-approximations when we restrict the LP to use only flow paths
of length bounded by a given parameter $\ell$. This fact will be useful in the sequel.

Let us also make explicit the notions of expansion used in this paper:

**Definition 1.** *Let $G$ be an uncapacitated $n$-vertex graph and for $U \subseteq V(G)$ let
$\partial U$ denote the set of edges of $G$ with exactly one end point in $U$. $G$ is said to*

---

[1] Notice that the LP, as given here, is path-based and has exponential size. To
solve it, we first solve a different, polynomial sized edge-based LP which uses flow-
conservation constraints, and then perform a simple path-decomposition on the solution.

*have expansion $\alpha$ if for all $U \subseteq V(G)$ we have $|U| \le n/2 \Rightarrow |\partial U| \ge \alpha|U|$. If $G$ is $\Delta$-regular, it is called a* strong expander *if for sufficiently small $\alpha$, arbitrary $\beta < 1 - \alpha$, $\gamma < \frac{1}{2}$, and $U \subseteq V(G)$, we have*

$$|U| \le \gamma n \Rightarrow |\partial U| \ge (1 - \alpha)\Delta|U| \text{ and } \gamma n < |U| \le n/2 \Rightarrow |\partial U| \ge \beta\Delta|U|.$$

## 3    Expansion-Based Approximation Bounds for UFP

Our approach is to show that the flows in any fractional solution to LPMAIN can be rerouted so as to yield a new fractional solution in which the flow paths use few edges of small capacity (we call such a solution *favorable*), where "few" is quantified using a parameter of the graph called its *flow number* [15]. The rerouting reduces the profit of the solution by a factor that can be made arbitrarily close to 1. Next, we show that a favorable fractional solution can be rounded efficiently to an integral solution with good profit.

Our rerouting is similar to that in [14,15] and has been obtained independently following [14]; our contribution is to restrict the notion of "short" to edges of low capacity. Note that the rerouting need not be algorithmically efficient since it is just used to establish a lower bound on the profit of a favorable fractional solution; the solution itself can be obtained by solving a modified LP.

We shall denote by $F$ the flow number of our input graph. For a definition of flow number we refer the reader to Kolman and Scheideler [15]. We shall use the facts, proven in [15], that $F$ can be computed in polynomial time and that $F = O(\Delta\alpha^{-1}\log n)$. We remark that we shall only be concerned with the flow number of *uncapacitated* graphs. The paper [15] in fact defines flow number for capacitated graphs as well, and proves a similar bound for $F$ in this case, with a "capacitated" definition of $\Delta$. Recall that we normalize the capacities and demands such that $c_{\min} = 1$ and $\rho_{\max} \le 1$.

**Definition 2.** *A fractional solution to* LPMAIN *is said to be $(c, d)$-favorable if every flow path used by the solution has at most $d$ edges of capacity at most $c$.*

**Theorem 1.** *For $0 < \varepsilon \le 1$ and $c \ge 1$, given a fractional solution to* LPMAIN *with profit $W$, there exists a $(c, 4cF/\varepsilon)$-favorable fractional solution with profit at least $W/(1 + \varepsilon)$. Alternatively, there exists a $(c, 4F/\varepsilon)$-favorable fractional solution with profit at least $W/(c(1 + \varepsilon))$.*

**Corollary 1.** *Suppose, as in UCUFP, that $c_e = 1$ for all edges $e$. For any $0 < \varepsilon \le 1$, given a fractional solution to* LPMAIN *with profit $W$, there exists a $(1, 4F/\varepsilon)$-favorable fractional solution with profit at least $W/(1 + \varepsilon)$.*

**Theorem 2.** *Given a $(\log n, d)$-favorable fractional solution to* LPMAIN *with profit $W$, we can efficiently compute a (random) integral solution with expected profit $W'$ such that, for large enough $d$,*

1. *$W' = \Omega(W/d)$.*
2. *If, additionally, $\rho_{\max} \le 1/B$, for integer $B \ge 2$, then $W' = \Omega\left(W/d^{1/(B-1)}\right)$.*

3. *If, additionally, each $\rho_i = 1/B$, for integer $B \geq 1$, then $W' = \Omega\left(W/d^{1/B}\right)$.*

**Corollary 2.** *For graphs with expansion $\alpha$ and maximum degree $\Delta$, there is an $O(\Delta\alpha^{-1}\log^2 n)$ approximation for UFP, and an $O(\Delta\alpha^{-1}\log n)$ approximation for UCUFP. For integer $B$, if $\rho_{\max} \leq 1/B$, then the approximations improve to $O\left((\Delta\alpha^{-1}\log^2 n)^{1/B}\right)$ and $O\left((\Delta\alpha^{-1}\log n)^{1/B}\right)$ for UFP and UCUFP respectively.*

*Remark.* It is worth noting that the $O(\Delta\alpha^{-1}\log n)$-approximation for UFP given in [15] uses the aforementioned "capacitated" definition of $\Delta$; their $\Delta$ can be as large as $c_{\max}/c_{\min}$ times the maximum degree. Clearly, the two definitions coincide for UCUFP.

*Proof.* For the first result on UFP, using $c = \log n$ and $\varepsilon = 1$ in Theorem 1, we have a $(\log n, O(F \log n))$-favorable fractional solution. Applying part 1 of Theorem 2 and using $F = O(\Delta\alpha^{-1}\log n)$ gives us the desired approximation guarantee. For the next result, by Corollary 1 we have a $(1, O(F))$-favorable solution which, for UCUFP, is the same as a $(\log n, O(F))$-favorable solution. Applying part 1 of Theorem 2 completes the proof.

Now suppose we have an instance $\mathcal{I}$ of UFP with $\rho_{\max} \leq \frac{1}{B}$, where $B \geq 2$. Let us create two new UFP instances, $\mathcal{I}_1$ and $\mathcal{I}_2$, with the same underlying graph as $\mathcal{I}$ but with source-sink pairs restricted to those with demands at most $\frac{1}{B+1}$ and those with demands greater than $\frac{1}{B+1}$, respectively. Part 2 of Theorem 2 gives us an $O(d^{1/B})$-approximation for $\mathcal{I}_1$ from a $(\log n, d)$-favorable fractional solution to the LP relaxation of $\mathcal{I}_1$. As for $\mathcal{I}_2$, note that each demand in $\mathcal{I}_2$ can be set equal to $\frac{1}{B}$ without affecting the space of feasible solutions. After making this change, part 3 of Theorem 2 gives us an $O(d^{1/B})$-approximation for $\mathcal{I}_2$. Either $\mathcal{I}_1$ or $\mathcal{I}_2$ has optimal profit at least half that of $\mathcal{I}$, so we can simply pick the better of the two approximate solutions. Finally, by Theorem 1, we can take $d = O(F \log n)$. This proves the third result. The last result, for UCUFP, follows similarly; we use Corollary 1 to take $d = O(F)$. □

For strong expanders we use a counting argument and the UCUFP approximation with $B = 2$ in Corollary 2 to obtain the following theorem.

**Theorem 3.** *For sufficiently large constant $\Delta$, there is an approximation algorithm with ratio $O(\sqrt{\log n})$ for UCUFP on $\Delta$-regular strong expanders.*

In the remainder of this section we prove the above theorems.

## 3.1   Producing a Favorable Solution

In this section we prove Theorem 1. We shall need the concept of a *balanced multicommodity flow problem* (BMFP), from Kolman and Scheideler [15]. For our purposes, a BMFP shall consist of an uncapacitated graph, a set of vertex pairs $\{(u_i, v_i)\}$, and demands $0 \leq \rho_i \leq 1$, one for each vertex pair. The total demand entering or leaving a vertex is required to equal its degree.

Suppose, as in the statement of Theorem 1, that we are given a fractional solution to LPMAIN with profit $W$. Let $\mathcal{P}$ be the set of all flow paths used by this solution. Set $L = 2cF/\varepsilon$. Let $\mathcal{P}'$ denote the subset of $\mathcal{P}$ consisting of paths with at least $2L$ edges of capacity at most $c$. We shall define a BMFP on the underlying uncapacitated graph $G$. For each flow path $\pi \in \mathcal{P}'$, orient it in the direction in which it carries flow. For a vertex $u$ on $\pi$, let $\mathrm{pred}_\pi(u)$ denote the vertex which is the predecessor of $u$ on $\pi$. We call $u$ a *good* vertex if $\mathrm{pred}_\pi(u)$ exists and the edge $(\mathrm{pred}_\pi(u), u)$ has capacity at most $c$. Let $u_1, u_2, \ldots, u_L$ be the first $L$ good vertices on $\pi$ and let $v_1, v_2, \ldots, v_L$ be the last $L$ good vertices. We add the pairs $\{(u_i, v_i) : 1 \le i \le L\}$, each with demand $f_\pi/c$, to the BMFP. Once we do this for all flow paths in $\mathcal{P}'$, the total demand entering or leaving any vertex is clearly *at most* its degree; we then add dummy demands, if required, to satisfy the definition of a BMFP. We shall need the following proposition, which follows from the work in [15].

**Proposition 1 (Kolman and Scheideler).** *A $\frac{1}{2F}$ fraction of all the demands in this BMFP can be concurrently satisfied on the underlying uncapacitated graph $G$, using a family of flow paths of length at most $2F$ each.*

Let $\mathcal{Q}$ be a family of flow paths guaranteed by Prop. 1. We reroute flow going through paths in $\mathcal{P}'$, through $\mathcal{Q}$. Notice that a path $\pi \in \mathcal{P}'$ is associated with $L$ paths in $\mathcal{Q}$ each of which "shortcuts" $\pi$. We send $f_\pi/L$ flow through each of these shortcuts, adjusting the flow on edges in $\pi$ appropriately. When we do this for all paths, we obtain a candidate fractional solution with profit $W$ that uses paths with at most $\max(L + 2F, 2L)$ edges of capacity at most $c$. Notice that $L + 2F \le 2L$, if $\varepsilon \le 1$. Thus, the flow paths in this candidate solution have at most $2L = 4cF/\varepsilon$ edges with capacity at most $c$.

This candidate solution could violate some edge capacities. However, by Prop. 1, had we sent $f_\pi/(2cF)$ flow through each shortcut for $\pi \in \mathcal{P}'$ we would have had a total flow of at most 1 on each edge. Since $L = 2cF/\varepsilon$, we in fact have a total flow of at most $\varepsilon$ on each edge due to the shortcuts. Thus, throwing in the flow paths in $\mathcal{P} \setminus \mathcal{P}'$ and scaling each flow value and each $x_i$ by $1/(1+\varepsilon)$, we will have a feasible solution. The new profit after scaling is clearly $W/(1+\varepsilon)$, which proves the first part of Theorem 1.

We adapt the above argument in a simple way to obtain a $(c, 4F/\varepsilon)$–favorable solution of profit $W/(c(1 + \varepsilon))$. Given a fractional solution with profit $W$ we scale each $x_i$ and $f_\pi$ by $1/c$, reducing the profit to $W/c$. For each edge in $G$ with capacity at most $c$, we reset the capacity to 1; the other capacities stay intact. This keeps the solution feasible, since $\rho_{\max} \le 1$. We apply the above argument to this modified solution with $L = 2F/\varepsilon$ to obtain the claimed result.

## 3.2   Rounding a Favorable Solution

We now turn to the proof of Theorem 2.

Srinivasan [19] and Baveja and Srinivasan [5] show that in a UCUFP instance, randomized rounding yields an $O(d)$ approximation if all flow path lengths are bounded by $d$. However, the proof is complicated and is based on the FKG inequality. In our case we have a fractional solution to LPMAIN that is $(\log n, d)$-favorable for some large enough $d$, and has profit $W$. It is conceivable that the analysis in [19,5] can be generalized to obtain an $O(d)$ approximation for our case but we believe it will be complex also. Instead, following the work of Srinivasan [21], we use randomized rounding followed by alteration to obtain an $O(d)$ approximation when rounding $(\log n, d)$-favorable fractional solutions. The proof is simple and transparent. Similar arguments yield an $O(1)$ approximation if $c_{min}$ is $\Omega(\log F)$ where $F$ is the flow number of the graph. For this latter case our proof is substantially simpler than the one in [15] which uses the Lovász Local Lemma.

Consider the following procedure: Fix an $i$ and randomly select exactly one of the paths in $\mathcal{P}_i$; for a path $\pi \in \mathcal{P}_i$, we select it with probability $\frac{f_\pi}{16d}$. Since $\sum_{\pi \in \mathcal{P}_i} f_\pi = x_i$, we will have selected *some* path in $\mathcal{P}_i$ with probability $\frac{x_i}{16d}$. Repeat this procedure for all $i \in \{1, \dots, k\}$. We shall refer to this as the *selection phase*.

Associate a path $\pi \in \mathcal{P}_i$ with the demand $\rho_i$; we shall write this demand as $\rho(\pi)$. Order all paths in $\bigcup_{i=1}^{k} \mathcal{P}_i$ in such a way that paths with $\rho(\pi) \geq \frac{1}{2}$ precede paths with $\rho(\pi) < \frac{1}{2}$. Starting from an empty set, add to it the paths picked in the selection phase, one by one, according to the above order; it is understood that when $\pi$ is added, the demand $\rho(\pi)$ is to be routed along $\pi$. If a path can be added to the current set of paths without violating any edge capacity, add it. Otherwise, discard it. We shall refer to this as the *pruning phase* or the *alteration phase*. It is clear that at the end of this phase we have a feasible integral solution.

**Lemma 1.** *The resulting random integral solution has expected profit* $\Omega(W/d)$.

Before we prove this lemma, we need a proposition, whose proof, based on Chernoff-Hoeffding bounds, we omit.

**Proposition 2.** *Let* $a_1, \dots, a_m, y_1, \dots, y_m \in [0, 1]$ *be such that* $\forall i < j : \left(a_i < \frac{1}{2} \Rightarrow a_j < \frac{1}{2}\right)$ *and furthermore* $\sum_{i=1}^{m} a_i y_i \leq 1$. *Let* $0 < \theta < 1$ *and let* $0\text{-}1$ *random variables* $Y_1, \dots, Y_m$ *and* $Z_1, \dots, Z_m$ *be defined as follows:*

$$Y_i = \begin{cases} 1, \text{ with probability } \theta y_i \\ 0, \text{ otherwise.} \end{cases} \quad ; \quad Z_i = \begin{cases} Y_i, \text{ if } \sum_{j<i} a_j Z_j \leq 1 - a_i \\ 0, \text{ otherwise.} \end{cases}$$

*Then, for each* $i$, $\Pr[Z_i = 0 \mid Y_i = 1] \leq (2 + 2e)\theta$.

*Proof (of Lemma 1).* Consider an edge $e \in E(G)$ and let $\pi_1, \dots, \pi_m$ be all the paths in the fractional solution that pass through $e$, arranged in the order that they are considered in the pruning phase. Let $X_i$ and $Z_i(e)$ be 0-1 random variables such that $X_i = 1$ iff $\pi_i$ was picked in the selection phase and $Z_i(e) = 1$ iff the inclusion of $\pi_i$ in the pruning phase did *not* violate the capacity of edge $e$. Then $X_i$ and $Z_i(e)$ are equivalent to $Y_i$ and $Z_i$ (respectively) of Prop. 2 with $y_i = f_{\pi_i}, a_i = \frac{\rho(\pi_i)}{c_e}$, and $\theta = \frac{1}{16d}$.

Note that $a_i \leq 1$ since $\rho(\pi_i) \leq 1$ and $c_e \geq 1$. Further, $\sum_{i=1}^{m} a_i y_i \leq 1$ because of the constraints in LPMAIN. Therefore Prop. 2 applies and we have

$$\Pr[Z_i(e) = 0 \mid X_i = 1] \leq \frac{2+2e}{16d}. \tag{1}$$

For those edges that satisfy $c_e > \log n$, we can say something stronger. Since $0 \le a_i \le 1/c_e$, the random variables $\{c_e a_i X_i\}_{i=1}^m$ are distributed in $[0,1]$ and their sum $X$ satisfies $\mathrm{E}[X] \le \frac{c_e}{16d}$. Now, for any $i$,

$$\Pr[Z_i(e) = 0 \mid X_i = 1] \le \Pr[\textstyle\sum_{j=1}^{i-1} a_j X_j > 1 - a_i] \le$$
$$\Pr[X > c_e - a_i c_e] \le \Pr[X > c_e - 1].$$

Let $\beta = \frac{c_e - 1}{\mathrm{E}[X]}$. The Chernoff-Hoeffding bound, after some routine algebra, gives

$$\Pr[Z_i(e) = 0 \mid X_i = 1] \le \left(\frac{e^\beta}{\beta^\beta}\right)^{\mathrm{E}[X]} \le \left(\frac{e}{\beta}\right)^{c_e - 1}$$

$$\le \left(\frac{e c_e}{16d(c_e - 1)}\right)^{c_e - 1} \le 2^{-3c_e} \le \frac{1}{n^3}, \qquad (2)$$

for large enough $d$. Finally, recall that we started with a fractional solution that was $(\log n, d)$-favorable. Let $Z_i$ be a 0-1 random variable with $Z_i = 1$ iff $\pi_i$ survives the pruning phase. Then (1) and (2) yield

$$\Pr[Z_i = 0 \mid X_i = 1] \le \textstyle\sum_{e \in \pi} \Pr[Z_i(e) = 0 \mid X_i = 1] \le d \cdot \frac{2 + 2e}{16d} + n^2 \cdot \frac{1}{n^3} \le \frac{1}{2},$$

for large enough $n$, because there at most $n^2$ edges on $\pi$ and at most $d$ of them have $c_e \le \log n$. This means that $\Pr[Z_i = 1] \ge (1 - \frac{1}{2}) \Pr[X_i = 1] = \frac{f_\pi}{32d}$. Therefore the expected profit of the solution after the pruning phase is at least $\frac{W}{32d} = \Omega(W/d)$.    $\square$

The first part of Theorem 2 follows immediately from Lemma 1. For the other two parts we need the following proposition which replaces Prop. 2 and is proven in a very similar manner.

**Proposition 3.** *Let* $\{a_i\}$, $\{y_i\}$, $\theta$, $\{Y_i\}$ *and* $\{Z_i\}$ *be exactly as in Prop. 2. Let* $B$ *be an integer. Then*

1. *If each* $a_i \in [0, \frac{1}{B}]$ *with* $B \ge 2$, *then* $\forall i$, $\Pr[Z_i = 0 \mid Y_i = 1] \le e^B \theta^{B-1}$.
2. *If each* $a_i = \frac{1}{B}$ *with* $B \ge 1$, *then* $\forall i$, $\Pr[Z_i = 0 \mid Y_i = 1] \le e^B \theta^B$.

### 3.3  An $O(\sqrt{\log n})$-Approximation for Strong Expanders

We now sketch the proof of Theorem 3. We shall need a result due to Frieze [10].

**Theorem 4 (Frieze).** *For an $n$-vertex $\Delta$-regular strong expander, with $\Delta$ sufficiently large, there exist constants $k_1, k_2$ such that any $(k_1 \Delta n / \log n)$ pairs of vertices, with no vertex appearing in more than $k_2 \Delta$ pairs, can be connected by disjoint paths of length $O(\log n)$ in polynomial time.*

*Proof (of Theorem 3).* Suppose we have an instance $\mathcal{I}$ of UCUFP. Fix an optimal integral solution $\mathcal{O}$ for $\mathcal{I}$ and partition the terminals pairs of $\mathcal{O}$ into three parts as follows:

- $\mathcal{O}_1$ includes exactly those pairs with demand at most $\frac{1}{2}$.
- $\mathcal{O}_2$ includes exactly those pairs not in $\mathcal{O}_1$, and routed by $\mathcal{O}$ on paths of length at most $\sqrt{\log n}$.
- $\mathcal{O}_3$ includes the rest of the pairs.

Our algorithm partitions $\mathcal{I}$ into two instances: $\mathcal{I}_1$, which is $\mathcal{I}$ restricted to demands $\rho_i \leq \frac{1}{2}$, and $\mathcal{I}_2$, which is $\mathcal{I} \setminus \mathcal{I}_1$. By Corollary 2, applied with $B = 2$, we can find a solution to $\mathcal{I}_1$ that $O(\Delta\beta^{-1} \log^{1/2} n)$-approximates the optimum (here $\beta$ is as in Definition 1). Since $\mathcal{O}_1$ is a feasible solution for $\mathcal{I}_1$, our solution is within the same factor of $\mathcal{O}_1$. Now we solve an LP relaxation of $\mathcal{I}_2$ with the added restriction that flow path lengths are at most $\sqrt{\log n}$. By part 1 of Theorem 2, the fractional solution can be rounded to give an $O(\sqrt{\log n})$-approximation to the LP optimum. Since $\mathcal{O}_2$ is feasible for the LP relaxation, we obtain an $O(\sqrt{\log n})$-approximation to the value of $\mathcal{O}_2$.

Finally, we pick the $(k_1 \Delta n / \log n)$ most profitable demands in $\mathcal{I}$, where $k_1$ is the constant in Theorem 4. By that theorem, we can find disjoint paths for all of them. Since each demand in $\mathcal{O}_3$ is more than $\frac{1}{2}$, any two flow paths in a feasible solution must be edge-disjoint, and thus the number of pairs in $\mathcal{O}_3$ is at most $|E(G)|/\sqrt{\log n} = O(\Delta n/\sqrt{\log n})$. It follows that simply accepting the most profitable pairs gives us an $O(\sqrt{\log n})$-approximation to the optimum of $\mathcal{O}_3$. Since one of $\mathcal{O}_1, \mathcal{O}_2, \mathcal{O}_3$ has at least a third of the profit of $\mathcal{O}$ and we approximated each within an $O(\sqrt{\log n})$ factor, we get the desired result.    $\square$

## 4    UFP on Line and Ring Networks

In this section we consider UFP restricted to the line network. We handle the ring network in a very similar fashion; we give the relevant details at the end of the section. Before we proceed, let us fix some notation. The terminal pairs now form intervals $I_1, I_2, \ldots, I_m$ on the line $[1, n]$, with $I_j$ having demand $\rho_j$ and weight (or profit) $w_j$. Edge $e$ on the line has capacity $c_e$, which we shall henceforth write as $c(e)$. For an edge $e$, let $\mathcal{I}(e)$ be the set of all demands (intervals) that contain $e$. Recall that we are working under the no-bottleneck assumption: $\rho_{\max} \leq 1 = \min_e c(e)$.

The UCUFP on the line is equivalent to a problem in resource allocation that has been studied recently [17,4,6,9]; however, we will not use the resource allocation terminology. Constant factor approximation algorithms for the resource allocation problem, and consequently UCUFP on the line, have been obtained via several different techniques — LP rounding [17], the local-ratio method [4,6], and primal-dual algorithms [4,6]. Most of these techniques do not seem to extend to UFP on the line where capacities are non-uniform. There is, however, one exception: a recent algorithm of Calinescu et al. [9] which gives constant factor approximations for UCUFP on the line. We extend their algorithm and analysis to non-uniform capacities. Their algorithm is the following: the demands are divided into two sets, one set containing demands which are "large" compared to the (common) capacity, say 1, and the other containing the rest. Dynamic programming is then invoked to find the optimal solution on the set of large demands. For the "small" demands, the algorithm solves the LP and then randomly rounds the solution (after scaling it by a constant $\alpha < 1$). The resulting set of demands has the right weight in expectation, but it may not be feasible. The alteration step then looks at the randomly chosen demands in order of their left end points, accepting a demand in the final output if adding it maintains feasibility. Since all edges have capacity 1, a demand $I_j$ is rejected in this step if demands sharing an edge with it and that have been inserted earlier add up to $1 - \rho_j$. However, these demands are small and their expected sum is at most $\alpha$, so applying a Chernoff bound shows the probability that a demand is chosen randomly and later rejected is small.

Our algorithm for UFP is very similar to that in [9], but the analysis requires new ideas. One difficulty is the following: in the alteration step, a demand $\rho_j$ which spans edges $e_1, e_2, \ldots, e_k$ in the left-to-right order is rejected if for some edge $e_i$, the demands already accepted that are using edge $e_i$ sum up to more than $c(e_k) - \rho_j$. In the uniform capacity case it is sufficient to just look at the edge $e_1$ for the rejection probability. In the non-uniform case, taking a union bound for the rejection probability over edges $e_1, \ldots, e_k$ is too weak to give a constant factor approximation and we need a more careful analysis.

Another idea is needed in defining small and large demands so that dynamic programming is still feasible for the large demands, and the small demands are still small enough to allow us to make the concentration arguments. To this end, we define the *bottleneck capacity* $b_j$ of a demand $I_j$ to be the capacity of the lowest capacity edge on this demand. Now a demand $I_j$ is $\delta$-*small* if $\rho_j \leq \delta b_j$, else it is $\delta$-*large*.

In the sequel, we show how to find the optimal solution for the $\delta$-large demands, and a constant factor approximation for the set of $\delta$-small demands, for some appropriate choice of $\delta$. We then output the better of the two solutions.

**Large Demands:** The following lemma is key to invoking dynamic programming to find an optimal solution for the $\delta$-large demands in $n^{O(1/\delta^2)}$ time ; the remaining details are routine and we omit them for lack of space.

**Lemma 2.** *The number of $\delta$-large demands that cross an edge in any feasible solution is at most $2\lfloor 1/\delta^2 \rfloor$.*

*Proof.* Fix a feasible solution $\mathcal{S}$, and consider an edge $e$. Let $S_e$ be the set of all $\delta$-large demands in $\mathcal{S}$ that cross $e$. We partition $S_e$ in to two sets $S_\ell$ and $S_r$ as follows: a demand in $S_e$ is in $S_\ell$ if it has a bottleneck capacity edge to the left of $e$ (including $e$), otherwise the demand is in $S_r$. We show that $|S_\ell| \leq \lfloor 1/\delta^2 \rfloor$, and a similar argument works for $|S_r|$.

Let $A$ be the set of bottleneck edges for demands in $S_\ell$ and let $e'$ be the rightmost edge in $A$. Since $e'$ is the bottleneck edge for some $\delta$-large demand $I_j \in S_\ell$, by definition, $\rho_j \geq \delta c(e')$. Since $\rho_j \leq c_{\min}$, it follows that $c(e') \leq c_{\min}/\delta$. Because $e'$ is the rightmost edge in $A$, all demands in $S_\ell$ pass through $e'$. But each demand $I_k$ in $S_\ell$ is $\delta$-large, which implies that $\rho_k \geq \delta b_k \geq \delta c_{\min}$. It follows that $|S_\ell| \leq \lfloor c(e')/(\delta c_{\min}) \rfloor \leq \lfloor 1/\delta^2 \rfloor$.                $\square$

**Small Demands:** When all demands are $\delta$-small, we give a 12-approximation for some small enough $\delta$. In the interest of clarity, we have not tried to optimize the constants too much. We first solve the LP LPMAIN for the problem. Let $x_j$ be the fractional value assigned to demand $I_j$. We define two $\{0,1\}$-random variables $X_j$ and $Y_j$ thus:

1. Let $\alpha < 1$ be a constant, to be chosen later. Let $X_j$ be set to 1 independently with probability $\alpha x_j$.
2. Sort the demands corresponding to $X_j = 1$ in order of their left end points (breaking ties arbitrarily). Look at them in this order, adding the current demand to the output if adding it does not violate any edge capacity. Set $Y_j = 1$ if demand $I_j$ is output.

By construction, this procedure produces a feasible solution. Clearly, $\mathrm{E}[X_j] = \Pr[X_j = 1] = \alpha x_j$, and the expected weight of the final solution is $\mathrm{E}[Y_j] = \Pr[Y_j = 1] = \alpha x_j \cdot \Pr[Y_j = 1|X_j = 1]$. The rest of the argument shows that $\Pr[Y_j = 0 \mid X_j = 1]$, the chance of rejection, is at most a half (for $\alpha = \frac{1}{6}$). This in turn shows that the expected weight is at least $\sum w_j x_j / 12$, giving us the claimed 12-approximation.

Let us focus on a particular demand $I_j$ with $X_j = 1$, and let $E_j = \langle e_1, \ldots, e_k \rangle$ be the edges on $I_j$ from left to right. The crucial idea is the following: when considering $I_j$, its probability of rejection depends on whether there is "enough room" on all these edges. Instead of taking a union bound over all edges, we choose a subsequence of edges such that the capacity of each edge drops by half, and such that for a demand to be rejected, a "bad" event happens at one of these chosen edges. Now a union bound on these bad events at these edges suffices. We show that this union bound gives us a geometric sum, and thus the chance of rejection is a constant times the probability of rejection on some edge $e_i$. Finally, arguments similar to that in [9] complete the proof.

Formally, create a subsequence $E'_j = \langle e_{i_1}, e_{i_2}, \ldots, e_{i_h} \rangle$ of $E_j$ as follows: set $i_1 = 1$, and hence $e_{i_1} = e_1$. For $\ell > 1$, set $i_\ell = \min\{t : t > i_{\ell-1}$ and $c(e_t) < c(e_{i_{\ell-1}})/2\}$. In other words $e_{i_\ell}$ is the first edge to the right of $e_{i_{\ell-1}}$ with capacity at most half the capacity of $e_{i_{\ell-1}}$. If there is no such edge we stop the construction of the sequence. For $1 \le a \le h$, let $\mathcal{E}_a$ denote the (bad) event that the random demands chosen in step 1 use at least $\frac{1}{2}c(e_{i_a}) - \delta b_j$ capacity in the edge $e_{i_a}$. The following lemma shows that it is enough to bound the chance that no bad event occurs on these chosen edges.

**Lemma 3.** $\Pr[Y_j = 0 \mid X_j = 1] \le \sum_{a=1}^{h} \Pr[\mathcal{E}_a]$.

*Proof.* If $Y_j = 0$ and $X_j = 1$ then some edge $e_g \in E_j$ had a capacity violation when $I_j$ was considered for insertion. Let $e_{i_a}$ be the edge in $E'_j$ to the left of $e_g$ and closest to it. (Here, an edge is considered to be "to the left" of itself.) Note that such an edge always exists since $e_{i_1} = e_1$, and $e_1$ is the left most edge in $I_j$. By the construction of the subsequence, $c(e_g) \ge \frac{1}{2}c(e_{i_a})$. If the capacity of $e_g$ was violated while trying to insert $I_j$, it must be that the capacity of demands already accepted that cross $e_g$ is at least $c(e_g) - \rho_j$ which is lower bounded by

$\frac{1}{2}c(e_{i_a}) - \delta b_j$: we use the fact that $I_j$ is small which implies that $\rho_j \leq \delta b_j$ and the fact that $c(e_g) \geq \frac{1}{2}c(e_{i_a})$. However, any interval that is accepted before $I_j$ and crosses $e_g$, must also cross $e_{i_a}$, and thus event $\mathcal{E}_a$ occurs. Applying the trivial union bound, we have $\Pr[Y_j = 0 \mid X_j = 1] \leq \sum_a \Pr[\mathcal{E}_a]$.     $\square$

We now show that the sequence $\Pr[\mathcal{E}_a]$ is *geometric*, which makes the union bound work:

**Lemma 4.** *For* $\alpha = \frac{1}{6}$ *and* $\delta = \frac{3}{100}$, *we have* $\Pr[\mathcal{E}_a] \leq \left(\frac{1}{3}\right)^{c(e_{i_a})}$, *and therefore,* $\sum_{a=1}^{h} \Pr[\mathcal{E}_a] \leq \frac{1}{2}$.

*Proof.* Let $Q_a = \sum_{I_s \in \mathcal{I}(e_{i_a})} \rho_s X_s$ be the random variable that gives the sum of demands that edge $a$ intersects *and* that are chosen in step 1. Since each $\rho_s \leq 1$, the independent variables $\{\rho_s X_s\}$ are distributed in $[0, 1]$. We have $\Pr[\mathcal{E}_a] = \Pr[Q_a \geq \frac{1}{2}c(e_{i_a}) - \delta b_j]$. Setting $\beta = (1/2 - \delta - \alpha)/\alpha$, and using the fact that $b_j \leq c(e_{i_a})$ gives $\Pr[\mathcal{E}_a] = \Pr[Q_a \geq \frac{1}{2}c(e_{i_a}) - \delta b_j] \leq \Pr[Q_a \geq (1 + \beta)\alpha c(e_{i_a})]$. Also, $\mathbb{E}[Q_a] = \sum_{I_s \in \mathcal{I}(e_{i_a})} \rho_s \mathbb{E}[X_s] = \sum_{I_s \in \mathcal{I}(e_{i_a})} \alpha \rho_s x_s \leq \alpha c(e_{i_a})$, where the last inequality follows from the feasibility of the LP solution. Since $Q_a$ is a sum of independent random variables distributed in $[0, 1]$ we apply a Chernoff-Hoeffding bound to get the following: $\Pr[Q_a \geq (1 + \beta)\alpha c(e_{i_a})] \leq \left(e^\beta / (1 + \beta)^{1+\beta}\right)^{\alpha c(e_{i_a})} \leq (1/3)^{c(e_{i_a})}$, where the final inequality follows by plugging in the constants we chose for $\alpha$ and $\delta$. Since $c(e_{i_a}) < c(e_{i_{a-1}})/2$ and each $c(e_{i_a}) \geq 1$, we now get: $\sum_{a=1}^{h} \Pr[\mathcal{E}_a] \leq \sum_a (1/3)^{c(e_{i_a})} \leq \sum_{i \geq 0} (1/3)^{2^i} \leq \sum_{j \geq 1} (1/3)^j = \frac{1}{2}$.     $\square$

Note that this implies that $\Pr[Y_j = 0 \mid X_j = 1] \leq \frac{1}{2}$, and hence gives us the claimed 12-approximation for the $\delta$-small demands. Combining this with the fact that we can optimally solve for the set of $\delta$-large demands, we get our main result of this section. We have not optimized the values of $\delta$ and $\alpha$ to obtain the best constant in the theorem below.

**Theorem 5.** *There is a 13-approximation for UFP on the line if* $\rho_{\max} \leq c_{\min}$.

**Corollary 3.** *There is a constant factor approximation for UFP on the line when* $\rho_{\max}/\rho_{\min}$ *is bounded even without the no-bottleneck assumption. Hence, for arbitrary demands we get an* $O(\log \rho_{\max})$ *approximation.*

*Proof.* Since the analysis for the $\delta$-small demands does not use the fact that $\rho_{\max} \leq c_{\min}$, we need to only consider the large demands. For the $\delta$-large demands, an argument similar to that in Lemma 2 works when $\rho_{\max}/\rho_{\min}$ is bounded.     $\square$

The performance of our algorithm matches the integrality gap of the LP to within a constant factor; we defer the proof to the full version.

**Theorem 6.** *The integrality gap of the natural LP is* $O(1)$ *when* $\rho_{\max} \leq c_{\min}$. *For arbitrary demands the integrality gap is* $\Theta(\log \rho_{\max})$ *which can be* $\Omega(n)$.

## 4.1 UFP on a Ring Network

Finally, we consider UFP on the ring network. Unlike the line network, this gives us a choice of one of two paths for each demand. However, we can reduce the problem on the ring to that on a line network with a slight loss in the approximation factor as follows. Let $e$ be any edge on the ring with $c(e) = c_{min}$. Consider any integral optimal solution $\mathcal{O}$ to the problem. The demands routed in $\mathcal{O}$ can be partitioned into two sets $\mathcal{O}_1$ and $\mathcal{O}_2$ where those in $\mathcal{O}_1$ use $e$ and those in $\mathcal{O}_2$ do not. We remove $e$ and solve the problem approximately on the resulting line network. This clearly approximates the value of $\mathcal{O}_2$. To approximate the solution for $\mathcal{O}_1$, for each demand we choose the path that uses $e$ and solve a knapsack problem to find the maximum weight set of demands that can be routed with capacity bounded by $c_e$. Since $c(e) = c_{min}$ any solution feasible at $e$ will be feasible for the entire network. Thus we obtain:

**Theorem 7.** *For UFP on the ring there is a $(1 + \alpha)$ approximation where $\alpha$ is the approximation factor for the problem on the line.*

## 5 Concluding Remarks

We also note that an online $O(F \log n)$-approximation for UFP can be obtained by combining the bounded greedy algorithm [13], and the online algorithm of Awerbuch, Azar and Plotkin [2]; we defer the details to the full version of the paper.

### Acknowledgments

We are grateful to Bruce Shepherd for suggesting the unsplittable flow problem on the line and for several discussions.

## References

1. N. Alon and J. Spencer. *The Probabilistic Method*. Wiley Interscience, New York, 1992.
2. B. Awerbuch, Y. Azar, and S. Plotkin. Throughput-competitive online routing. In *Proceedings of the 34th Annual IEEE Symposium on Foundations of Computer Science*, pp. 32–40. 1993.
3. Y. Azar and O. Regev. Strongly polynomial algorithms for the unsplittable flow problem. In *Proceedings of the 8th Integer Programming and Combinatorial Optimization Conference*. 2001.
4. A. Bar-Noy, R. Bar-Yehuda, A. Freund, J. S. Naor, and B. Scheiber. A unified approach to approximating resource allocation and scheduling. In *Proceedings of the 32nd Annual ACM Symposium on Theory of Computing*, pp. 735–744, 2000.
5. A. Baveja and A. Srinivasan. Approximation algorithms for disjoint paths and related routing and packing problems. *Math. Oper. Res.*, 25(2):255–280, 2000.

6. P. Berman and B. DasGupta. Improvements in throughput maximization for real-time scheduling. In *Proceedings of the 32nd Annual ACM Symposium on Theory of Computing*, pp. 680–687, 2000.

7. T. Bohman and A. M. Frieze. Arc-disjoint paths in expander digraphs. In *Proceedings of the 42nd Annual IEEE Symposium on Foundations of Computer Science*. 2001.

8. A. Z. Broder, A. M. Frieze, and E. Upfal. Existence and construction of edge-disjoint paths on expander graphs. *SIAM Journal on Computing*, 23(5):976–989, 1994.

9. G. Calinescu, A. Chakrabarti, H. Karloff, and Y. Rabani. Improved approximation algorithms for resource allocation. In *Proceedings of the 9th Integer Programming and Combinatorial Optimization Conference*, 2002.

10. A. M. Frieze. Edge-disjoint paths on expander graphs. *SIAM Journal on Computing*, 30(6):1790–1801, 2001.

11. V. Guruswami, S. Khanna, R. Rajaraman, F. B. Shepherd, and M. Yannakakis. Near-optimal hardness results and approximation algorithms for edge-disjoint paths and related problems. In *Proceedings of the 31st Annual ACM Symposium on Theory of Computing*, pp. 19–28. 1999.

12. J. M. Kleinberg. *Approximation Algorithms for Disjoint Paths Problems*. Ph.D. thesis, MIT, 1996.

13. J. M. Kleinberg and R. Rubinfeld. Short paths in expander graphs. In *Proceedings of the 37th Annual IEEE Symposium on Foundations of Computer Science*, pp. 86–95. 1996.

14. P. Kolman and S. Scheideler. Simple on-line algorithms for the maximum disjoint paths problem. In *Proceedings of 13th ACM Symposium on Parallel Algorithms and Architectures*. 2001.

15. P. Kolman and S. Scheideler. Improved bounds for the unsplittable flow problem. In *Proceedings of the 13th Annual ACM-SIAM Symposium on Discrete Algorithms*. 2002.

16. F. T. Leighton and S. B. Rao. Multicommodity max-flow min-cut theorems and their use in designing approximation algorithms. *Journal of the ACM*, 46(6):787–832, 1999. (Preliminary version in *29th Annual Symposium on Foundations of Computer Science*, pages 422–431, 1988).

17. C. A. Phillips, R. N. Uma, and J. Wein. Off-line admission control for general scheduling problems. In *Proceedings of the 11th Annual ACM-SIAM Symposium on Discrete Algorithms*, pp. 879–888. 2000.

18. P. Raghavan and C. D. Thompson. Randomized rounding: a technique for provably good algorithms and algorithmic proofs. *Combinatorica*, 7(4):365–374, 1987.

19. A. Srinivasan. Improved approximations for edge-disjoint paths, unsplittable flow, and related routing problems. In *Proceedings of the 38th Annual IEEE Symposium on Foundations of Computer Science*, pp. 416–425. 1997.

20. A. Srinivasan. Improved approximation guarantees for packing and covering integer programs. *SIAM J. Comput.*, 29(2):648–670, 1999.

21. A. Srinivasan. New approaches to covering and packing problems. In *Proceedings of the 12th Annual ACM-SIAM Symposium on Discrete Algorithms*, pp. 567–576. 2001.

# 1.5-Approximation for Treewidth of Graphs Excluding a Graph with One Crossing as a Minor*

Erik D. Demaine[1], MohammadTaghi Hajiaghayi[1], and Dimitrios M. Thilikos[2]

[1] Laboratory for Computer Science, Massachusetts Institute of Technology,
200 Technology Square, Cambridge, MA 02139, U.S.A.
[2] Departament de Llenguatges i Sistemes Informàtics,
Universitat Politècnica de Catalunya,
Campus Nord – Mòdul C5, Desp. 211b, c/Jordi Girona Salgado, 1-3,
E-08034, Barcelona, Spain

**Abstract.** We give polynomial-time constant-factor approximation algorithms for the treewidth and branchwidth of any $H$-minor-free graph for a given graph $H$ with crossing number at most 1. The approximation factors are 1.5 for treewidth and 2.25 for branchwidth. In particular, our result directly applies to classes of nonplanar graphs such as $K_5$-minor-free graphs and $K_{3,3}$-minor-free graphs. Along the way, we present a polynomial-time algorithm to decompose $H$-minor-free graphs into planar graphs and graphs of treewidth at most $c_H$ (a constant dependent on $H$) using clique sums. This result has several applications in designing fully polynomial-time approximation schemes and fixed-parameter algorithms for many NP-complete problems on these graphs.

## 1 Introduction

Treewidth plays an important role in the complexity of several problems in graph theory. The notion was first defined by Robertson and Seymour in [RS84] and served as one of the cornerstones of their lengthy proof of the Wagner conjecture, now known as the Graph Minors Theorem. (For a survey, see [RS85].) Treewidth also has several applications in algorithmic graph theory. In particular, a wide range of otherwise-intractable combinatorial problems are polynomially solvable, often linearly solvable, when restricted to graphs of bounded treewidth [ACP87,Bod93].

Roughly speaking, the *treewidth* of a graph is the minimum $k$ such that the graph can be "decomposed" into a tree structure of bags, with each vertex of graph spread over a connected subtree of bags, so that edges only connect two

---

* The work of the third author was supported by the IST Programme of the EU under contract number IST-1999-14186 (ALCOM-FT), the Spanish CICYT project TIC2000-1970-CE, and the Ministry of Education and Culture of Spain (Resolución 31/7/00 – BOE 16/8/00). Emails: {hajiagha, edemaine}@theory.lcs.mit.edu, and sedthilk@lsi.upc.es

K. Jansen et al. (Eds.): APPROX 2002, LNCS 2462, pp. 67–80, 2002.
© Springer-Verlag Berlin Heidelberg 2002

vertices occupying a common bag, and at most $k + 1$ vertices occupy each bag. (For the precise definition, see Section 2.)

Much research has been done on computing and approximating the treewidth of a graph. Computing treewidth is NP-complete even if we restrict the input graph to graphs of bounded degree [BT97], cocomparability graphs [ACP87, HM94], bipartite graphs [Klo93], or the complements of bipartite graphs [ACP87]. On the other hand, treewidth can be computed exactly in polynomial time for chordal graphs, permutation graphs [BKK95], circular-arc graphs [SSR94], circle graphs [Klo93], and distance-hereditary graphs [BDK00].

From the approximation viewpoint, Bodlaender et al. [BGHK95] gave an $O(\log n)$-approximation algorithm for treewidth on general graphs. A famous open problem is whether treewidth can be approximated within constant factor. Treewidth can be approximated within constant factor on AT-free graphs [BT01] (see also [BKMT]) and on planar graphs. The approximation for planar graphs is a consequence of the polynomial-time algorithm given by Seymour and Thomas [ST94] for computing the parameter *branchwidth*, whose value approximates treewidth within a factor of 1.5. To our knowledge, until now it remained open whether treewidth could be approximated within a constant factor for other kinds of graphs.

In this paper, we make a significant step in this direction. We prove that, if $H$ is a graph that can be drawn in the plane with a single crossing (a *single-crossing* graph), then there is a polynomial-time algorithm that computes the treewidth of any $H$-minor-free graph. The two simplest examples of such graph classes are $K_5$-minor-free graphs and $K_{3,3}$-minor-free graphs. Our result is based on a structural characterization of the graphs excluding a single-crossing graph as a minor. This characterization allows us to decompose such a graph into planar graphs and graphs of small treewidth according to clique sums. This decomposition theorem is a generalization of the current decomposition results for graphs excluding special single-crossing graph such as $K_{3,3}$ [Asa85] and $K_5$ [KM92]. We also show how this decomposition can be computed in polynomial time.

Our decomposition theorem has two main applications. First, we show how the tree decomposition and treewidth of each component in the decomposition can be combined in order to obtain an approximation for the whole input graph. Second, we show how the constructive decomposition can be applied to obtain fully polynomial-time schemes and fixed-parameter algorithms for a wide variety of NP-complete problems on these graphs.

This paper is organized as follows. First, in Section 2, we introduce the terminology used throughout the paper, and formally define the parameters treewidth and branchwidth. In Section 3, we introduce the concept of clique-sum graphs and prove several results on the structure of graphs excluding single-crossing graphs as minors. The main approximation algorithm is described in Section 4. In Section 5, we present several applications of clique-sum decompositions in designing algorithms for these graphs. Finally, in Section 6, we conclude with some remarks and open problems.

## 2   Background

### 2.1   Preliminaries

All the graphs in this paper are undirected without loops or multiple edges. The reader is referred to standard references for appropriate background [BM76].

Our graph terminology is as follows. A graph $G$ is represented by $G = (V, E)$, where $V$ (or $V(G)$) is the set of vertices and $E$ (or $E(G)$) is the set of edges. We denote an edge $e$ between $u$ and $v$ by $\{u, v\}$. We define $n$ to be the number of vertices of a graph when this is clear from context.

The *(disjoint) union* of two disjoint graphs $G_1$ and $G_2$, $G_1 \cup G_2$, is the graph $G$ with merged vertex and edge sets: $V(G) = V(G_1) \cup V(G_2)$ and $E(G) = E(G_1) \cup E(G_2)$.

One way of describing classes of graphs is by using *minors*, introduced as follows. *Contracting* an edge $e = \{u, v\}$ is the operation of replacing both $u$ and $v$ by a single vertex $w$ whose neighbors are all vertices that were neighbors of $u$ or $v$, except $u$ and $v$ themselves. A graph $G$ is a *minor* of a graph $H$ if $H$ can be obtained from a subgraph of $G$ by contracting edges. A graph class $\mathcal{C}$ is a *minor-closed* class if any minor of any graph in $\mathcal{C}$ is also a member of $\mathcal{C}$. A minor-closed graph class $\mathcal{C}$ is $H$-*minor-free* if $H \notin \mathcal{C}$.

For example, a planar graph is a graph excluding both $K_{3,3}$ and $K_5$ as minors.

### 2.2   Treewidth

The notion of treewidth was introduced by Robertson and Seymour [RS86] and plays an important role in their fundamental work on graph minors. To define this notion, first we consider the representation of a graph as a tree, which is the basis of our algorithms in this paper. A *tree decomposition* of a graph $G = (V, E)$, denoted by $TD(G)$, is a pair $(\chi, T)$ in which $T = (I, F)$ is a tree and $\chi = \{\chi_i | i \in I\}$ is a family of subsets of $V(G)$ such that: (1) $\bigcup_{i \in I} \chi_i = V$; (2) for each edge $e = \{u, v\} \in E$ there exists an $i \in I$ such that both $u$ and $v$ belong to $\chi_i$; and (3) for all $v \in V$, the set of nodes $\{i \in I | v \in \chi_i\}$ forms a connected subtree of $T$. To distinguish between vertices of the original graph $G$ and vertices of $T$ in $TD(G)$, we call vertices of $T$ *nodes* and their corresponding $\chi_i$'s *bags*. The maximum size of a bag in $TD(G)$ minus one is called the *width* of the tree decomposition. The *treewidth* of a graph $G$ (tw($G$)) is the minimum width over all possible tree decompositions of $G$.

### 2.3   Branchwidth

A *branch decomposition* of a graph $G$ is a pair $(T, \tau)$, where $T$ is a tree with vertices of degree 1 or 3 and $\tau$ is a bijection from the set of leaves of $T$ to $E(G)$. The *order* of an edge $e$ in $T$ is the number of vertices $v \in V(G)$ such that there are leaves $t_1, t_2$ in $T$ in different components of $T(V(T), E(T) - e)$ with $\tau(t_1)$ and $\tau(t_2)$ both containing $v$ as an endpoint. The *width* of $(T, \tau)$ is the maximum order over all edges of $T$, and the *branchwidth* of $G$ is the minimum width over all branch decompositions of $G$. The following result implies that branchwidth is a 1.5-approximation on treewidth:

**Theorem 1 ([RS91], Section 5).** *For any graph $G$ with $m$ edges, there exists an $O(m^2)$-time algorithm that*

1. *given a branch decomposition $(T, \tau)$ of $G$ of width $\leq k+1$, constructs a tree decomposition $(\chi, T)$ of $G$ that has width $\leq \frac{3}{2}k$; and*
2. *given a tree decomposition $(\chi, T)$ of $G$ that has treewidth $k+1$, constructs a branch decomposition $(T, \tau)$ of $G$ of width $\leq k$.*

While the complexity of treewidth on planar graphs remains open, the branchwidth of a planar graph can be computed in polynomial time:

**Theorem 2 ([ST94], Sections 7 and 9).** *One can construct an algorithm that, given a planar graph $G$,*

1. *computes in $O(n^3)$ time the branchwidth of $G$; and*
2. *computes in $O(n^5)$ time a branch decomposition of $G$ with optimal width.*

Combining Theorems 1 and 2, we obtain a polynomial-time 1.5-approximation for treewidth in planar graphs:

**Theorem 3.** *One can construct an algorithm that, given a planar graph $G$,*

1. *computes in $O(n^3)$ time a value $k$ with $k \leq \mathrm{tw}(G) + 1 \leq \frac{3}{2}k$; and*
2. *computes in $O(n^5)$ time a tree decomposition of $G$ with width $k$.*

This approximation algorithm will be one of two "base cases" in our development in Section 4 of a 1.5-approximation algorithm for nonplanar graphs excluding a single-crossing graph as a minor.

# 3  Computing Clique-Sum Decompositions for Graphs Excluding a Single-Crossing-Graph Minor

This section describes the general framework of our results, using the key tool of *clique-sums*; see [HNRT01,Haj01].

## 3.1  Clique Sums

Suppose $G_1$ and $G_2$ are graphs with disjoint vertex-sets and $k \geq 0$ is an integer. For $i = 1, 2$, let $W_i \subseteq V(G_i)$ form a clique of size $k$ and let $G'_i$ ($i = 1, 2$) be obtained from $G_i$ by deleting some (possibly no) edges from $G_i[W_i]$ with both endpoints in $W_i$. Consider a bijection $h : W_1 \to W_2$. We define a $k$-sum $G$ of $G_1$ and $G_2$, denoted by $G = G_1 \oplus_k G_2$ or simply by $G = G_1 \oplus G_2$, to be the graph obtained from the union of $G'_1$ and $G'_2$ by identifying $w$ with $h(w)$ for all $w \in W_1$. The images of the vertices of $W_1$ and $W_2$ in $G_1 \oplus_k G_2$ form the *join set*.

In the rest of this section, when we refer to a vertex $v$ of $G$ in $G_1$ or $G_2$, we mean the corresponding vertex of $v$ in $G_1$ or $G_2$ (or both). It is worth mentioning that $\oplus$ is not a well-defined operator and it can have a set of possible results. See Figure 1 for an example of a 5-sum operation.

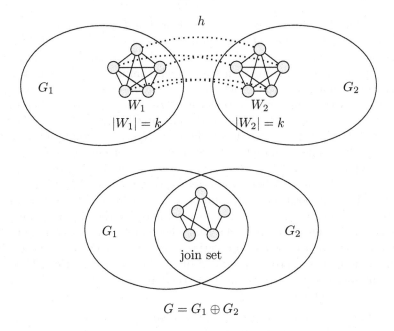

$$G = G_1 \oplus G_2$$

**Fig. 1.** Example of 5-sum of two graphs.

## 3.2  Connection to Treewidth

The following lemma shows how the treewidth changes when we apply a clique-sum operation, which will play an important role in our approximation algorithms in Section 4. This lemma is mentioned in [HNRT01] without proof. For the sake of completeness, we present the proof here.

**Lemma 1 ([HNRT01]).** *For any two graphs $G$ and $H$, $\mathrm{tw}(G \oplus H) \le \max\{\mathrm{tw}(G), \mathrm{tw}(H)\}$.*

*Proof.* Let $W$ be the set of vertices of $G$ and $H$ identified during the $\oplus$ operation. Since $W$ is a clique in $G$, in every tree decomposition of $G$, there exists a node $\alpha$ such that $W$ is a subset of $\chi_\alpha$ [BM93]. Similarly, the same is true for $W$ and a node $\alpha'$ of each tree decomposition of $H$. Hence, we can construct a tree decomposition of $G$ and a tree decomposition of $H$ and add an edge between $\alpha$ and $\alpha'$.                                                    □

## 3.3  Computing Clique-Sum Decompositions

The main theorem of this section is an algorithmic version of the following existential theorem of Robertson and Seymour:

**Theorem 4 ([RS93]).** *For any single-crossing graph $H$, there is an integer $c_H \ge 4$ (depending only on $H$) such that every graph with no minor isomorphic to $H$ can be obtained by 0-, 1-, 2- or 3-sum of planar graphs and graphs of treewidth at most $c_H$.*

We use Theorem 4 in our constructive algorithm. For a graph $G = (V, E)$, we call a subset $S$ of $V(G)$ a $k$-*cut* if the induced subgraph $G[V - S]$ is disconnected and $|S| = k$. A $k$-cut is *strong* if $G - S$ has more than two connected components, or it has two connected components and each component has more than one vertex. This definition is less strict than the notion of strong cuts introduced in [KM92], where a similar (but consequently weaker) version of Lemma 2 is obtained. Let $S \subseteq V(G)$ be a cut that separates $G$ into $h \geq 2$ components $G_1, \cdots, G_h$. For $1 \leq i \leq h$, we denote by $G_i \cup K(C)$ the graph obtained from $G[V(G_i) \cup C]$ by adding an edge between any pair of nonadjacent vertices in $C$. The graphs $G_i \cup K(C)$, $1 \leq i \leq k$, are called the *augmented components* induced by $C$. Theorem 5 below describes a constructive algorithm to obtain a clique-sum decomposition of a graph $G$ as in Theorem 4 with the additional property that the decomposition graphs are minors of the original graph $G$. In this sense, the result is even stronger than Theorem 4. This additional property, that each graph in the clique-sum series is a minor of the original graph, is crucial for designing approximation algorithms in the next section; Theorem 4 alone would not suffice. First we illustrate the important influence of strong cuts on augmented components:

**Lemma 2.** *Let $C$ be a strong 3-cut of a 3-connected graph $G = (V, E)$, and let $G_1, G_2, \cdots, G_h$ denote the $h$ induced components of $G[V - C]$. Then each augmented component of $G$ induced by $C$, $G_i \cup K(C)$, is a minor of $G$.*

*Proof.* Suppose $C = \{x, y, z\}$. First consider the case that $C$ disconnects the graph into at least 3 components. By symmetry, it suffices show that $G_1 \cup K(C)$ is a minor of $G$. Contract all edges of $G_2$ and $G_3$ to obtain super-vertices $y'$ and $z'$. Because $G$ is 3-connected, both $y'$ and $z'$ are adjacent to all vertices in $C$. Now contract edges $\{y, y'\}$ and $\{z, z'\}$ to obtain super-vertices $y''$ and $z''$, respectively. Then $x$, $y''$, and $z''$ form a clique, so we have arrived at the augmented component $G_1 \cup K(C)$ via contractions. Next consider the case in which $G[V - C]$ has only two components $G_1$ and $G_2$, and both have at least two vertices. Again it suffices to show that $G_1 \cup K(C)$ is a minor of $G$. First suppose that $G_2$ is a tree. Because $G$ is 3-connected, there is a vertex $x'$ in $G_2$ that neighbors $x$, and similarly a vertex $z'$ in $G_2$ that neighbors $z$. Now contract every other vertex of $G_2$ arbitrarily to either $x'$ or $z'$, to obtain super-vertices $x''$ and $z''$. Because $G_2$ is connected, there is an edge between $x''$ and $z''$. Because $G_2$ has no cycle, there cannot be more than one edge between the components corresponding to super-vertices $x''$ and $z''$. Because $G$ is 3-connected, there is an edge between $y$ and either $x''$ or $z''$, say $x''$. Again because $G$ is 3-connected, $z''$ is connected to a vertex of $C$ other than $z$. If $z''$ is adjacent to $y$, contract the edges $\{x'', x\}$ and $\{z'', z\}$, and if $z''$ is adjacent to $x$, contract the edges $\{x'', y\}$ and $\{z'', z\}$, to form a clique on the vertices of $C$. Finally suppose that $G_2$ has a cycle $C'$. We claim that there are three vertex-disjoint paths connecting three vertices of $C'$ to three vertices of $C$ in $G_2$. By contracting these paths and then contracting edges of $C'$ to form a triangle, we have a clique on the vertices of $C$ as desired. To prove the claim, augment the graph $G$ by adding a vertex $v_1$ connected to every vertex in $C$, and by adding a vertex $v_2$ connected to every vertex in $C'$. Because $|C| = 3$ and $|C'| \geq 3$, the augmented graph is still vertex 3-connected. Therefore there exist at least three vertex-disjoint paths from $v_1$ to $v_2$. Each of these paths must be in $G_2$, begin by entering a vertex of $C$, and end by leaving a vertex of $C'$, and these vertices of $C$ and $C'$ must be different among the three paths (because they are vertex-disjoint). Thus, if we remove the first vertex $v_1$ and last vertex $v_2$ from each path, we obtain the desired paths.    □

**Theorem 5.** *For any graph $G$ excluding a single-crossing graph $H$ as a minor, we can construct in $O(n^4)$ time a series of clique-sum operations $G = G_1 \oplus G_2 \oplus \cdots \oplus G_m$ where each $G_i$, $1 \leq i \leq m$, is a minor of $G$ and is either a planar graph or a graph of treewidth at most $c_H$. Here each $\oplus$ is a 0-, 1-, 2- or 3-sum.*

*Proof.* The algorithm works basically as follows. Given a graph $G$, compute its connectivity. If it is disconnected, consider each of its connected components separately. If it has a 1-cut or 2-cut, recursively apply the algorithm on the augmented components induced by that 1-cut or 2-cut. If its connectivity is at least three, find a strong 3-cut and recursively apply the algorithm on the augmented components induced by that strong 3-cut. If the graph is 3-connected but has no strong 3-cut, then we claim that it is either planar or has treewidth at most $c_H$.

We first prove the correctness of the algorithm above, and later fill in the algorithmic details and analyze the running time. If $G$ has a 1-cut or 2-cut, then each augmented component is a minor of $G$, and thus by Theorem 4 we can recurse on each augmented component. The same holds for strong 3-cuts if $G$ is 3-connected, because Theorem 2 implies that the property of excluding graph $H$ as a minor is inherited by all its augmented components. Now suppose that the graph $G$ is 3-connected yet it has no strong 3-cut. It remains to show that either the treewidth of $G$ is greater than $c_H$ or that $G$ is planar. Suppose for contradiction that neither of these properties hold. By Theorem 4, $G$ can be obtained by 3-sums of a sequence of elementary graphs $\mathcal{C} = (J_1, \ldots, J_r)$. (Because $G$ is 3-connected, we have no $k$-sums for $k \leq 2$.) We claim that one of the graphs in $\mathcal{C}$ must be a planar graph with at least five vertices. If this were not the case, then all the graphs in $\mathcal{C}$ would have treewidth at most $c_H$ so, by Lemma 1, $G$ would also have treewidth $\leq c_H$, which is a contradiction. Notice also that we can not have more than one graph in $\mathcal{C}$ with at least five vertices because we do not have strong 3-cuts. Therefore, $\mathcal{C}$ contains a planar graph with at least 5 vertices and all the other graphs in $\mathcal{C}$ are $K_4$'s. We claim that $G$ itself must be planar, establishing a contradiction. Suppose to the contrary that, during the clique-sum operations forming $G$, there is a 3-sum $G'' = G' \oplus K_4$ with join set $C$ such that $G'$ is planar but $G''$ is not planar. Consider a planar embedding of $G'$. Because $C$ is a triangle in $G'$ and $G' \oplus K_4$ is not planar, there are some vertices inside triangle $C$ and some vertices outside triangle $C$. Thus $G'' - C$ has at least three components so $C$ is a strong 3-cut in $G''$. Because $G''$ is a graph in the clique-sum sequence of $G$, $C$ is also a strong 3-cut in $G$, which is again a contradiction.

To analyze the running time of the algorithm, first we claim that, for a $H$-minor-free graph $G$ where $H$ is single-crossing, we have $|E(G)| = O(|V(G)|)$. This claim follows because the number of edges in planar graphs and graphs of treewidth at most $c_H$ is a linear function in the number of vertices, and the total number of vertices of graphs in a clique-sum sequence forming $G$ is linear in $|V(G)|$ (we have linear number of $k$-sums and $k \leq 3$). In linear time we can obtain all 1-cuts [Tar72] and we can obtain all 2-cuts using the algorithms of Hopcroft and Tarjan [HT73] or Miller and Ramachandran [MR92]. The number of 3-cuts in a 3-connected graph is $O(n^2)$ and we can obtain all 3-cuts in $O(n^2)$ time [KR91]. We can check whether each 3-cut is strong in $O(n)$ time using a depth-first search. All other operations including checking planarity and having treewidth at most $c_H$ can be performed in linear time [Wil84,Bod96]. Now, if the algorithm makes no recursive calls, the running time of the algorithm, $T(n)$, is $O(n)$. If it makes recursive calls for a 1-cut, we have that $T(n) = T(n_1) + T(n - n_1 + 1) + O(n)$, $n_1 \geq 2$, where $n_1$ and $n - n_1 + 1$ are the sizes of the two augmented components. (We only split the graph into two 2-connected components at once, possibly leaving the same 1-cut for the recursive calls.) Similarly, for recursive calls for a 2-cut, we have $T(n) = T(n_1) + T(n - n_1 + 2) + O(n)$, $n_1 \geq 3$. For recursive calls for a strong 3-cut with exactly two components, we have $T(n) = T(n_1) + T(n - n_1 + 3) + O(n^3)$, $n_1 \geq 4$. Finally, if we have recursive calls for a strong 3-cut with at least three components, we have that $T(n) = T(n_1) + T(n_2) + T(n - n_1 - n_2 + 6) + O(n^3)$, $4 \leq n_1, n_2, n - n_1 - n_2 + 6 \leq n - 2$, where $n_1$, $n_2$, and $n - n_1 - n_2 + 6$ are the sizes of the augmented components. (Again, we only split the graph into three 3-connected components, possibly leaving the same 3-cut for the recursive calls.) Solving this recurrence concludes a worst-case running time of $O(n^4)$.                                  □

We can also parallelize this algorithm to run in $O(\log^2 n)$ time using an approach similar to that described by Kezdy and McGuinness [KM92]. The details are omitted from this paper.

## 3.4   Related Work

Theorems 2 and 5 generalize a characterization of $K_{3,3}$-minor-free graphs and $K_5$-minor-free graphs by Wagner [Wag37]. He proved that a graph has no minor isomorphic to $K_{3,3}$ if and only if it can be obtained from planar graphs and $K_5$ by 0-, 1-, and 2-sums. He also showed that a graph has no minor isomorphic to $K_5$ if and only if it can be obtained from planar graphs and $V_8$ by 0-, 1-, 2-, and 3-sums. Here $V_8$ denotes the graph obtained from a cycle of length 8 by joining each pair of diagonally opposite vertices by an edge. We note that both $K_5$ and $V_8$ have treewidth 4, i.e., $c_H = 4$. Constructive algorithms for obtaining such clique-sum series have also been developed. Asano [Asa85] showed how to construct in $O(n)$ time a series of clique-sum operations for $K_{3,3}$-minor-free graphs. Kézdy and McGuinness [KM92] presented an $O(n^2)$-time algorithm to construct such a clique-sum series for $K_5$-minor-free graphs.

## 4    Approximating Treewidth

We are now ready to prove our final result, a 1.5-approximation algorithm on treewidth:

**Theorem 6.** *For any single-crossing graph $H$, we can construct an algorithm that, given an $H$-minor-free graph as input, outputs in $O(n^5)$ time a tree decomposition of $G$ of width $k$ where $\mathrm{tw}(G) \leq k+1 \leq \frac{3}{2}\mathrm{tw}(G)$.*

*Proof.* The algorithm consists of the following four steps:

**Step 1:** Let $G$ be a graph excluding a single-crossing graph $H$. By Theorem 5, we can obtain a clique-sum decomposition $G = G_1 \oplus G_2 \oplus \cdots \oplus G_m$ where each $G_i$, $1 \leq i \leq m$, is a minor of $G$ and is either a planar graph or a graph of treewidth at most $c_H$. According to the same theorem, this step requires $O(n^4)$ time. Let $B$ be the set of bounded treewidth components and $P$ be the set of planar components: $B = \{i \mid 1 \leq i \leq m, \ \mathrm{tw}(G_i) \leq c_H\}$, $P = \{1, \ldots, m\} - B$.

**Step 2:** By Theorem 3, we can construct, for any $i \in P$, a tree decomposition $D_i$ of $G_i$ with width $k_i$ and such that

$$k_i \leq \mathrm{tw}(G_i) + 1 \leq \frac{3}{2}k_i \quad \text{for all } i \in P. \tag{1}$$

The construction of each of these tree decompositions requires $O(|V(G_i)|^5)$ time. As $m = O(n)$ and $\sum_{1 \leq i \leq m} |V(G_i)| = O(n)$, the total time for this step is $O(n^5)$.

**Step 3:** Using Bodlaender's algorithm in [Bod96], for any $i \in B$, we can obtain a tree decomposition of $G_i$ with minimum width $k_i$, in linear time where the hidden constant depends only on $c_H$. Combining (1) with the fact that $\mathrm{tw}(G_i) = k_i$ for each $i \in B$, we obtain

$$k_i \leq \mathrm{tw}(G_i) + 1 \leq \frac{3}{2}k_i \quad \text{for all } i \in \{1, \ldots, m\}. \tag{2}$$

**Step 4:** Now that we have tree decompositions $D_i$ of each $G_i$, we glue them together using the construction given in the proof of Lemma 1. In this way, we obtain a tree decomposition of $G$ that has size $k = \max\{k_i \mid 1 \leq i \leq m\}$. Combining this equality with (2), we have

$$k \leq \max\{\mathrm{tw}(G_i) \mid i = 1, \ldots, m\} + 1 \leq \frac{3}{2}k. \tag{3}$$

Finally, we prove that the algorithm is a 1.5-approximation. By Lemma 1, we have that $\mathrm{tw}(G) \leq \max\{\mathrm{tw}(G_i) \mid i = 1, \ldots, m\}$. By Theorem 5, each $G_i$ is a minor of $G$ and therefore $\mathrm{tw}(G_i) \leq \mathrm{tw}(G)$. Thus, $\mathrm{tw}(G) = \max\{\mathrm{tw}(G_i) \mid i = 1, \ldots, m\}$ and from (3) we conclude that $k \leq \mathrm{tw}(G) + 1 \leq \frac{3}{2}k$ and the theorem follows. □

Notice that, in the theorem above, if we just want to output the value $k$ without the corresponding tree decomposition, then we just make use of Theorem 3(1) in Step 2 and skip Step 4, and the overall running time drops to $O(n^4)$. Using the same approach as Theorem 6, one can prove a potentially stronger theorem:

**Theorem 7.** *If we can compute the treewidth of any planar graph in polynomial time, then we can compute the treewidth of any H-minor-free graph, where H is single-crossing, in polynomial time.*

*Proof.* We just use the polynomial-time algorithm for computing treewidth of planar graphs in Step 2 of the algorithm described in the proof of Theorem 6. □

# 5  Other Applications of Constructing Clique-Sum Decompositions

In this section, we show that the constructive algorithm described in Section 5 has many other important applications in algorithm design for the class of graphs excluding a single-crossing graph as a minor. Roughly speaking, because both planar graphs and graphs of bounded treewidth have good algorithmic properties, clique-sum decompositions into these graphs enable the design of efficient algorithms for many NP-complete problems.

## 5.1  Polynomial-Time Approximation Schemes (PTASs)

Much work designs PTASs for NP-complete problems restricted to certain special graphs. Lipton and Tarjan [LT80] were the first who proved various NP-optimization problems have PTASs over planar graphs. Alon et al. [AST90] generalized Lipton and Tarjan's ideas to graphs excluding a fixed minor. Because these PTASs were impractical [CNS82], Baker [Bak94] developed practical PTASs for the problems considered by Lipton and Tarjan and Alon et al. Eppstein [Epp00] showed that Baker's technique can be extended by replacing "bounded outerplanarity" with "bounded local treewidth." Intuitively, a graph has *bounded local treewidth* if the treewidth of an $r$-neighborhood (all vertices of distance at most $r$) of each vertex $v \in V(G)$ is a function of $r$, $r \in \mathbb{N}$, and not the number of vertices. Unfortunately, Eppstein's algorithms are impractical for nonplanar graphs. Hajiaghayi et al. [HNRT01,Haj01] designed practical PTASs for both minimization and maximization problems on graphs excluding one of $K_5$ or $K_{3,3}$ as a minor, which is a special class of graphs with bounded local treewidth. Indeed, they proved the following more general theorem:

**Theorem 8 ([HNRT01,Haj01]).** *Given the clique-sum series of an H-minor-free graph G, where H is a single-crossing graph, there are PTASs with approximation ratio $1 + 1/k$ (or $1 + 2/k$) running in $O(c^k n)$ time (c is a small constant) on graph G for hereditary maximization problems (see [Yan78] for exact definitions) such as maximum independent set and other problems such as maximum triangle matching, maximum H-matching, maximum tile salvage, minimum vertex cover, minimum dominating set, minimum edge-dominating set, and subgraph isomorphism for a fixed pattern.*

Applying Theorem 5, we obtain the following corollary:

**Corollary 1.** *There are PTASs with approximation ratio $1 + 1/k$ (or $1 + 2/k$) with running time $O(c^k n + n^4)$ for all problems mentioned in Theorem 8 on any H-minor-free graph H where H is single-crossing.*

## 5.2  Fixed-Parameter Algorithms (FPTs)

Developing fast algorithms for NP-hard problems is an important issue. Recently, Downey and Fellows [DF99] introduced a new approach to cope with this NP-hardness, called *fixed-parameter tractability*. For many NP-complete problems, the inherent combinatorial explosion can be attributed to a certain aspect of the problem, a *parameter*. The parameter is often an integer and small in practice. The running times of simple algorithms may be exponential in the parameter but polynomial in the rest of the problem size. For example, it has been shown that $k$-vertex cover (finding a vertex cover of size $k$) has an algorithm with running time $O(kn + 1.271^k)$ [CKJ99] and hence this problem is fixed-parameter tractable. Alber et al. [ABFN00] demonstrated a solution to the planar $k$-dominating set in time $O(4^{6\sqrt{34k}} n)$. This result was the first nontrivial results for the parameterized version of an NP-hard problem where the exponent of the exponential term grows sublinearly in the parameter (see also [KP02] for a recent improvement of the time bound of [ABFN00] to $O(2^{27\sqrt{k}} n)$). Using this result, others could obtain exponential speedup of fixed parameter algorithms for many NP-complete problems on planar graphs (see e.g. [KC00,CKL01,KLL01,AFN01]). Recently, Demaine et al. [DHT02] extended these results to many NP-complete problems on graphs excluding either $K_5$ or $K_{3,3}$ as a minor. In fact, they proved the following general theorem:

**Theorem 9 ([DHT02]).** *Given the clique-sum series of an H-minor-free graph G, where H is a single-crossing graph, there are algorithms that in $O(2^{27\sqrt{k}} n)$ time decide whether graph G has a subset of size k dominating set, dominating set with property P, vertex cover, edge-dominating set, minimum maximal matching, maximum independent set, clique-transversal set, kernel, feedback vertex set and a series of vertex removal properties (see [DHT02] for exact definitions).*

Again, applying Theorem 5, we obtain the following corollary:

**Corollary 2.** *There are algorithms that in $O(2^{27\sqrt{k}} n + n^4)$ time decide whether any H-minor-free graph G, where H is single-crossing, has a subset of size k with one of the properties mentioned in Theorem 9.*

## 6  Conclusions and Future Work

In this paper, we obtained a polynomial-time algorithm to construct a clique-sum decomposition for $H$-minor-free graphs, where $H$ is a single-crossing graph. As mentioned above, this polynomial-time algorithm has many applications in designing approximation algorithms and fixed-parameter algorithms for these kinds of graphs [Haj01,HNRT01,DHT02]. Also, using this result, we obtained a 1.5-approximation algorithm for treewidth on these graphs. Here we present several open problems that are possible extensions to this paper.

One topic of interest is finding characterization of other kinds of graphs such as graphs excluding a double-crossing graph (or a graph with a bounded number of crossings) as a minor. We suspect that we can obtain such characterizations using $k$-sums for $k > 3$. Designing polynomial-time algorithms to construct such decompositions would be instructive.

It would also be interesting to find other problems than those mentioned by Hajiaghayi et al. [Haj01,HNRT01,DHT02] for which the technique of obtaining clique-sum decomposition can be applied. We think that this approach can be applied for many other NP-complete problems that have good (approximation) algorithms for planar graphs and graphs of bounded treewidth.

From Theorem 2, the treewidth is a 1.5-approximation on the branchwidth. A direct consequence of this fact and our result is the existence of a 2.25-approximation for the branchwidth of the graphs excluding a single-crossing graph. One open problem is how one can use clique-sum decomposition to obtain a better approximation or an exact algorithm for the branchwidth of this graph class.

## Acknowledgments

We thank Prabhakar Ragde and Naomi Nishimura for their encouragement and help on this paper.

# References

ABFN00.  Jochen Alber, Hans L. Bodlaender, Henning Fernau, and Rolf Niedermeier. Fixed parameter algorithms for planar dominating set and related problems. In *Algorithm theory—Scandinavian Workshop on Algorithm Theory 2000 (Bergen, 2000)*, pages 97–110. Springer, Berlin, 2000.

ACP87.  Stefan Arnborg, Derek G. Corneil, and Andrzej Proskurowski. Complexity of finding embeddings in a $k$-tree. *SIAM J. Algebraic Discrete Methods*, 8(2):277–284, 1987.

AFN01.  Jochen Alber, Henning Fernau, and Rolf Niedermeier. Parameterized complexity: Exponential speed-up for planar graph problems. In *Electronic Colloquium on Computational Complexity (ECCC)*. Germany, 2001.

Asa85.  Takao Asano. An approach to the subgraph homeomorphism problem. *Theoret. Comput. Sci.*, 38(2-3):249–267, 1985.

AST90.  Noga Alon, Paul Seymour, and Robin Thomas. A separator theorem for for graphs with excluded minor and its applications. In *Proceedings of the 22nd Annual ACM Symposium on Theory of Computing (Baltimore, MD, 1990)*, pages 293–299, 1990.

Bak94.  Brenda S. Baker. Approximation algorithms for NP-complete problems on planar graphs. *J. Assoc. Comput. Mach.*, 41(1):153–180, 1994.

BDK00.  Hajo J. Broersma, Elias Dahlhaus, and Ton Kloks. A linear time algorithm for minimum fill-in and treewidth for distance hereditary graphs. *Discrete Appl. Math.*, 99(1-3):367–400, 2000.

BGHK95.   Hans L. Bodlaender, John R. Gilbert, Hjálmtýr Hafsteinsson, and Ton
          Kloks. Approximating treewidth, pathwidth, frontsize, and shortest elimi-
          nation tree. *J. Algorithms*, 18(2):238–255, 1995.
BKK95.    Hans L. Bodlaender, Ton Kloks, and Dieter Kratsch. Treewidth and path-
          width of permutation graphs. *SIAM J. Discrete Math.*, 8(4):606–616, 1995.
BKMT.     Vincent Bouchitté, Dieter Kratsch, Haiko Müller, and Ioan Todinca. On
          treewidth approximations. In *Cologne-Twente Workshop on Graphs and
          Combinatorial Optimization (CTW'01)*.
BM76.     John A. Bondy and U. S. R. Murty. *Graph Theory with Applications*.
          American Elsevier Publishing Co., Inc., New York, 1976.
BM93.     Hans L. Bodlaender and Rolf H. Möhring. The pathwidth and treewidth
          of cographs. *SIAM J. Discrete Math.*, 6(2):181–188, 1993.
Bod93.    Hans L. Bodlaender. A tourist guide through treewidth. *Acta Cybernetica*,
          11:1–23, 1993.
Bod96.    Hans L. Bodlaender.   A linear-time algorithm for finding tree-
          decompositions of small treewidth. *SIAM J. Comput.*, 25(6):1305–1317,
          1996.
BT97.     Hans L. Bodlaender and Dimitrios M. Thilikos. Treewidth for graphs with
          small chordality. *Discrete Appl. Math.*, 79(1-3):45–61, 1997.
BT01.     Vincent Bouchitté and Ioan Todinca. Treewidth and minimum fill-in:
          grouping the minimal separators. *SIAM J. Comput.*, 31(1):212–232 (elec-
          tronic), 2001.
CKJ99.    Jianer Chen, Iyad A. Kanj, and Weijia Jia. Vertex cover: further observa-
          tions and further improvements. In *Graph-theoretic concepts in computer
          science (Ascona, 1999)*, pages 313–324. Springer, Berlin, 1999.
CKL01.    Maw-Shang Chang, Ton Kloks, and Chuan-Min Lee. Maximum clique
          transversals. In *Proceedings of the 27th International Workshop on Graph-
          Theoretic Concepts in Computer Science*, pages 300–310. Mathematical
          Programming Society, Boltenhagen, Germany, 2001.
CNS82.    Norishige Chiba, Takao Nishizeki, and Nobuji Saito. An approximation
          algorithm for the maximum independent set problem on planar graphs.
          *SIAM J. Comput.*, 11(4):663–675, 1982.
DF99.     Rodney G. Downey and Michael R. Fellows. *Parameterized Complexity*.
          Springer-Verlag, New York, 1999.
DHT02.    Erik D. Demaine, Mohammadtaghi Hajiaghayi, and Dimitrios M. Thilikos.
          Exponential speedup of fixed parameter algorithms on $K_{3,3}$-minor-free or
          $K_5 - minor - free$ graphs. Technical Report MIT-LCS-TR-838, M.I.T,
          March 2002.
Epp00.    David Eppstein. Diameter and treewidth in minor-closed graph families.
          *Algorithmica*, 27(3-4):275–291, 2000.
Haj01.    MohammadTaghi Hajiaghayi. Algorithms for Graphs of (Locally) Bounded
          Treewidth. Master's thesis, University of Waterloo, September 2001.
HM94.     Michel Habib and Rolf H. Möhring. Treewidth of cocomparability graphs
          and a new order-theoretic parameter. *ORDER*, 1:47–60, 1994.
HNRT01.   Mohammadtaghi Hajiaghayi, Naomi Nishimura, Prabhakar Ragde, and
          Dimitrios M. Thilikos. Fast approximation schemes for $K_{3,3}$-minor-free or
          $K_5$-minor-free graphs. In *Euroconference on Combinatorics, Graph Theory
          and Applications 2001 (Barcelona, 2001)*. 2001.
HT73.     J. E. Hopcroft and R. E. Tarjan. Dividing a graph into triconnected com-
          ponents. *SIAM J. Comput.*, 2:135–158, 1973.

KC00.      Tom Kloks and Leizhen Cai. Parameterized tractability of some (efficient) $Y$-domination variants for planar graphs and t-degenerate graphs. In *International Computer Symposium (ICS)*. Taiwan, 2000.

KLL01.     Tom Kloks, C.M. Lee, and Jim Liu. Feedback vertex sets and disjoint cycles in planar (di)graphs. In *Optimization Online*. Mathematical Programming Society, Philadelphia, 2001.

Klo93.     Ton Kloks. Treewidth of circle graphs. In *Algorithms and computation (Hong Kong, 1993)*, pages 108–117. Springer, Berlin, 1993.

KM92.      André Kézdy and Patrick McGuinness. Sequential and parallel algorithms to find a $K_5$ minor. In *Proceedings of the Third Annual ACM-SIAM Symposium on Discrete Algorithms (Orlando, FL, 1992)*, pages 345–356, 1992.

KP02.      Iyad A. Kanj and Ljubomir Perkovic. Improved parameterized algorithms for planar dominating set. In *27th International Symposium on Mathematical Foundations of Computer Science, MFCS 2002*. Warszawa - Otwock, Poland, August 26-30, 2002. To appear.

KR91.      Arkady Kanevsky and Vijaya Ramachandran. Improved algorithms for graph four-connectivity. *J. Comput. System Sci.*, 42(3):288–306, 1991. Twenty-Eighth IEEE Symposium on Foundations of Computer Science (Los Angeles, CA, 1987).

LT80.      Richard J. Lipton and Robert Endre Tarjan. Applications of a planar separator theorem. *SIAM J. Comput.*, 9(3):615–627, 1980.

MR92.      Gary L. Miller and Vijaya Ramachandran. A new graph triconnectivity algorithm and its parallelization. *Combinatorica*, 12(1):53–76, 1992.

RS84.      Neil Robertson and Paul D. Seymour. Graph minors. III. Planar tree-width. *Journal of Combinatorial Theory Series B*, 36:49–64, 1984.

RS85.      Neil Robertson and Paul D. Seymour. Graph minors — a survey. In I. Anderson, editor, *Surveys in Combinatorics*, pages 153–171. Cambridge Univ. Press, 1985.

RS86.      Neil Robertson and Paul D. Seymour. Graph minors. II. Algorithmic aspects of tree-width. *J. Algorithms*, 7(3):309–322, 1986.

RS91.      Neil Robertson and Paul D. Seymour. Graph minors. X. Obstructions to tree-decomposition. *Journal of Combinatorial Theory Series B*, 52:153–190, 1991.

RS93.      Neil Robertson and Paul Seymour. Excluding a graph with one crossing. In *Graph structure theory (Seattle, WA, 1991)*, pages 669–675. Amer. Math. Soc., Providence, RI, 1993.

SSR94.     Ravi Sundaram, Karan S. Singh, and Pandu C. Rangan. Treewidth of circular-arc graphs. *SIAM J. Discrete Math.*, 7(4):647–655, 1994.

ST94.      Paul D. Seymour and Robin Thomas. Call routing and the ratcatcher. *Combinatorica*, 14(2):217–241, 1994.

Tar72.     Robert Tarjan. Depth-first search and linear graph algorithms. *SIAM J. Comput.*, 1(2):146–160, 1972.

Wag37.     Kehrer Wagner. Über eine Eigenschaft der eben Komplexe. *Deutsche Math.*, 2:280–285, 1937.

Wil84.     S. G. Williamson. Depth-first search and Kuratowski subgraphs. *J. Assoc. Comput. Mach.*, 31(4):681–693, 1984.

Yan78.     Mihalis Yannakakis. Node- and edge-deletion NP-complete problems. In *Conference Record of the Tenth Annual ACM Symposium on Theory of Computing (San Diego, CA, 1978)*, pages 253–264. ACM press, New York, 1978.

# Typical Rounding Problems

Benjamin Doerr

Mathematisches Seminar II, Christian-Albrechts-Universität zu Kiel,
Ludewig-Meyn-Str. 4, D-24098 Kiel, Germany,
bed@numerik.uni-kiel.de,
http://www.numerik.uni-kiel.de/

**Abstract.** The linear discrepancy problem is to round a given $[0,1]$–vector $x$ to a binary vector $y$ such that the rounding error with respect to a linear form is small, i.e., such that $\|A(x-y)\|_\infty$ is small for some given matrix $A$. The discrepancy problem is the special case of $x = (\frac{1}{2}, \ldots, \frac{1}{2})$. A famous result of Beck and Spencer (1984) as well as Lovász, Spencer and Vesztergombi (1986) shows that the linear discrepancy problem is not much harder than this special case: Any linear discrepancy problem can be solved with at most twice the maximum rounding error among the discrepancy problems of the submatrices of $A$.

In this paper we strengthen this result for the common situation that the discrepancy of submatrices having $n_0$ columns is bounded by $Cn_0^\alpha$ for some $C > 0, \alpha \in (0,1]$. In this case, we improve the constant by which the general problem is harder than the discrepancy one, down to $2(\frac{2}{3})^\alpha$. We also find that a random vector $x$ has expected linear discrepancy $2(\frac{1}{2})^\alpha Cn^\alpha$ only. Hence in the typical situation that the discrepancy is decreasing for smaller matrices, the linear discrepancy problem is even less difficult compared to the discrepancy one than assured by the results of Beck and Spencer and Lovász, Spencer and Vesztergombi.

**Keywords:** rounding, discrepancy, games.

## 1 Introduction

In this paper we deal with rounding problems, and in particular with the question how much easier it is to round a vector with all entries $\frac{1}{2}$ compared to the general case of $[0,1]$–vectors. A famous result of Beck and Spencer [4] and Lovász, Spencer and Vesztergombi [8] shows that the general problem can be reduced to the $\frac{1}{2}$–case. In this paper we refine their result for the typical case that the rounding problem for smaller matrices can be solved better than for larger ones.

Let us be more precise: For a given matrix $A \in \mathbb{R}^{m \times n}$ and a vector $x \in \mathbb{R}^n$ we are interested in finding a vector $y \in \mathbb{Z}^n$ such that (1) $\|x - y\|_\infty \leq 1$ and (2) the rounding error $\|A(x - y)\|_\infty$ is small. $y$ is sometimes called *approximate integer solution* for the linear system $Ay = Ax$.

It is easy to see from the problem statement that only the fractional part of $x$ is important. Therefore we usually assume $x \in [0,1]^n$ and consequently have $y \in \{0,1\}$. It is also clear that rescaling $A$ does not chance the problem

K. Jansen et al. (Eds.): APPROX 2002, LNCS 2462, pp. 81–93, 2002.

substantially: If we replace $A$ by $\lambda A$ for some $\lambda > 0$, the set of optimal solutions is not changed and their rounding error just changes by a factor of $\lambda$. Thus we lose nothing by assuming $A \in [-1, 1]^{m \times n}$.

In discrepancy theory, this problem is known under the term *linear discrepancy problem*:

$$\text{lindisc}(A, x) = \min_{y \in \{0,1\}^n} \|A(x - y)\|_\infty,$$

$$\text{lindisc}(A) = \max_{x \in [0,1]^n} \text{lindisc}(A, x).$$

The special case that $x = \frac{1}{2}\mathbf{1}_n$ is called *combinatorial discrepancy problem*. It can be seen as the problem to partition the columns of $A$ into two groups such that the row sums within each group are similar. We write

$$\text{disc}(A) := \min_{y \in \{0,1\}^n} \|A(\tfrac{1}{2}\mathbf{1}_n - y)\|_\infty = \tfrac{1}{2} \min_{y \in \{-1,1\}^n} \|Ay\|_\infty.^1$$

Note that already the combinatorial discrepancy problem is far from being easy: It is $NP$–hard to decide whether a $0, 1$ matrix has discrepancy zero or not. On the other hand, a number of results and algorithms are known:

- If all column vectors have $l_1$–norm at most $t$, then $\text{disc}(A) \leq t$ (Beck, Fiala [2]).
- A $y \in \{-1,1\}^n$ such that $\|Ay\|_\infty \leq \sqrt{2n \ln(2m)}$ can be computed in time polynomial in $n$ and $m$. In particular, $\text{disc}(A) \leq \sqrt{\frac{1}{2}n \ln(2m)}$ (Alon, Spencer [1]).
- If $m \geq n$, then $\text{disc}(A) \leq 3\sqrt{n \ln(2m/n)}$ (Spencer [12]).
- If $A$ is the incidence matrix of a hypergraph $\mathcal{H}$ having primal shatter function $\pi_{\mathcal{H}} = O(n^d)$, then $\text{disc}(A) = O(n^{\frac{1}{2} - \frac{1}{2d}})$ (Matoušek [10]). Hence this bound in particular holds if $\mathcal{H}$ has VC-dimension $d$.
- If the dual shatter function satisfies $\pi_{\mathcal{H}}^* = O(n^d)$, then the discrepancy is $\text{disc}(A) = O(n^{\frac{1}{2} - \frac{1}{2d}} \sqrt{\log(n)})$ (Matoušek, Welzl, Wernisch [11]).

We refer to the chapter Beck and Sós [3] and the book Matoušek [9] for further discrepancy results. For our purposes a result of Beck and Spencer [4] and Lovász, Spencer and Vesztergombi [8] is crucial: It shows that the linear discrepancy problem is not much harder than the combinatorial one:

---

[1] Note that some papers define the linear and combinatorial discrepancy to be twice our values. This is motivated by the notion of hypergraph discrepancy: The discrepancy of a hypergraph is the least $k \in \mathbb{N}_0$ such that there is a 2–coloring of the vertex set such in each hyperedge the number of vertices in one color deviates from that in the other by at most $k$. If a hypergraph has discrepancy $k$, its incidence matrix has discrepancy $\frac{1}{2}k$ (in our notation), and vice versa. This motivates to define the discrepancy of a matrix A by $\min_{y \in \{-1,1\}} \|Ay\|_\infty$. On the other hand, from the viewpoint of rounding problems, our notation seems more appropriate.

**Theorem 1.** *For any $A \in [-1,1]^{m \times n}$ and $x \in [0,1]^n$, there is a $y \in \{0,1\}^n$
such that*

$$\|A(x-y)\|_\infty \le 2 \max_{A_0 \le A} \operatorname{disc}(A_0).^2$$

*A $y \in \{0,1\}^n$ such that $\|A(x-y)\|_\infty \le 2D + O(2^{-k}n)$ can be computed by
$k$ times solving a combinatorial discrepancy problem for a submatrix of $A$ with
discrepancy at most $D$.*

The constant of 2 in Theorem 1 cannot be improved in general: For arbitrary $n \in \mathbb{N}$, Lovász, Spencer and Vesztergombi [8] provide an example $A \in \{0,1\}^{(n+1) \times n}$, $x \in [0,1]^n$ such that any $y \in \{0,1\}^n$ fulfills

$$\|A(x-y)\|_\infty = 2(1 - \tfrac{1}{n+1}) \max_{A_0 \le A} \operatorname{disc}(A_0).$$

On the other hand, Theorem 1 is known to be not sharp: The factor of 2 can
be replaced by $2(1 - \tfrac{1}{2m})$ as shown in [5]. In between these two results little
seems to be known. Spencer conjectures that $2(1 - \tfrac{1}{n+1})$ is the right constant in
Theorem 1. This has been proven for totally unimodular matrices in [6].

Before explaining our results, we would like to point out that Theorem 1
requires understanding not only the discrepancy problem for $A$, but also for
all submatrices of $A$. This is known under the term 'hereditary discrepancy
problem', the corresponding notion is the *hereditary discrepancy* of $A$ defined by

$$\operatorname{herdisc}(A) := \max_{A_0 \le A} \operatorname{disc}(A_0).$$

It is not difficult to construct examples where the discrepancy of a submatrix
(and thus the hereditary discrepancy) is much larger than the discrepancy of
the matrix itself (which might even be zero). However, all these examples have
the flavor of being artificially designed for this purpose. The situation usually
encountered (both when looking at examples or results like the ones above) is
that the discrepancy or the upper bound given by a result does not deviate
significantly from the respective maximum taken over all submatrices.

In many cases the discrepancy behavior is even more regular: Smaller matrices tend to have lower discrepancies. To formalize this we introduce the
(hereditary) discrepancy function of $A$: Define $h_A(n_0)$ to be the largest discrepancy among all submatrices of $A$ having at most $n_0$ columns (to save some
floors, we regard $h_A$ as a function on the non-negative real numbers). Then most
of the results above show $h_A(n_0) \le Cn_0^\alpha$ for some $C > 0$ and $\alpha \le 1$.

The main result of this paper is that this stronger discrepancy assumption
can be exploited for the linear discrepancy problem. This reduces the constant
of 2 in Theorem 1, the factor by which the general problem can be harder than
the combinatorial one, down to $2(\tfrac{2}{3})^\alpha$ (e.g., 1.63 for $h_A = O(\sqrt{n_0})$):

**Theorem 2.** *If $h_A(n_0) \le Cn_0^\alpha$ for all $n_0 \in \{1, \ldots, n\}$, then*

$$\operatorname{lindisc}(A) \le 2\left(\tfrac{2}{3}\right)^\alpha Cn^\alpha.$$

---

$^2$ *We write $A_0 \le A$ to denote that $A_0$ is a submatrix of $A$.*

A more detailed analysis yields bounds for $\text{lindisc}(A, x)$ that take into account the vector $x$. We present a function $w : [0, 1] \to [0, \frac{2}{3}]$ such that

$$\text{lindisc}(A, x) \le 2 \left( \sum_{i=1}^{n} w(x_i) \right)^{\alpha} Cn^{\alpha}$$

holds for all $x \in [0, 1]^n$. This allows an average case analysis showing that an $x$ picked uniformly at random has expected $\text{lindisc}(A, x)$ at most $2(\frac{1}{2})^{\alpha} Cn^{\alpha}$. It also shows that 'small' $x$ have lower linear discrepancies: We prove $\text{lindisc}(A, x) \le 2(2\|x\|_1(-\log_2(\frac{1}{n}\|x\|_1) + 1))^{\alpha} Cn^{\alpha}$. This might seem natural at first, but recall that in the example $(A, x)$ such that $\text{lindisc}(A, x) = 2(1 - \frac{1}{n+1})$ herdisc$(A)$ in [8] we have $x = (\frac{1}{n+1}, \ldots, \frac{1}{n+1})$.

All our results are constructive in the following sense: Let $A$ be given. Assume that we can solve discrepancy problems for submatrices of $A$ having $n_0$ columns with rounding error at most $Cn_0^{\alpha}$. Then for any $x \in [0, 1]^n$ we can compute a $y \in \{0, 1\}^n$ such that $\|A(x-y)\|_\infty \le 2\left(\sum_{i=1}^{n} w(x_i)\right)^{\alpha} Cn^{\alpha} + O(2^{-k}n)$ by solving $k$ discrepancy problems for submatrices of $A$.

## 2    Reduction to Game Theory

Our proofs are based on the proof of Theorem 1, which we state here in a language suitable for our further work. Here and in the remainder we use the shorthand $[n]$ to denote the set $\{1, \ldots, n\}$.

*Proof (of Theorem 1).* Let $x \in [0, 1]^n$. We construct a $y \in \{0, 1\}^n$ such that $\|A(x - y)\|_\infty$ is small. As

$$x \mapsto \min_{y \in \{0,1\}^n} \|A(x - y)\|_\infty$$

is a continuous function and $\{\sum_{i=1}^{n} b_i 2^{-i} \mid n \in \mathbb{N}, b_1, \ldots, b_n \in \{0, 1\}\}$ is dense in $[0, 1]$, we may assume that $x$ has finite binary expansion of length $k$, i.e., there is a $k \in \mathbb{N}$ such that $2^k x$ is integral.

Set $a^{(0)} := x$. We define a series of intermediate 'roundings' $a^{(l)}, l = 1, \ldots, k$ having binary length at most $k - l$. Suppose that for $l \in \{1, \ldots, k\}$, $a^{(l-1)}$ is already defined and satisfies $a^{(l-1)} 2^{k-l+1} \in \mathbb{Z}^n$. Set

$$X^{(l)} := \{j \in [n] \mid a_j^{(l-1)} 2^{k-l+1} \text{ odd}\},$$

the set of all $j$ such that the binary expansion of $a_j^{(l-1)} 2^{k-l+1}$ ends in 1. By the definition of combinatorial discrepancy, there is an $\varepsilon^{(l)} : X^{(l)} \to \{-1, +1\}$ such that

$$d_i^{(l)} := \tfrac{1}{2} \sum_{j \in X^{(l)}} a_{ij} \varepsilon^{(l)}(j)$$

satisfies $|d_i^{(l)}| \leq h_A(|X^{(l)}|)$ for all $i \in [m]$. Define

$$a_j^{(l)} := \begin{cases} a_j^{(l-1)} - 2^{-(k-l+1)}\varepsilon^{(l)}(j) & \text{if } j \in X^{(l)} \\ a_j^{(l-1)} & \text{otherwise.} \end{cases}$$

Then $a^{(l)}2^{k-l} \in \mathbb{Z}^n$ and $A(a^{(l-1)} - a^{(l)}) = 2^{-(k-l)}d^{(l)}$. Having defined $a^{(l)}$ for all $l \in \{0, \ldots, k\}$, we put $y = a^{(k)}$ and compute

$$\|A(x - y)\|_\infty = \left\| A\left(\sum_{l=1}^{k}(a^{(l-1)} - a^{(l)})\right) \right\|_\infty$$

$$= \left\| \sum_{l=1}^{k} 2^{-(k-l)}d^{(l)} \right\|_\infty$$

$$\leq \sum_{l=1}^{k} 2^{-(k-l)}h_A(|X^{(l)}|).$$

From $h_A(|X^{(l)}|) \leq \operatorname{herdisc}(A)$ for all $l \in [k]$ we get the original result $\|A(x - y)\|_\infty \leq 2\operatorname{herdisc}(A)$. □

There is one option we did not use in the above algorithm: At any time during the above rounding process, we may replace $\varepsilon^{(l)}$ by $-\varepsilon^{(l)}$. This changes the resulting $a^{(l)}$, but does not violate our discrepancy guarantee as we just replace $d^{(l)}$ by $-d^{(l)}$. By choosing signs for the $\varepsilon^{(l)}$, $l \in [k]$ in a clever way, we try to keep the sets $X^{(l)}, l \in [k]$ small and thus improve the discrepancy bound.

Note that if we change the sign of one $\varepsilon^{(l)}$, this does not only change the last digit of the binary expansion of the $a^{(l)}$, but may change any digit. Furthermore, it is very difficult get suitable information about the $\varepsilon^{(l)}$ in the general case. We therefore regard the sign-choosing problem as an on-line problem, i.e., we analyze what can be achieved by choosing the sign of $\varepsilon^{(l)}$ without knowing the possible colorings $\varepsilon^{(l+1)}$.

Worst-case analyses of on-line problems naturally lead to games. One player represents the on-line algorithm and the other one the data not known to the algorithm. Our problem is modeled by the following two-player game. For obvious reasons we call the players 'Pusher' and 'Chooser'. Let $f$ be any real function with domain containing $\{0, \ldots, n\}$.

**The Game $G(a^{(0)}, f)$:** The starting position of the game is a vector $a^{(0)} \in [0,1]^n$ having a finite binary expansion of length at most $k$, i.e., $2^k a^{(0)}$ is integral. The game then consist of $k$ rounds of the following structure:

*Round $l$:*
- Set $X^{(l)} := \{j \in [n] \mid 2^{k-l+1}a_j^{(l-1)} \text{ odd}\}$.
- Pusher selects a partition $S^{(l)} \dot\cup T^{(l)} = X^{(l)}$.
- Chooser chooses one partition class $Y^{(l)} \in \{S^{(l)}, T^{(l)}\}$.

– The position is updated according to

$$a_j^{(l)} := \begin{cases} a_j^{(l-1)} - 2^{-(k-l+1)} & \text{if } j \in X^{(l)} \setminus Y^{(l)} \\ a_j^{(l-1)} + 2^{-(k-l+1)} & \text{if } j \in Y^{(l)} \\ a_j^{(l-1)} & \text{otherwise.} \end{cases}$$

*Objective of the game:* We call the value $\sum_{l=1}^{k} 2^{-k+l} f(|X^{(l)}|)$ the pay-off (for Pusher). As the name suggests, it is Pusher's aim to maximize this value (and Chooser's, to keep it small). The maximum pay-off Pusher can enforce in a game started in position $a^{(0)}$ is the value $v(a^{(0)}, f)$ of this game.

From the discussion above the following connection between the game $G(x, f)$ and our rounding problem is obvious:

**Lemma 1.** lindisc$(A, x) \leq v(x, h_A)$.

We complete the proof of our main results by estimating the values of the corresponding games. Note that we may always replace $h_A$ by a pointwise not smaller function $f$: Then $v(x, h_A) \leq v(x, f)$ implies lindisc$(A, x) \leq v(x, f)$. We call such an $f$ an upper bound for $h_A$. The results cited in the introduction also indicate that we lose little by assuming $f$ to be concave, non-decreasing and non-negative.

## 3    Worst-Case Analysis

In this section we prove an upper bound on the game values, which by Lemma 1 yields an upper bound on the linear discrepancy of $A$.

**Lemma 2.** *Let $f : [0, n] \rightarrow \mathbb{R}$ be concave and non-decreasing. Then $v(a^{(0)}, f) \leq 2f(\frac{2}{3}n)$ holds for all starting positions $a^{(0)}$.*

*Proof.* To give an upper bound on $v(a^{(0)}, f)$, we have to show that Chooser has a strategy such that no matter what partitions Pusher selects, the pay-off will never exceed this bound. We analyze the following strategy.

*Chooser's strategy:* Assume all notation given as in the definition of the game. We may assume that $k$ is even. In an even numbered round $l$, Chooser chooses the partition class arbitrarily. If $l$ is odd, this is in particular not the last round, Chooser proceeds like this: He chooses that one of the two alternatives, that minimizes the size of $X^{(l+1)}$.

*Analysis:* Let $l \in [k]$ be odd. Let $X^{(l)} = S^{(l)} \dot{\cup} T^{(l)}$ be the partition given by Pusher. Denote by $X^{(l+1)} \circ Y^{(l)}$ the value of $X^{(l+1)}$ resulting from Chooser's move $Y^{(l)}$. Now we easily see that $(X^{(l+1)} \circ S^{(l)}) \cap X^{(l)}$ and $(X^{(l+1)} \circ T^{(l)}) \cap X^{(l)}$ form a partition of $X^{(l)}$. On the other hand, $(X^{(l+1)} \circ S^{(l)}) \setminus X^{(l)} = (X^{(l+1)} \circ T^{(l)}) \setminus X^{(l)}$, that is, the complement of $X^{(l)}$ is not affected by Chooser's move. Hence Chooser has to choose $Y^{(l)}$ in such a way that $(X^{(l+1)} \circ Y^{(l)}) \cap X^{(l)}$ is minimized. Then $|(X^{(l+1)} \circ Y^{(l)}) \cap X^{(l)}| \leq \frac{1}{2}|X^{(l)}|$, and thus

$$
\begin{aligned}
|X^{(l+1)} \circ Y^{(l)}| &= |(X^{(l+1)} \circ Y^{(l)}) \setminus X^{(l)}| + |(X^{(l+1)} \circ Y^{(l)}) \cap X^{(l)}| \\
&\leq |[n] \setminus X^{(l)}| + \frac{1}{2}|X^{(l)}| \\
&= n - \frac{1}{2}|X^{(l)}|.
\end{aligned}
$$

We conclude that if Chooser follows the strategy proposed above, we have $|X^{(l+1)}| \leq n - \frac{1}{2}|X^{(l)}|$ for all odd $l \in [k]$. Let

$$
\overline{f} : [0, n] \to \mathbb{R}; x \mapsto 2f(n - \tfrac{1}{2}x) + f(x).
$$

Since $f$ is concave, we have $\overline{f}(x) = 3(\frac{2}{3}f(n - \frac{1}{2}x) + \frac{1}{3}f(x)) \leq 3f(\frac{2}{3}n)$ for all $x \in [0, n]$. By definition, $\overline{f}(\frac{2}{3}n) = 3f(\frac{2}{3}n)$. Hence $\frac{2}{3}n$ is a global maximum of $\overline{f}$. Using this we bound the pay-off:

$$
\begin{aligned}
\sum_{l=1}^{k} 2^{-k+l} f(|X^{(l)}|) &= \sum_{\substack{l \in [k] \\ l \text{ odd}}} 2^{-k+l} \left( 2f(|X^{(l+1)}|) + f(|X^{(l)}|) \right) \\
&\leq \sum_{\substack{l \in [k] \\ l \text{ odd}}} 2^{-k+l} \left( 2f(n - \tfrac{1}{2}|X^{(l)}|) + f(|X^{(l)}|) \right) \\
&\leq \sum_{\substack{l \in [k] \\ l \text{ odd}}} 2^{-k+l} \left( 2f(n - \tfrac{1}{3}n) + f(\tfrac{2}{3}n) \right) \\
&\leq \sum_{l=1}^{k} 2^{-k+l} f(\tfrac{2}{3}n) \leq 2f(\tfrac{2}{3}n).
\end{aligned}
$$

$\square$

Lemma 1 and 2 give the following theorem, a slight generalization of Theorem 2 in the introduction.

**Theorem 3.** *If $f$ is a concave and non-decreasing upper bound for $h_A$, then*

$$
\mathrm{lindisc}(A) \leq 2f(\tfrac{2}{3}n).
$$

It may seem that our on-line strategy is very simple: Every second decision is chosen arbitrarily, the remaining ones only take into account the next move. Nevertheless, the game-theoretic analysis is tight in the worst-case:

**Lemma 3.** *For any $f$ and any $k \in \mathbb{N}$ there is a starting position $a^{(0)}$ such that Pusher can enforce a pay-off of $2(1 - 2^{-k})f(\frac{2}{3}n)$ in a $k$-round game of $G(a^{(0)}, f)$.*

*Proof.* Let $n$ be a multiple of 3. Put $x_k := \sum_{i=1}^{\lfloor k/2 \rfloor} 2^{-2i}$. If $k$ is odd, let $a^{(0)}$ be such that one third of its components equal $x_k$ and two thirds are $x_k + 2^{-k}$. If $k$ is even, two thirds of the $a_j^{(0)}, j \in [n]$ shall equal $x_k$, and one third $x_{k-2} + 2^{-(k+2)}$.

If $k$ is odd, $X^{(1)} := \{j \in [n] \mid 2^k a_j^{(0)} \text{ odd}\}$ has cardinality $\frac{2}{3}n$ by definition of $a^{(0)}$. Let $S^{(1)} \dot\cup T^{(1)}$ be any partition of $X^{(1)}$ such that $|S^{(1)}| = |T^{(1)}| = \frac{1}{3}n$. Regardless of Chooser's choice, half of the $x_k + 2^{-k}$–values (and thus a total of $\frac{1}{3}n$) are rounded up to $x_{k-2} + 2^{-(k+2)}$, the remaining ones are rounded down to $x_k$. Hence we end up with the starting position for the game lasting $k - 1$ rounds.

Similarly, if $k$ is even, we have $|X^{(1)}| = \frac{2}{3}n$, and partitioning $X^{(1)}$ into equally sized classes proceeds the game to the starting position for the game lasting $k-1$ rounds.

Hence by induction we conclude that this strategy ensures $|X^{(l)}| = \frac{2}{3}n$ for all $l \in [k]$, and thus a pay-off of $2(1 - 2^{-k})h_A(\frac{2}{3}n)$.  $\square$

The strategy described in the proof of Lemma 2 can also be applied to non-concave functions $f$. The following corollary shows that an example $A \in \{0,1\}^{m \times n}$, $x \in [0,1]^n$ having $\mathrm{lindisc}(A, x) > 2(1 - \frac{1}{n+1})\mathrm{herdisc}(A)$ (thus being stronger than the currently best known ones of Lovász, Spencer and Vesztergombi) must have constant discrepancy function on $[\frac{2}{3}n, n]$. Hence such examples — should they exist — do not display the regular discrepancy behavior we investigate in this paper, and thus must have a rather particular structure.

**Corollary 1.** *If $A \in \{0,1\}^{m \times n}$ and $h_A(\frac{2}{3}n) < \mathrm{herdisc}(A)$, then*

$$\mathrm{lindisc}(A, x) \le 2(1 - \tfrac{1}{n+1})\mathrm{herdisc}(A).$$

*Proof.* If Chooser follows the strategy proposed in the proof of Lemma 2, he ensures that $|X^{(l)}| \le \frac{2}{3}n$ or $|X^{(l-1)}| \le \frac{2}{3}n$ holds for all $l \in [k]$. Hence

$$\mathrm{lindisc}(A) \le \sum_{l=0}^{\infty} 2^{-2l}(\mathrm{herdisc}(A) + \tfrac{1}{2}h_A(\tfrac{2}{3}n))$$

follows along the lines of the proof of Lemma 2. If $h_A(\frac{2}{3}n) < \mathrm{herdisc}(A)$, then $h_A(\frac{2}{3}n) \le \mathrm{herdisc}(A) - \frac{1}{2}$. This yields $\mathrm{lindisc}(A) \le 2\,\mathrm{herdisc}(A) - \frac{1}{3}$, proving the claim for the case that $\mathrm{herdisc}(A) < \frac{1}{6}(n + 1)$. The case $\mathrm{herdisc}(A) \ge \frac{1}{6}(n + 1)$ is trivial, since $\mathrm{lindisc}(A) \le \frac{1}{4}(n + 1)$ holds for any $A \in \{0,1\}^{m \times n}$.  $\square$

## 4   Improved Strategies and Average Case

As Lemma 3 shows, the on-line sign-choosing algorithm presented in the proof of Lemma 2 is optimal in the worst case (given by a particular starting position). In the following we present a more complicated on-line strategy, that is optimal in the worst-case as well, but yields tighter bounds for other starting positions. It is a potential function strategy, that is, we define a potential function for all positions and Chooser's strategy is to minimize this potential.

For a finite binary sequence $b = (b_1, \ldots, b_k) \in \{0,1\}^k$ recursively define

$$w(b, k) := b_k,$$

$$w(b, i) := \begin{cases} \frac{1}{2}w(b, i+1) & \text{if } b_i = b_{i+1} \\ 1 - \frac{1}{2}w(b, i+1) & \text{otherwise.} \end{cases}$$

Put $w(b) = \sum_{i=1}^{k} 2^{-i} w(b, i)$. For a number $a \in [0, 1[$ having finite binary expansion $a = \sum_{i=1}^{k} 2^{-i} b_i$, we write $w(a) := w((b_1, \ldots, b_k))$. Put $w(1) = 0$. We have

**Lemma 4.** *Let* $a \in [0, 1]$ *having finite binary expansion* $a = \sum_{i=1}^{k} 2^{-i} b_i$ *such that* $b_k = 1$.

(i) $\sum_{b \in \{0,1\}^k} w(b) = 2^{k-1} - \frac{1}{2}$.
(ii) $w(a) = w(1 - a)$.
(iii) $w(a) \leq \frac{2}{3}$.
(iv) $w(a) \leq 2a(-\log_2(a) + 1)$.
(v) $w(a) = 2^{-k} + \frac{1}{2}(w(a + 2^{-k}) + w(a - 2^{-k}))$.

*Proof.* Ad (i): We use induction on $k$. For $k = 1$ we compute

$$\sum_{b \in \{0,1\}^k} w(b) = w((0)) + w((1)) = 0 + \frac{1}{2}.$$

Let $k \geq 2$. For $b = (b_1, \ldots, b_k) \in \{0,1\}^k$ put $\bar{b} = (1 - b_1, \ldots, b_k)$. Then $w(b, i) = w(\bar{b}, i)$ for $i \geq 2$ and $w(b, 1) + w(\bar{b}, 1) = 1$. Further, we have $w(b) = \frac{1}{2}w(b, 1) + \frac{1}{2}w((b_2, \ldots, b_k))$. Thus

$$w(b) + w(\bar{b}) = \frac{1}{2}(w(b, 1) + w(\bar{b}, 1)) + w((b_2, \ldots, b_k)) = \frac{1}{2} + w((b_2, \ldots, b_k)).$$

Hence

$$\sum_{b \in \{0,1\}^k} w(b) = \sum_{\substack{b \in \{0,1\}^k \\ b_1 = 0}} (w(b) + w(\bar{b}))$$

$$= \sum_{\substack{b \in \{0,1\}^k \\ b_1 = 0}} (\tfrac{1}{2} + w((b_2, \ldots, b_k)))$$

$$= 2^{k-2} + \sum_{b \in \{0,1\}^{k-1}} w(b) = 2^{k-1} - \frac{1}{2}.$$

Ad (ii): If $a = 0$, then (ii) is satisfied by definition. Hence assume $a \neq 0$ and $k \geq 1$. Let $b = (b_1, \ldots, b_k)$. Define $\tilde{b} \in \{0,1\}^k$ by $\tilde{b}_k = 1$ and $\tilde{b}_i = 1 - b_i$ for $i = 1, \ldots, k-1$. Then $1 - a = \sum_{i=1}^{k} 2^{-i} \tilde{b}_i$. Now (ii) follows from $w(b, i) = w(\tilde{b}, i)$ for $i = 1, \ldots, k$.

Ad (iii): By (ii), we may assume $b_1 = 0$. If $b_2 = 0$, then

$$w(b) = \tfrac{1}{4}w(b, 2) + \tfrac{1}{4}w(b, 2) + \tfrac{1}{4}w((b_3, \ldots, b_k)) \leq \frac{1}{4} + \frac{1}{4} \cdot \frac{2}{3} = \frac{2}{3}$$

by induction on $k$. If $b_2 = 1$, then

$$w(b) = \tfrac{1}{2}(1 - \tfrac{1}{2}w(b, 2)) + \tfrac{1}{4}w(b, 2) + \tfrac{1}{4}w((b_3, \ldots, b_k)) \leq \frac{1}{2} + \frac{1}{4} \cdot \frac{2}{3} = \frac{2}{3}$$

again by induction.

Ad (iv): If $2^{-(\ell+1)} \le a < 2^{-\ell}$, then $b_1 = \ldots = b_\ell = 0$ and $b_{\ell+1} = 1$. Thus

$$w(a) = \ell 2^{-\ell} w(b, \ell) + 2^{-\ell} w((b_{\ell+1}, \ldots, b_k)) \le (\ell + 1) 2^{-\ell} < 2a(-\log_2 a + 1).$$

Ad (v): Assume for simplicity that $a + 2^{-k} \ne 1$ (this case is easily solved separately). Let $b^+ \in \{0, 1\}^{k-1}$ such that

$$a + 2^{-k} = \sum_{i=1}^{k-1} 2^{-i} b_i^+.$$

Note that $a - 2^{-k} = \sum_{i=1}^{k-1} 2^{-i} b_i$. Let $\ell \in [k-1]$ be minimal subject to $b_{\ell+1} = \ldots = b_k = 1$. Then $b_i^+ = 0$ for $\ell+1 \le i < k$, $b_\ell^+ = 1$ and $b_i^+ = b_i$ for all $i < \ell$. Thus we have $w((b_1, \ldots, b_{k-1}), i) + w(b^+, i) = 2^{-k+i} + 0 = 2w(b, i)$ for $\ell + 1 \le i < k$. We also compute $w((b_1, \ldots, b_{k-1}), \ell) + w(b^+, \ell) = 1 - 2^{-k+\ell} + 1 = 2w(b, \ell)$. Since $b_i^+ = b_i$ for all $i < \ell$, an easy induction yields $w((b_1, \ldots, b_{k-1}), i) + w(b^+, i) = 2w(b, i)$ also for the remaining $i \in [k-1]$. Now (v) follows from the definition of $w$.     □

A reader familiar with probabilistic game analysis (cf. Spencer [13]) might prefer this randomized view: Given $a$, we repeat rounding the last non-zero digit of its binary expansion up or down with equal probabilities $\frac{1}{2}$. Then $w(b, i)$ is the probability that $b_i = 1$ when all higher bits are already rounded. Thus $w(x_i)$ is the expected contribution of a single entry of $x$ to the pay-off $\sum_{l=1}^{k} 2^{-k+l} |X(l)|$ of the game $G(x, \mathrm{id})$, if Chooser plays randomly.

By (v) of the lemma above, $w$ is continuous on the set of numbers having finite binary expansion. Hence there is a unique continuation on $[0, 1]$, which we denote by $w$ as well. Note that (ii) to (iv) of Lemma 4 now hold for arbitrary $a \in [0, 1]$. The inequality (iii) is sharp as shown by $a = \frac{1}{3}$ and $a = \frac{2}{3}$.

For a game position $x \in [0, 1]^n$ we put $w(x) = \sum_{i=1}^{n} w(x_i)$. Then

**Lemma 5.** *Let $f$ be a concave, non-decreasing and non-negative function. Let $x \in [0, 1]^n$ be a starting position of a $k$-round game. If Chooser plays the strategy to minimize $w$, then the pay-off in the game $G(x, f)$ is at most $2f(w(x))$. Consequently, if $f$ is an upper bound on $h_A$, then*

$$\mathrm{lindisc}(A, x) \le 2f(w(x))$$

*holds for all $x \in [0, 1]^n$.*

*Proof.* Assume first that $f = \mathrm{id}_{\mathbb{R}}$. We proceed by induction on the length $k$ of the binary expansion of $a$. If $k = 0$, the game ends before it started, and the pay-off is $0 = w(a)$. Hence let $k \ge 1$. Let $X = \{j \in [n] \mid 2^k a_j \text{ odd}\}$, and let $S \dot\cup T = X$ denote Pusher's move. Let $s, t$ denote the positions that arise if Chooser chooses $S, T$. If Chooser takes $S$, by induction the pay-off is bounded by $2^{-k+1}|X| + 2w(s)$ (and an analogous statement holds for $T$). Let $y \in \{s, t\}$ be such that $w(y) = \min\{w(s), w(t)\}$. Then the pay-off resulting from choosing $y$ is bounded by $2^{-k+1}|X| + 2w(y) \le 2^{-k+1}|X| + w(s) + w(t) \le w(a)$, where the latter inequality follows from Lemma 4.

In the notation of the definition of the game, we just showed that Chooser's strategy yields $\sum_{l=1}^{k} 2^{-k+l}|X^{(l)}| < 2w(a)$. Now let $f$ be an arbitrary concave, non-decreasing and non-negative function. Then the pay-off of $G(a, f)$ is bounded by

$$\sum_{l=1}^{k} 2^{-k+l} f(|X^{(l)}|) \leq 2f\left(\sum_{l=1}^{k} 2^{-k+l-1}|X^{(l)}|\right) \leq 2f(w(a)).$$

$\square$

Lemma 5 allows an average case analysis of the linear discrepancy problem.

**Theorem 4.** *Let $f$ be a concave and non-decreasing upper bound for $h_A$. For an $x$ chosen uniformly at random from $[0, 1]^n$, the expected linear discrepancy satisfies*

$$E(\text{lindisc}(A, x)) \leq 2f(\tfrac{1}{2}n).$$

*Proof.* Let $k \in \mathbb{N}$ and $B = \{\sum_{l=1}^{k} 2^{-l} b_l \mid b_1, \ldots, b_k \in \{0, 1\}\}$. For a number $r \in [0, 1]$ denote by $\tilde{r}$ the largest element of $B$ not exceeding $r$. Put $\tilde{x} = (\tilde{x}_1, \ldots, \tilde{x}_n)$. Then

$$\begin{aligned}
E(\text{lindisc}(A, x)) &\leq E(\text{lindisc}(A, \tilde{x})) + n2^{-k} \\
&= \sum_{\tilde{x} \in B^n} 2^{-nk} \text{lindisc}(A, \tilde{x}) + n2^{-k} \\
&\leq 2 \sum_{\tilde{x} \in B^n} 2^{-nk} f(w(\tilde{x})) + n2^{-k} \\
&\leq 2f\left(2^{-nk} \sum_{\tilde{x} \in B^n} w(\tilde{x})\right) + n2^{-k} \\
&\leq 2f(\tfrac{1}{2}n) + n2^{-k},
\end{aligned}$$

where the latter inequality follows from

$$\begin{aligned}
\sum_{\tilde{x} \in B^n} w(\tilde{x}) &= \sum_{\tilde{x} \in B^n} \sum_{i=1}^{n} w(\tilde{x}_i) \\
&= \sum_{i=1}^{n} \sum_{\tilde{x} \in B^n} w(\tilde{x}_i) \\
&= \sum_{i=1}^{n} 2^{(n-1)k} \sum_{b \in B} w(b) \\
&= n2^{(n-1)k} \sum_{b \in B} w(b)
\end{aligned}$$

and Lemma 4.

$\square$

Another consequence of Lemma 5 is that 'small' $x$ have lower linear discrepancy:

**Lemma 6.** *Let $f$ be a concave and non-decreasing upper bound for $h_A$. Let $x \in [0,1]^n$ and $\overline{x} := \frac{1}{n} \sum_{i=1}^{n} x_i$. Then*

$$\mathrm{lindisc}(A, x) \le 2f(2\|x\|_1(-\log_2(\overline{x}) + 1)).$$

*Proof.* From Lemma 4 and the concavity of $a \mapsto 2a(-\log_2(a) + 1)$, we conclude $w(x) \le 2\|x\|_1(-\log_2(\overline{x}) + 1)$. Thus Lemma 5 proves the claim. □

## 5   Conclusion

In this paper we investigated the relation between the linear discrepancy problem (rounding arbitrary vectors) and the combinatorial discrepancy problem (rounding vectors with entries $\frac{1}{2}$ only). We assumed that the discrepancy problem can be solved better for submatrices having fewer columns. This assumption is justified by the fact that most results are of this type. We showed that the classical results of Beck and Spencer and Lovász, Spencer and Vesztergombi on the relation of both rounding problems can be strengthened in this situation. We analyzed both the worst- and average case. Like in [7], our results indicate that the assumption of decreasing discrepancies is both natural and powerful.

We have to leave it as an open problem how tight our bounds are. Another open problem is for which vectors $x$ the rounding problem is hardest. 'Small' vectors cause lower errors, and as $w(a) = \frac{2}{3}$ for some $a \in [0,1]$ implies $a \in \{\frac{1}{3}, \frac{2}{3}\}$, our bound $\mathrm{lindisc}(A, x) \le 2f(\frac{2}{3}n)$ can only be tight if $x_i \in \{\frac{1}{3}, \frac{2}{3}\}$ for all $i \in [n]$. On the other hand, most examples seem to indicate that $x \approx \frac{1}{2}\mathbf{1}_n$ is the most difficult instance.

## References

1. N. Alon and J. Spencer. *The Probabilistic Method.* John Wiley & Sons, Inc., 2nd edition, 2000.
2. J. Beck and T. Fiala. "Integer making" theorems. *Discrete Applied Mathematics*, 3:1–8, 1981.
3. J. Beck and V. T. Sós. Discrepancy theory. In R. Graham, M. Grötschel, and L. Lovász, editors, *Handbook of Combinatorics*. 1995.
4. J. Beck and J. Spencer. Integral approximation sequences. *Math. Programming*, 30:88–98, 1984.
5. B. Doerr. Linear and hereditary discrepancy. *Combinatorics, Probability and Computing*, 9:349–354, 2000.
6. B. Doerr. Lattice approximation and linear discrepancy of totally unimodular matrices. In *Proceedings of the 12th Annual ACM-SIAM Symposium on Discrete Algorithms*, pages 119–125, 2001.
7. B. Doerr and A. Srivastav. Recursive randomized coloring beats fair dice random colorings. In A. Ferreira and H. Reichel, editors, *Proceedings of the 18th Annual Symposium on Theoretical Aspects of Computer Science (STACS) 2001*, volume 2010 of *Lecture Notes in Computer Science*, pages 183–194, Berlin–Heidelberg, 2001. Springer Verlag.

8. L. Lovász, J. Spencer, and K. Vesztergombi. Discrepancies of set-systems and matrices. *Europ. J. Combin.*, 7:151–160, 1986.
9. J. Matoušek. *Geometric Discrepancy*. Springer-Verlag, Berlin, 1999.
10. J. Matoušek. Tight upper bound for the discrepancy of half-spaces. *Discr. & Comput. Geom.*, 13:593–601, 1995.
11. J. Matoušek, E. Welzl, and L. Wernisch. Discrepancy and approximations for bounded VC–dimension. *Combinatorica*, 13:455–466, 1984.
12. J. Spencer. Six standard deviations suffice. *Trans. Amer. Math. Soc.*, 289:679–706, 1985.
13. J. Spencer. Randomization, derandomization and antirandomization: Three games. *Theor. Comput. Sci.*, 131:415–429, 1994.

# Approximating Min-sum Set Cover

Uriel Feige[1], László Lovász[2], and Prasad Tetali[3,*]

[1] Department of Computer Science and Applied Mathematics,
the Weizmann Institute, Rehovot 76100, Israel
`feige@wisdom.weizmann.ac.il`
[2] Microsoft Research, One Microsoft Way,
Redmond, WA 98052
`lovasz@microsoft.com`
[3] School of Mathematics and College of Computing,
Georgia Institute of Technology,
Atlanta, GA 30332-0160
`tetali@math.gatech.edu`

**Abstract.** The input to the *min sum set cover* problem is a collection of $n$ sets that jointly cover $m$ elements. The output is a linear order on the sets, namely, in every time step from 1 to $n$ exactly one set is chosen. For every element, this induces a first time step by which it is covered. The objective is to find a linear arrangement of the sets that minimizes the sum of these first time steps over all elements.

We show that a greedy algorithm approximates min sum set cover within a ratio of 4. This result was implicit in work of Bar-Noy, Bellare, Halldorsson, Shachnai and Tamir (1998) on *chromatic sums*, but we present a simpler proof. We also show that for every $\epsilon > 0$, achieving an approximation ratio of $4 - \epsilon$ is NP-hard. For the min sum *vertex* cover version of the problem, we show that it can be approximated within a ratio of 2, and is NP-hard to approximate within some constant $\rho > 1$.

## 1 Introduction

The *min sum set cover* (**mssc**) problem is a problem related both to the classical *min set cover* problem, and to the linear arrangement problems.

There are two equivalent ways by which we describe our problems. In one of them, $S$ is a set of points, and $\mathcal{F} = \{S_1, S_2, \ldots, S_s\}$ is a collection of subsets of $S$. An equivalent representation is via a hypergraph with vertex set $V$ and hyperedge set $E$. The hyperedges of the hypergraph correspond to the points in the set system, and the vertices of the hypergraph correspond to the subsets. A hypergraph is *r-uniform* if every hyperedge contains exactly $r$ vertices. Likewise, we call a set system $r$-uniform if every point appears in exactly $r$ subsets. A hypergraph is *d-regular* if every vertex has degree $d$, namely, is contained in exactly $d$ hyperedges. Likewise, we call a set system $d$-regular if every subset is of cardinality $d$.

* research in part supported by the NSF grant DMS-0100298

K. Jansen et al. (Eds.): APPROX 2002, LNCS 2462, pp. 94–108, 2002.

**Min Sum Set Cover (mssc).** Viewing the input as a hypergraph $H(V, E)$, a linear ordering is a bijection $f$ from $V$ to $\{1, \ldots, |V|\}$. For a hyperedge $e$ and linear ordering $f$, we define $f(e)$ as the minimum of $f(v)$ over all $v \in e$. The goal is to find a linear ordering that minimizes $\sum_e f(e)$.

We note that minimizing the sum of $f(e)$ is equivalent to minimizing the average of $f(e)$. So another way of viewing **mssc** is as that of seeking a linear arrangment of the vertices of a hypergraph that minimizes the average cover time for the hyperedges.

An important special case of **mssc** is the following.

**Min Sum Vertex Cover (msvc).** The hypergraph is a graph $G(V, E)$. (Equivalently, in the set system representation, every point belongs to exactly two subsets.) Hence one seeks a linear arrangement of the vertices of a graph that minimizes the average cover time of the edges. Linear arrangement problems on graphs often come up as heuristics for speeding up matrix computation. And indeed, **msvc** came up in ([4], Section 4) in the context of designing efficient algorithms for solving semidefinite programs, and was one of the motivations for our work.

Another problem that in a sense is a special case of **mssc** is the following.

**Min Sum Coloring.** The input to this problem is a graph. The output is linear ordering of its independent sets, or equivalently, a legal coloring of its vertices by natural numbers. The objective is to find such a coloring that minimizes the sum of color-numbers assigned to vertices. Given an input graph $G'(V', E')$, one can cast the min sum coloring problem as an **mssc** problem as follows. The vertices of the hypergraph $H$ are the independent sets of $G'$, and the hyperedges of the hypergraph $H$ are the vertices $V'$. Note however that the size of the hypergraph $H$ would typically be exponential in the size of the graph $G'$. Min sum coloring has been extensively studied in the past and many of the results carry over to **mssc**. We shall later mention the results of [2,3].

We note here that min sum vertex cover can be viewed as a special case of min sum coloring, by taking $G'$ to be the complement of the line graph of $G$.

All the above problems are NP-hard, and we shall study their approximability.

## 1.1   Related Work

We are not aware of previous work on the min sum set cover problem. Regarding min sum vertex cover, this problem was suggested to us by the authors of [4]. They use a greedy algorithm that repeatedly takes the vertex of largest degree in the remaining graph as a heuristic for **msvc** . The problem **msvc** itself is used as a heuristic for speeding up a solver for semidefinite programs.

Min sum coloring was studied extensively. It models the issue of minimizing average response time in distributed resource allocation problems. The vertices of the underlying graph (the so-called *conflict graph*) represent tasks that need to be performed, and an edge between two vertices represents a conflict – the corresponding tasks cannot be scheduled together. Part of the difficulty of the min sum coloring problem is that of identifying the independent sets in the conflict graph, which makes it more difficult than **mssc** (where the underlying hypergraph is given explicitly). In [2] it is observed that min sum coloring is hard to approximate within a ratio of $n^{1-\epsilon}$ for every $\epsilon > 0$, due to the hardness of distinguishing between graphs that have no independent sets of size $n^\epsilon$ and graphs that have chromatic number below $n^\epsilon$ (which is shown in [7]). This hardness result does not apply to **mssc**.

In [2] it is shown that the greedy algorithm that iteratively picks (and removes) the largest independent set in the graph approximates min sum coloring within a factor 4. This algorithm can be applied for certain families of graphs (such as perfect graphs), and also in the case of **mssc** (where of course we iteratively pick the vertex with largest degree in the remaining hypergraph). We observe that the proof in [2] of the factor 4 approximation applies also to **mssc** (and not just to the special case of min sum coloring). Hence **mssc** is approximable within a factor of 4.

In [3] examples are shown where the greedy algorithm does not approximate min sum coloring within ratios better than 4, showing the optimality of the analysis in [2]. We observe that the proof given there also applies to the use of the greedy algorithm for min sum vertex cover (which is the algorithm used in [4]).

Min sum coloring was studied also on some special families of graphs (such as interval graphs, bipartite graphs, line graphs), often obtaining ratios of approximation better than 4. As noted earlier, min sum vertex cover is a special case of min sum coloring (on complements of line graphs), but we are not aware of explicit treatment of this special case.

There are close connections between **mssc** and set-cover. For problems related to set cover, tight approximation thresholds (up to low order terms) are often known. Examples include $\ln n$ for min set cover and $(1 - 1/e)$ for max $k$-cover [5], $\ln n$ for the Domatic Number [6], roughly $\sqrt{n}$ for maximum disjoint packing of sets (a result published in the context of auction design). This is some indication that one may be able to find a tight approximation threshold for **mssc**. However, let us point out a major difference between **mssc** and other problems related to set cover. Given an instance of **mssc** which is composed of two disjoint instances, the optimal solution is not necessarily a combination of the optimal solutions to each of the sub-instances. This makes it more difficult to design and analyse algorithms for **mssc**. In particular, we do not even know if there is a polynomial time algorithm for min sum vertex cover when the underlying graph is a tree (whereas min vertex cover is polynomial time solvable on trees.) As we shall later see, the hardest instances for **mssc** (in terms of approximation ratio) have different properties than the hardest instances for min set cover. A major difference (already manifested in [3]) is that they are not regular.

## 1.2   New Results

The main result regarding the approximation of min sum set cover is the following.

**Theorem 1.**   *1. The greedy algorithm approximates min sum set cover within a ratio no worse than 4.*
*2. For every $\epsilon > 0$, it is NP-hard to approximate min sum set cover within a ratio of $4 - \epsilon$.*

As noted earlier, the first part of Theorem 1 was essentially already proved in [2]. However, we present an alternative proof. Our proof was inspired by the primal-dual approach for approximation algorithms based on linear programming. Using an intuitive interpretation of the dual, we present our proof without the need to presenting the underlying linear program. (Note however that it is implicit in our proof that the underlying linear program approximates **mssc** within a ratio of 4.) Using our approach we avoid the evaluations of summations that are involved in the proof in [2].

The second part of Theorem 1 is proved by modifying a reduction of [5], and combining it with ideas from [3].

For min sum vertex cover, we observe that the results of [3] imply that the greedy algorithm does not approximate it within a ratio better than 4. We then show:

**Theorem 2.**   *1. An approximation algorithm based on linear programming approximates min sum vertex cover within a ratio of 2.*
*2. There exists a constant $\rho > 1$ such that min sum vertex cover is NP-hard to approximate within a ratio better than $\rho$.*

The first part of Theorem 2 is proved by using a linear programming relaxation for **msvc** , and rounding it using a randomized rounding technique. We are currently working on understanding the integrality gap of this linear program better. It appears that we have examples showing that the integrality gap of this LP is at least 16/9, and that we have rounding techniques that give approximation ratios better than $c$ for some $c < 2$. As this work is not complete yet, it is omitted from this manuscript.

Our last set of results relate to the special case of **mssc** instances on $r$-uniform $d$-regular instances. We observe that on such instances **mssc** can be approximated within a ratio of $2r/(r+1)$. For large values of $r$, this approximation ratio tends to 2. For **msvc** (where $r = 2$), this approximation ratio is $4/3$. Our main extensions of these results are as follows:

**Theorem 3.**   *1. For every $\epsilon > 0$, there exist $r, d$ such that it is NP-hard to approximate min sum set cover within a ratio better that $2 - \epsilon$ on $r$-uniform $d$-regular hypergraphs.*
*2. For some $\rho < 4/3$ and every $d$, min sum vertex cover can be approximated within a ratio of $\rho$ on $d$-regular graphs.*

The first part of Theorem 3 is obtained as part of the proof of the second part of Theorem 1. The proof of the second part of Theorem 3 uses semidefinite programming.

## 2   The Greedy Algorithm

Let $H(V, E)$ be a hypergraph on which we wish to approximate min sum set cover. The greedy algorithm produces a sequence of vertices that cover all hyperedges as follows.

1. Initialize $i = 1$.
2. While hypergraph $H$ has an edge do
   (a) Take $v_i$ to be a vertex of maximum degree in $H$.
   (b) Update $H$ by removing $v_i$ and all hyperedges incident with it from $H$.
   (c) Increment $i$.

**Theorem 4.** *The greedy algorithm approximates min sum set cover within a ratio no worse than 4.*

Theorem 4 is proved in [2]. We present a different proof. A major difference between these proofs is that in [2] the solution of the greedy algorithm is compared directly with the optimal solution, whereas in our proof it is compared with a lower bound on the optimal solution, thereby avoiding some computations that are involved in the proof of [2].

*Proof.* Let **opt** denote the optimal value of the min sum set cover problem. Let **greedy** denote the value returned by the greedy algorithm.

For $i = 1, 2, \ldots$, let $X_i$ denote the set of edges first covered in step $i$ by the greedy algorithm. Let $R_i = \sum_{j=i}^{\infty} |X_i|$ be the number of edges that are uncovered at the beginning of step $i$. In the following proposition and throughout this section, the range of summations is from $i = 1$ until the largest value of $i$ for which the objects of the summation are still defined.

**Proposition 1.** *Using the notation above, we have that*

$$\mathbf{greedy} = \sum_i i|X_i| = \sum_i R_i \qquad (1)$$

*Proof.* Our notation is such that the greedy algorithm covers exactly $|X_i|$ new edges in step $i$, and hence the first equality in equation (1) follows by summing the contribution to **greedy** over sets $X_i$. The second equality is obtained by summing the contribution to **greedy** over time steps rather than over edges.

Clearly, **greedy** $\geq$ **opt**. We wish to show that **greedy** $\leq$ 4**opt**. For this we shall introduce a lower bound on **opt**, that we shall denote by **dual**. That is, **dual** $\leq$ **opt**. Then we shall show that **greedy** $\leq$ 4**dual**, proving the theorem.

The lower bound **dual** was derived by us by formulating the problem of min sum set cover as an integer program, relaxing it to a linear program, and taking the dual of this linear program. This dual has an intuitive interpretation that we explain next, avoiding the linear programming machinery. We shall retain the name dual for this lower bound.

The dual is expressed in terms of a *profit* (denoted by **profit**) and *tax* (denoted by **tax**). The profit component can be chosen arbitrarily. Then the tax is computed on it, and the net profit is **profit** − **tax**. The value of **dual** is the maximum possible net profit. As will be seen, the tax is computed in such a way that the net profit is not a monotone function of the profit: increasing the profit may well cause the net profit to decrease.

With each edge $e$ of the hypergraph, associate an arbitrary nonnegative profit $p_e$. Then the profit is defined as **profit** $= \sum_e p_e$. Given the profit per edge, the tax is computed as follows. There is tax for every time step $t = 1, 2 \ldots$. The tax at time $t$ is computed by selecting a vertex $v$, and taxing each edge incident with $v$ by $\max[0, p_e - t]$. That is, the tax associated with vertex $v$ at time $t$ is $\sum_{\{e | v \in e\}} \max[0, p_e - t]$. At each time step $t$, the profit is taxed according to the vertex that maximizes the tax. Finally **tax** is the sum of taxes over all time steps. We allow the same vertex to be used in taxing different time steps. Note that for $t \geq \max_e [p_e]$, the tax becomes 0 and need not be computed further.

**Proposition 2.** *For every way of assigning nonnegative profits to the edges,* **profit** − **tax** ≤ **opt**. *In particular,* **dual** ≤ **opt**.

*Proof.* Consider an arbitrary assignment of nonnegative profits to the edges. We show a particular choice of tax that brings the net profit to no more than **opt**. Hence under the maximum tax, this holds as well.

Let $v_1, v_2, \ldots$, be the sequence of vertices chosen by the optimum solution to the min sum set cover problem. Then we tax time step $i$ using vertex $v_i$ from this sequence. Let $e$ be an edge that was first covered by the optimal solution in step $t$. Then the contribution of $e$ to the net profit becomes $p_e - \max[0, p_e - t] \leq t$. Summing over all edges, we get that the net profit is at most the value of the optimal solution. (See also Figure 1 for an illustration that for any assignment of profits, **profit** − **tax** ≤ **greedy**.)

Having established that **dual** ≤ **opt**, we now show that **greedy** ≤ 4**dual**. This is done by exhibiting a particular assignment of profits, based on the solution of the greedy algorithm. For this assignment, we will show that the net profit is at least **greedy**/4, establishing the theorem.

Recall the notation of $X_i$ and $R_i$ for the greedy algorithm. Define $P_0 = 0$, and for every $i \geq 1$, $P_i = \max[P_{i-1}, \frac{R_i}{2|X_i|}]$. For every edge $e \in X_i$, let $p_e = P_i$. Recall that **profit** $= \sum_e p_e$.

**Proposition 3.** *Using the notation above, we have that*

$$\mathbf{profit} \geq \frac{1}{2} \sum_i R_i \tag{2}$$

$$\mathbf{profit} = \sum_i P_i (R_i - R_{i+1}) = \sum_i R_i (P_i - P_{i-1}) \tag{3}$$

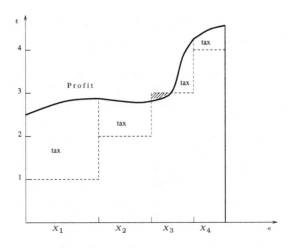

**Fig. 1.** Profit - tax $\leq$ greedy

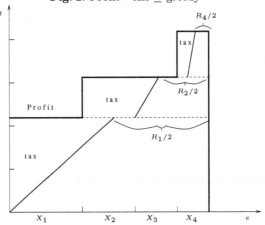

**Fig. 2.** tax $\leq \frac{1}{2}$ Profit

*Proof.* Inequality (2) is derived by summing the profit over sets $X_i$ :

$$\mathbf{profit} = \sum_i |X_i| P_i \geq \sum_i |X_i| \frac{R_i}{2|X_i|} = \sum_i R_i/2.$$

Equation (3) is obtained by summing the profit over time steps.

**Corollary 1.** *For the assignment of profits given above,* **greedy** $\leq 2$**profit**.

*Proof.* Follows by comparing equations (1) and (2).

To complete the proof of the theorem, we use the following proposition.

**Proposition 4.** *For the assignment of profits given above,* **tax** $\leq$ **profit**/2.

*Proof.* There is insufficient information in order to derive an exact expression for the tax. However, we derive an upper bound on the tax as follows.

The allocation of profits that we described above can be viewed as the end result of the following continuous process. Each edge has a bucket into which profit is poured at a uniform rate of one unit of profit per time step, and all buckets are filled in parallel. When the profit of an edge $e$ reaches the desired profit $p_e$, no more profit is poured into its bucket. Eventually, all edges reach their desired profit and the process ends.

We can identify different stages of this process. Initially, every edge has profit $P_0 = 0$. Stage 1 ends when all ($R_1$) edges have profit $P_1$. In stage 2, the profit of the edges in $X_1$ remains at $P_1$, whereas the profit of the edges of $R_2$ is increased until it reaches $P_2$. (This stage takes 0 time if $P_1 = P_2$). The general rule is that in stage i, the profit of the edges of $R_i$ is raised from $P_{i-1}$ to $P_i$, whereas the profit of the other edges remain unchanged. Observe that the total rate of increase of profit in stage i is exactly $R_i$ (giving equation (3)).

We now compute an upper bound on the tax. We think of tax collecting also as a continuous process. As the profit increases from time instance $t_1$ to time instance $t_2$, we collect additional tax. This additional tax is computed as the maximum possible tax collectable on the profit up to $t_2$, minus the maximum possible tax collectable on the profit up to $t_1$. (Thus, possibly the additional tax between $t_1$ and $t_2$ is larger than the additional profit.) Eventually, this gives the total tax.

Initially, the tax is 0. We now give an upper bound on the rate of collection of additional tax. Assume that we are at time instance $t$, and this time instance is within stage i. Then only edges in $R_i$ increase their profit. The cardinality of the largest set in $R_i$ is $|X_i|$. Recall that every (integer) time step up to $t$ has to be taxed. There are at most $t$ such time steps. Hence the rate of collection of additional tax is at most $t|X_i|$. But within stage i, $t \leq P_i \leq R_i/2|X_i|$. It follows that the rate of collection of tax in stage i is at most $R_i/2$, which is half the rate of profit increase. As this holds for all stages of the continuous process described above, the tax is at most half the profit. (See Figure 2.)

Summing up:    **greedy** $\leq 2$**profit** $\leq 4($**profit** $-$ **tax**$) \leq 4$**opt**,

where the inequalities follow from Corollary 1, Proposition 4 and Proposition 2 respectively. This completes the proof of Theorem 4.

## 3    Min-sum Vertex Cover

In [3] it is shown that for every $\epsilon > 0$, there are instances of min sum set cover for which the approximation ratio of the greedy algorithm is no better than $4 - \epsilon$. (Technically, this result is stated for min sum coloring, but it applies also to min sum set cover which is a more general problem.) We observe that the same negative result applies also to min sum vertex cover, by the following proposition. We omit the proof here due to space limitation, but we include it in the journal version of the paper.

**Proposition 5.** *Let I be an arbitrary instance of min sum set cover, and let* **opt** *be the value of its optimal solution and* **greedy** *be the value of its greedy solution. Then there there is an instance I' of min sum vertex cover with optimal value* **opt** *and for which the greedy solution has value* **greedy***.*

Hence the greedy algorithm does not approximate min sum vertex cover within a ratio better than 4. We now show a different algorithm that does approximate min sum vertex cover within a ratio better than 4.

Consider the following integer program for min sum vertex cover. The indices $i$ and $j$ run over all vertices. The index $t$ runs over all time steps. The variable $x_{it}$ is an indicator variable that indicates whether vertex $i$ is chosen at step $t$. $y_{ijt}$ is an indicator variable that indicates whether edge $(i, j)$ is still uncovered before step $t$.

**Minimize** $\sum_{(i,j) \in E} \sum_t y_{ijt}$ **subject to**

1. $x_{it} \in \{0, 1\}$. (Integrality constraint.)
2. $y_{ijt} \in \{0, 1\}$. (Integrality constraint.)
3. $\sum_i x_{it} \leq 1$. (In every time step, at most one vertex is chosen.)
4. $y_{ijt} \geq 1 - \sum_{t' < t}(x_{it'} + x_{jt'})$. (An edge is uncovered at the beginning of time $t$ unless one of its endpoints was covered at a previous time step.)

The integer program is relaxed to a linear program by relaxing the integrality constraints to $0 \leq x_{it} \leq 1$ and $0 \leq y_{ijt} \leq 1$. Clearly, the linear program (that is solvable in polynomial time) provides a lower bound for min sum vertex cover.

We propose a procedure for rounding a fractional solution of the linear program. The procedure is randomized and produces an integer solution with expected value at most twice that of the linear program. We note that the rounding technique can be made deterministic using the method of conditional expectation.

The rounding technique works in two stages. The first stage is performed independently for each vertex. Consider vertex $i$ and the fractional variables $x_{it}$ for $t \geq 1$. Let $t_i$ be that value of $t'$ for which $\sum_{t < t'} x_{it} < 1/2$ and $\sum_{t \leq t'} x_{it} \geq 1/2$. (If no such $t'$ exists, namely, $\sum_t x_{it} < 1/2$, then let $t_i = \infty$.) Now introduce new variables $z_{it}$, where $z_{it} = 2x_{it}$ for $t < t_i$, $z_{it_i} = 1 - \sum_{t < t_i} z_{it}$, and $z_{it} = 0$ for $t > t_i$. Note that $\sum_t z_{it} \leq 1$. Now randomly choose at most one value of $t$, where value $t$ is chosen with probability $z_{it}$. For the chosen $t$, $x_{it}$ is rounded to 1, and for all other values of $t$, $x_{it}$ is rounded to 0. Let $\bar{x}_{it}$ denote the rounded values obtained by this procedure.

The outcome of the first stage of the rounding technique satisfies the integrality constraints for the $\bar{x}_{it}$ (constraint 1) but may violate constraint 3. In the second stage of the rounding technique we scan the time steps one by one. For time step $t$, let $s_t = \sum_i \bar{x}_{it}$. Replace time step $t$ by $s_t$ time slots. Now allocate the vertices $i$ for which $\bar{x}_{it} = 1$ to these time slots in a random order. (The value of $t$ for which $\bar{x}_{it} = 1$ is shifted to the respective time slot.) Now constraint 3 is satisfied, because each time slot has exactly one vertex assigned to it.

Given values for $\bar{x}_{it}$ that satisfy constraints 1 and 3, a 0/1 assignment to the $y_{ijt}$ is derived in a straightforward way (ignoring the assignment originally given to them by the fractional solution). This completes the description of the rounding procedure.

**Lemma 1.** *The expected value of the rounded solution to the LP is at most twice the fractional value of the LP.*

*Proof.* Consider an arbitrary edge $(i, j)$ and an arbitrary time step $t$. The contribution of this to the fractional solution is $y_{ijt} \geq 1 - \sum_{t' < t}(x_{it'} + x_{jt'})$. We will compare this to the expected constribution of edge $(i, j)$ to time step $t$ in the rounded solution. This contribution is a product of two factors:

1. The probability that edge $(i, j)$ is not covered before time step $t$.
2. Conditioned on edge $(i, j)$ not being covered before time step $t$, the expected number of time slots within time step $t$. (Note that we will be comparing time step $t$ of the fractional solution to all time slots derived from it, rather than to time slot $t$.) Here there is subtlety involved. The number of time slots under consideration is not exactly $s_t$ (the number of time slots into which time step $t$ is transformed). There is the possibility that in the rounded solution edge $(i, j)$ was first covered in time step $t$. Then the particular time slot within time step $t$ in which $(i, j)$ was covered is random, and later time slots need not be counted.

For the first factor, we compute the probability that edge $(i, j)$ is not covered by the rounded solution before time $t$. This probability is

$$\left(1 - \sum_{t' < t} z_{it'}\right)\left(1 - \sum_{t' < t} z_{jt'}\right) \leq y_{ijt}$$

where the inequality follows from the relation $\sum_{t' < t} z_{it'} = \min[1, 2\sum_{t' < t} x_{it'}]$, and from constraint 4.

For the second factor (the value of $s_t$, and the random order within the time slots) we introduce two new random variables, $r$ (for "rest") and $w$ (for "waiting time"). $r$ counts the number of vertices other than $i$ and $j$ that are rounded to 1 at time step $t$. $w$ counts the number of relevant time slots within time step $t$ (those at the beginning of which edge $(i, j)$ is not yet covered). We are interested in the expectation of $w$. The value of this expectation can be expressed as a function of $r$, taking into acount also the random order of time slots within a time step. We obtain:

- If $\bar{x}_{it} = \bar{x}_{jt} = 0$, then $r = s_t$ and $w = r$.
- If $\bar{x}_{it} = 0$ and $\bar{x}_{jt} = 1$, or $\bar{x}_{it} = 1$ and $\bar{x}_{jt} = 0$, then $r = s_t - 1$ and $E[w] = 1 + E[r]/2$.
- If $\bar{x}_{it} = \bar{x}_{jt} = 1$, then $r = s_t - 2$ and $E[w] = 1 + E[r]/3$.

Now $E[r] = \sum_{k \neq i,j} E(\bar{x}_{kt}) = \sum_{k \neq i,j} z_{kt} \leq 2\sum_{k \neq i,j} x_{kt} \leq 2$, due to constraint 3. It follows that in all cases $E[w] \leq 2$. Hence, altogether the contribution of $y_{ijt}$ to the rounded solution is at most $2y_{ijt}$.

Using the linearity of expectation (over all $y_{ijt}$), the expected value of the rounded solution is at most twice that of the fractional solution.

The analysis of the rounding technique is essentially best possible. This can be verified by considering a graph composed of disjoint edges. The fractional solution can cover edge by edge by giving its two endpoints weight $1/2$. The rounded solution will then take both endpoints, paying twice as much. As mentioned in the introduction, we are currently working on a different rounding technique that gives an approximation ratio better than 2. We suspect that using semidefinite programming rather than linear programming can further improve the approximation ratio. This we can show for the special case of regular graphs. The integrality ratio of the LP is $4/3$ (on a clique), whereas semidefinite programming gives a better approximation ratio (see Theorem 6).

# 4   Regular Hypergraphs

Let $H$ be an $r$-uniform, $d$-regular hypergraph. That is, each hyperedge contains exactly $r$ vertices, and each vertex has degree $d$. Let $n$ denote the number of vertices and $m$ the number of hyperedges. (Clearly, $rm = dn$.)

For every such hypergraph, the optimal value of min sum set cover is at least $m\frac{n+r}{2r}$, because at most $d$ hyperedges are covered per step, and at this rate, it takes $\frac{m}{d} = \frac{n}{r}$ steps to cover all hyperedges. Hence the average number of steps until a hyperedge is covered is at least $\frac{n+r}{2r}$.

On the other hand, the optimal solution has value at most $m\frac{n+1}{r+1}$. This can be seen as follows. Consider a random permutation of all vertices. Then for every hyperedge, the expected step in which it is first covered is exactly $\frac{n+1}{r+1}$. (This last statement can be proven by considering a random cyclic permutation on $n+1$ elements, of which $r+1$ are special. Now select at random which of the $r+1$ special elements marks the start of the permutation on the rest of the $n$ elements, and the other $r$ special elements are the vertices that compose the hyperedge. Then over the choice of where the permutation starts, the expected number of steps until another special element is reached is exactly $(n+1)/(r+1)$.) Hence there is some ordering of the vertices for which the average time to cover a hyperedge is at most $\frac{n+1}{r+1}$. Moreover, the greedy algorithm produces such an ordering. (One way of seeing this is that the method of conditional expectations produces the greedy algorithm as a derandomization of the randomized algorithm.)

The above proves the following theorem.

**Theorem 5.** *For every $r$-uniform $d$-regular hypergraph, the approximation ratio of the greedy algorithm for min sum set cover is no worse than $\frac{2r}{r+1}$. In particular, the approximation ratio of the greedy algorithm for min sum vertex cover on regular graphs is no worse than $4/3$.*

As $r$ gets larger, the approximation ratio of the greedy algorithm on $r$-uniform regular hypergraphs approaches 2. This cannot be significantly improved unless P=NP, as we shall see in Theorem 7. However, for the special case of $r = 2$ (regular graphs), we can improve over the greedy algorithm.

**Theorem 6.** *There is some constant $1 < \rho < 4/3$ such that min sum vertex cover can be approximated within a ratio of $\rho$ on regular graphs.*

*Proof.* The central algorithmic tool used in our proof is semidefinite programming. The presentation of the algorithm is greatly simplified (perhaps at some loss in the approximation ratio) by using in a "blackbox" manner previous results regarding the use of semidefinite programming for the *max k-vertex cover* problem. The details can be found in the journal version of the paper.

## 5   Hardness of Approximation

**Theorem 7.** *For every $\epsilon > 0$, it is NP-hard to approximate min sum set cover within a ratio of $2 - \epsilon$ on uniform regular hypergraphs.*

Theorem 5 shows that Theorem 7 is essentially best possible. The proof of Theorem 7 is very similar to the proof given in [5] of the result that it is NP-hard to approximate the max $k$-coverage problem within a ratio better than $1 - 1/e + \epsilon$. We do not wish to reproduce here the proof already given in [5]; in stead, we provide a sketch of the proof of Theorem 7, using the terminology of [5].

**Proof sketch of Theorem 7:** In [5] a reduction from max 3SAT-5 to max $k$-coverage is described. We note that the resulting instance of max $k$-coverage (which is a hypergraph) is regular (each set contains the same number of points/ each vertex appears in the same number of hyperedges) but not uniform (some points are covered by more sets than others/ some hyperedges contain more vertices than others). To make the hypergraph also uniform, we change the starting point of the reduction. Rather than starting from a 3CNF formula in which each variable appears in exactly five clauses, we start from a 3CNF formula in which each literal appears in exactly three clauses (and each variable in six clauses). We call the satisfiability problem for such formulas 3SAT-6. We note that for some $\delta < 1$, it is NP-hard to distinguish between satisfiable 3SAT-6 formulas, and those in which at most a $\delta$-fraction of the clauses are satisfiable. (This can be proven by reduction from 3SAT-5, in which each variable appears 3 times in positive form and twice negated. For a 3SAT-5 formula with $n$ variables, join to it a satisfiable 2SAT-6 formula with $n$ clauses on a fresh set of variables, and to each of the 2CNF clauses add one of the original variables negated.) The (adaptation of the) reduction of [5] now gives a regular uniform hypergraph. This is a consequence of the following symmetries:

1. Every clause in the CNF formula contains the same number of literals (three in our case).
2. Every literal appears in the same number of clauses (three in our case).
3. Every code word (in the proof system of Section 2 in [5]) has exactly the same Hamming weight ($\ell/2$ when the Hadamard code is used).
4. In the partition system (proof of Theorem 12 in [5]) each part has exactly the same size ($m/k$, using the notation of [5]).

Now there are two cases:

1. If the original 3SAT-6 formula is satisfiable, then the reduction has the property that there is a collection of disjoint sets (and necessarily of equal cardinality) that covers all points. Let us denote by $t$ the number of sets used in such a cover. Hence for the min sum set cover problem, a hyperedge is covered by step $t/2$, on average.

2. If the original 3SAT-6 formula was only $\delta$-satisfiable (for $\delta < 1$), the reduction has the following property: for every constants $c_0 > 0$ and $\epsilon > 0$, it is possible to choose the parameter $\ell$ (number of repetitions in the proof system) to be a large enough constant so that for every $1 \leq c \leq c_0 t$, at most a fraction of $1 - (1 - 1/t)^c + \epsilon$ of the points can be covered by $c$ sets. The proof of this is an extension of the proof of Theorem 12 in [5] and is omitted. It follows (via simple calculations) that for the min sum set cover problem, a hyperedge is covered by step $(1 - O(\epsilon))t$, on average. (Essentially, for a random edge, it has probability of roughly $1/t$ of being covered at each time step, and hence the expected time until it is covered is roughly $t$.)

The gap between the two cases can be made arbitrarily close to a factor of 2. □

For nonregular instances of **mssc** we prove a stronger hardness of approximation result which matches the positive result of Theorem 4. The proof of the following theorem was inspired by [3].

**Theorem 8.** *For every $\epsilon > 0$, it is NP-hard to approximate min sum set cover within a ratio of $4 - \epsilon$ on uniform hypergraphs.*

*Proof.* Via a reduction from the regular uniform case (Theorem 7). Due to space limitations, we omit the proof here, but include it in the journal version.

We now prove hardness of approximation for **msvc** . The proof illustrates a property by which min sum set cover (and min sum vertex cover) differ from set cover. Namely, the optimal solution to a disjoint union of two input instances does not necessarily induce optimal solutions of the original two instances.

**Theorem 9.** *For some $\epsilon' > 0$, it is NP-hard to approximate min sum vertex cover within ratios better than $1 + \epsilon'$.*

*Proof.* For some univeral constant $d \geq 3$, let $G$ be a graph with $n$ vertices, $m$ edges, and degree at most $d$. It is known that min vertex cover is hard to approximate on graphs of bounded degree [1]. More specifically, for every $d \geq 3$ there is some $\delta > 1$ for which it is NP-hard to approximate min vertex cover with a ratio better than $\delta$.

We reduce the problem of approximating min vertex cover problem on bounded degree graphs to the problem of approximating min sum vertex cover.

Assume without loss of generality that $G$ does not have isolated vertices (as they can be removed from $G$ without changing the optimal solution of min vertex cover). Observe that at least $n/(d + 1)$ steps are needed to cover all edges of $G$, because each vertex covers at most all edges connected to itself and to its $d$

neighbors. It follows that it is NP-hard to approximate min vertex cover on $G$ within an additive factor of $\delta n/d$. Let us denote $\delta/d$ by $\epsilon$, and note that $\epsilon$ is a constant greater than 0.

Let $k = \frac{d}{2\epsilon}$. Construct a graph $G'$ that is the disjoint union of $G$ and $kn$ additional isolated edges (i.e., vertex disjoint union of $G$ and a matching of size $kn$). On $G'$ we wish to approximate min sum vertex cover. The optimal solution to min sum vertex cover on $G'$ may be assumed without loss of generality to first cover all edges of $G$, and only then cover the isolated edges, because the isolated edges can be covered at a rate of at most one at a time, and edges in $G$ can be covered at a rate of at least one at a time.

Let us consider the case that $G$ has a vertex cover with at most $t$ vertices. Then $G'$ has a min sum vertex cover of value at most $m\frac{t}{2} + kn(t + \frac{kn}{2})$.

Let us now consider the case that $G$ has no vertex cover with less than $t + \epsilon n$ vertices. Then it costs at least $kn(t + \epsilon n + \frac{kn}{2})$ to cover the $kn$ isolated edges, and we use this as a lower bound on the value of min sum vertex cover for $G'$.

The difference between the two cases is at least $\epsilon kn^2 - m\frac{t}{2}$. Using $m \leq dn/2$ and $t \leq n$, this difference is at least $n^2(\epsilon k - \frac{d}{4})$. Using $k = \frac{d}{2\epsilon}$, this difference is at least $\frac{\epsilon k}{2}n^2$. The optimal solution to min sum vertex cover on $G'$ is of value at most $m\frac{n}{2} + kn(n + \frac{kn}{2}) \leq k^2n^2$, where the last inequality follows from simple manipulations, using $d \geq 3$ and $\epsilon \leq 1/2$ (which are true in our context). Setting $\epsilon' = 1/2k = \epsilon/d$, it follows that if we could approximate min sum vertex cover in $G'$ within a ratio better than $1 + \epsilon'$, then we could approximate min vertex cover in $G$ within a ratio better than $\delta$.

We have not made an effort to find the best possible value of $\epsilon'$ for Theorem 9.

# References

1. S. Arora, C. Lund, R. Motwani, M. Sudan, M. Szegedy. "Proof verification and the hardness of approximation problems". *JACM* 45(3):501–555, 1998.
2. A. Bar-Noy, M. Bellare, M. Halldorsson, H. Shachnai, T. Tamir. "On chromatic sums and distributed resource allocation." *Information and Computation*, 140:183–202, 1998.
3. A. Bar-Noy, M. Halldorsson, G. Kortsarz. "A matched approximation bound for the sum of a greedy coloring". *Information Processing Letters*, 1999.
4. S. Burer and R. Monteiro. "A projected gradient algorithm for solving the maxcut SDP relaxation". *Optimization Methods and Software*, 15 (2001) 175-200.
5. U. Feige. "A threshold of ln $n$ for approximating set cover". *Journal of the ACM*, 45(4), 634–652, 1998.
6. U. Feige, M. Halldorsson, G. Kortsarz, A. Srinivasan. "Approximating the domatic number". Preliminary version in STOC 2000.
7. U. Feige and J. Kilian. "Zero knowledge and the chromatic number". *Journal of Computer and System Sciences*, 57(2):187–199, 1998.
8. U. Feige and M. Langberg. "Approximation algorithms for maximization problems arising in graph partitioning". *Journal of Algorithms* 41, 174–211 (2001).
9. E. Halperin and U. Zwick. "A unified framework for obtaining improved approximation algorithms for maximum graph bisection problems". In proceedings of IPCO, 2001.

# Approximating Maximum Edge
# Coloring in Multigraphs

Uriel Feige, Eran Ofek, and Udi Wieder

Weizmann Institute of Science, Rehovot 76100, Israel,
{feige,erano,uwieder}@wisdom.weizmann.ac.il

**Abstract.** We study the complexity of the following problem that we call *Max edge t-coloring*: given a multigraph $G$ and a parameter $t$, color as many edges as possible using $t$ colors, such that no two adjacent edges are colored with the same color. (Equivalently, find the largest edge induced subgraph of $G$ that has chromatic index at most $t$). We show that for every fixed $t \geq 2$ there is some $\epsilon > 0$ such that it is NP-hard to approximate *Max edge t-coloring* within a ratio better than $1 - \epsilon$. We design approximation algorithms for the problem with constant factor approximation ratios. An interesting feature of our algorithms is that they allow us to estimate the value of the optimum solution up to a multiplicative factor that tends to 1 as $t$ grows. Our study was motivated by call admittance issues in satellite based telecommunication networks.

## 1 Introduction

A *multigraph* is a graph that may contain multiple edges between any two vertices, but cannot contain self loops. Given a multigraph $G = (V, E)$, an *edge t-coloring* of the graph is an assignment $c : E \mapsto [t]$, where adjacent edges are mapped to distinct colors. The smallest $t$ for which there exists an edge $t$-coloring, is called the *chromatic index* of the graph and is denoted as $\chi'(G)$. We study a related problem:

*Problem 1.* Given a multigraph $G$ and a number $t$, legally color as many edges as possible using $t$ colors. We call this problem *Max edge t-coloring*.

We use the *Max edge t-coloring* to model a satellite based communication network. In this network, a ground station serves those clients that are geographically close to it. To establish a session between two clients that are far apart, their respective ground stations establish a connection via a satellite. A connection through the satellite requires a dedicated cell in an FTDM (frequency division multiplexing combined with time division multiplexing) matrix, namely, a matrix with $t$ time slots and $f$ frequencies. Each ground station contains a modem that can handle at most one connection per time slot, and use an arbitrary frequency for this connection. Thus to establish a session between two distant clients, one needs to find a free cell in the FTDM matrix, and moreover, this cell must belong to a time slot that is not in use by the respective modems.

K. Jansen et al. (Eds.): APPROX 2002, LNCS 2462, pp. 108–121, 2002.
© Springer-Verlag Berlin Heidelberg 2002

The *Max edge t-coloring* problem models the above setting in the offline case in which a collection of session requests is given in advance, and one has to simultaneously satisfy as many requests as possible. The vertices of the multigraph represent the base stations, the edges represent session requests, and the parameter $t$ specifies the number of time slots in the FTDM matrix. A solution to the resulting *Max edge t-coloring* problem can be used in order to decide how to allocate sessions in the FTDM matrix. Note that in our modelling we ignored the bound $f$ on the number of frequencies in the FTDM matrix. This is justified for the following reason. Given an arbitrary $t$-coloring of a multigraph, there is a simple balancing procedure that ensures that the cardinality of every two color classes in the solution differ by at most one. Thereafter, either no color class exceeds the bound on $f$, or all color classes contain at least $f$ edges. In the former case, we are done. In the latter case, we discard edges from the solution until all color classes are of cardinality exactly $f$. At this point we are making full use of the FTDM matrix, which is necessarily optimal.

## 1.1  Our Results

1. Hardness results: we show that for every $t \geq 2$ there exists an $\epsilon > 0$ such that it is NP-hard to approximate the *Max edge t-coloring* problem within a ratio better than $1 - \epsilon$ (Sect. 3).
2. We observe that a simple greedy algorithm achieves an approximation ratio of $1 - (1 - \frac{1}{t})^t$ (Sect. 4). The bulk of our work is concerned with improving over this ratio.
3. For $t = 2$ we give an improved algorithm which uses an LP relaxation and has an approximation ratio of at least $\frac{10}{13} \approx 0.77$ (Sect. 5.2).
4. The main result of this paper appears in Sects. 6, 7, 8.
   a. Denote by $\alpha$ the best approximation ratio for the chromatic index in multigraphs. We show an algorithm for the *Max edge t-coloring* whose approximation ratio tends to $\frac{1}{\alpha}$ as $t \to \infty$.
   b. We show an estimation for optimum value of *Max edge t-coloring* (without actually finding a coloring). The ratio between the estimation and the optimum tends to 1 as $t \to \infty$.

## 2  Related Work

An online version of *Max edge t-coloring* was investigated in [8]. Edges of the graph are given to the algorithm one by one. For each edge the algorithm must either color the edge with one of the $t$ colors, or reject it, before seeing the next edge. Once an edge has been colored the color cannot be altered and a rejected edge cannot be colored later. The aim is to color as many edges as possible. They showed that *any* algorithm for online *Max edge t-coloring* is at most $\frac{4}{7}$-competitive. Assume the $t$ colors are numbered by $\{1 \ldots t\}$. The *First Fit* algorithm upon receiving an edge, colors it with the lowest ranking color available. It is shown that *First Fit* is 0.48-competitive.

## 2.1   Edge-Coloring

*Max edge t-coloring* is closely related to the problem of finding the chromatic index of a graph. Approximating the chromatic index has received a large amount of attention.

**Lower Bounds on the Chromatic Index.** Clearly the maximum degree of a multigraph which is denoted by $\Delta$ is a lower bound. Another lower bound is the *odd density* of a graph. Let $S \subseteq V$ be a subset of the vertices of the graph. Every color class in a legal edge coloring of the edges inside $S$ forms a matching, thus it's size is bounded by $\lfloor \frac{|S|}{2} \rfloor$. It follows that the number of colors needed to color the subgraph induced by $S$ is at least $\left\lceil \frac{|E(S)|}{\lfloor \frac{|S|}{2} \rfloor} \right\rceil$. This number can be bigger than $\Delta$ if $|S|$ is odd. The *odd density* of a multigraph $G$ is defined as: $\rho(G) \stackrel{\triangle}{=} \max_{\substack{S \subseteq V, \\ |S|=2k+1}} \left\lceil \frac{|E(S)|}{k} \right\rceil$. We conclude that $\chi'(G) \geq \max\{\rho(G), \Delta(G)\}$. In general it is not known how to efficiently compute $\rho(G)$, though the value of $\max\{\rho(G), \Delta(G)\}$ can be computed in polynomial time.

The chromatic index can be viewed as the minimum number of matchings required to cover the graph. A natural relaxation would be to assign fractional weights to matchings such that each edge is covered by matchings with total weight of at least 1. The *fractional chromatic index* is the minimum total weight of matchings required to cover the graph. It is denoted by $\chi'^*(G)$. Clearly $\chi'^*(G) \leq \chi'(G)$. Edmonds proved that $\chi'^*(G) = \max\{\Delta(G), \rho(G)\}$ (see for example [17]). The value of $\chi'^*(G)$ together with a fractional coloring of $G$ can be computed efficiently (this is proved implicitly in Sect. 5).

**Upper Bounds for the Chromatic Index.** The *multiplicity* of a multigraph is the maximum number of edges between any specified two vertices.

**Theorem 1 (Vizing [24]).** *Let $G$ be a multigraph with multiplicity $d$, then:*

$$\Delta(G) \leq \chi'(G) \leq \Delta(G) + d$$

In particular for simple graphs, the chromatic index is at most $\Delta + 1$. Vizing's proof is constructive and yields an efficient algorithm. The current best bound which also yields an efficient algorithm is:

**Theorem 2 (Nishizeki and Kashiwagi [18]).** $\chi'(G) \leq \max\{\lfloor 1.1\Delta(G) + 0.8 \rfloor, \rho(G)\}$

The additive factor was later improved to 0.7 by [2]. The term $(1.1\Delta(G) + 0.7)$ is tight for graphs such as the Petersen graph for which $\Delta(G) = \rho(G) = 3$, but $\chi'(G) = 4$. The current state of affairs is that it is possible to approximate the chromatic index up to an *additive* factor of 1 in simple graphs (which is a tight result), and some *multiplicative* factor in multigraphs. It seems likely that a generalization of Vizing's theorem to multigraphs should hold:

*Conjecture 1.* For any multigraph $G$, $\chi'(G) \leq \chi'^*(G) + 1$.

This was proposed by Goldberg [10], Anderson [1] , and again by Seymour [23] in the stronger form: $\chi'(G) \leq \max\{\Delta(G)+1, \rho(G)\}$. While conjecture 1 remains unproven, an "asymptotic" version of it was proven:

**Theorem 3 (Kahn [15]).** *For every $\epsilon > 0$ there exists $D(\epsilon)$ so that for any multigraph $G$ with $\chi'^*(G) \geq D(\epsilon)$ we have $\chi'(G) \leq (1 + \epsilon)\chi'^*(G)$.*

Kahn's proof is not constructive. It is unknown whether there is a (probabilistic) polynomial time algorithm with performance guarantee that matches Theorem 3.

For some families of graphs it is possible to compute the chromatic index exactly. These include *bipartite graphs* for which the chromatic index always equals the maximum degree (see for instance [5]) and simple *planar graphs* with maximal degree larger than 8. See [19] for an overview.

## 2.2    Matchings, the Matching Polytope

As noted, when coloring the edges of a graph, each color class forms a *matching*. There is a known efficient algorithm that finds a matching with maximum cardinality in a graph. See [7] and [6]. Given a graph $G$ with edge set $E$, any matching $M$ can be associated with a point $x \in \{0,1\}^{|E|}$, where $x_e = 1$ iff $e \in M$. The *matching polytope* of $G$ is the convex hull of all the points in $\mathbb{R}^{|E|}$ which represent a matching in $G$. This polytope can be described as an intersection of half-spaces, each half space is represented by a constraint on the set of variables $x_e$. For a set of vertices $S \subseteq V$, we denote by $X(S)$ the total sum of edge variables in the subgraph induced by $S$.

**Theorem 4 (Edmonds [6]).** *The following linear constraints define the matching polytope of a graph:*

### The Matching LP

$$\forall v \in V \qquad\qquad \sum_{e|v \in e} x_e \leq 1 \text{ Degree constraint} \qquad (1)$$

$$\forall S \subseteq V, |S| \text{ is odd} \qquad X(S) \leq \left\lfloor \frac{|S|}{2} \right\rfloor \text{ Blossom constraint} \qquad (2)$$

$$\forall e \in E \qquad\qquad 0 \leq x_e \leq 1 \text{ Capacity constraint} \qquad (3)$$

The number of blossom constraints is exponentially large in the number of variables. In order to solve maximum weighted matching in an LP approach (using for instance the ellipsoid method) one needs a separation oracle for the blossom constraints. Padberg and Rao in [21] showed a polynomial time algorithm which finds a violated blossom constraint in the special case that degree constraints are not violated. This suffices for solving the LP.

**Maximum $b$-matching.** In the maximum $b$-matching problem the input is a multigraph $G$, and a vector $b \in \mathbb{N}^{|V|}$. The goal is to find the maximum (with respect to the number of edges) subgraph $H$, for which the degree of a vertex $v$ in $H$ is at most $b_v$. The name for this problem might be misleading: even if $b_v = t$ for all $v \in V$ this problem is different from the *Max edge t-coloring* problem (in which one has to find a maximum edge subgraph which can be covered by $t$ matchings). For example a triangle is a graph with all degrees bounded by 2, however it can not be covered by 2 matchings. In the special case that $b_v = 1$ for all $v \in V$ the $b-$matching problem is simply a maximum matching problem. A more natural name for this problem can be *the maximum degree constrained subgraph*, however we use the name $b$-matching for historical reasons. Edmonds and Johnson gave a generalization of the blossom constraints (see appendix B.1 of [9], [11]) which together with degree constraints induce a polytope with integer vertices; the vertices of this polytope are exactly all the edge subgraphs of $G$ which obey the degree constraints. A separation oracle for finding a violated constraint of this LP appears at [21]. This implies a polynomial time algorithm for solving the b-matching problem. More details can be found at [17] and [11].

## 3    Hardness of *Max Edge t-Coloring*

**Theorem 5.** *For every $t \geq 2$ there exists an $\epsilon_t$ such that it is NP-Hard to approximate Max edge t-coloring within a factor better than $1 - \epsilon_t$.*

Holyer showed [14] that deciding whether an input graph is 3 edge colorable is NP-hard. A similar result for the problem of deciding whether an input multigraph is $t$ edge colorable (for every $t > 3$) was given by Leven and Galil [16]. This implies that for every $t \geq 3$ the *Max edge t-coloring* problem is NP-hard. The reductions used by Holyer, Leven and Galil are from MAX 3 SAT. If the domain is limited to MAX 3SAT-3 instances (3SAT formulas in which every variable appears in at most 3 clauses) then these reductions are in fact $L$ reductions (in the terms of [22]). Therefore the reduction preserves the inapproximability of MAX 3SAT-3. Details are omitted from this manuscript. The case of *Max edge 2-coloring* is different since deciding whether the chromatic index of a graph is 2 can be done in polynomial time. The proof of Theorem 5 for $t = 2$ uses a modified version of a reduction attributed to Papadimitriou in [4]. The details appear in the full version of this paper [9].

## 4    The Greedy Approach

A simple greedy algorithm is based on the online approach in which the edges are colored one after the other (for instance *First Fit*). The following offline algorithm *Greedy* has a better performance guarantee: in each iteration (out of $t$) a maximum matching is selected, colored and removed from the graph.

The *Max edge t-coloring* problem is a special case of the *maximum coverage problem*. We wish to cover the maximum number of edges of $G$ with $t$ sets, where each set is a matching. The following Theorem is well known (see for example [13] Theorem 3.8):

**Theorem 6.** *The greedy algorithm for the maximum coverage problem yields a* $1 - \left(1 - \frac{1}{t}\right)^t$ *approximation.*

It follows that *Greedy* has an approximation ratio of at least $1 - \left(1 - \frac{1}{t}\right)^t$. We could not design for every value of $t$ examples that show that the approximation ratio of *Greedy* is no better than $1 - \left(1 - \frac{1}{t}\right)^t$. However, a simple example shows that for every even $t$, the approximation ratio is no better than $\frac{3}{4}$. Consider a line of four edges, each with multiplicity of $\frac{t}{2}$. All edges can be colored with $t$ colors. However, *Greedy* might take the first $\frac{t}{2}$ matchings to consist of the first and fourth edges, and then it will cover only $\frac{3}{4}$ of the edges.

## 5    A Linear Program

An optimal solution to *Max edge t-coloring* has a maximum degree and odd density bounded by $t$. Therefore a natural approach for solving *Max edge t-coloring* may be to solve the following problem as an intermediate step:

*Problem 2.* Given a multigraph $G$ and a parameter $t$, find a largest (in edges) subgraph of $G$ whose degree and odd density are bounded by $t$. We call this problem *Max Sparse t-Matching*.

The following is an LP relaxation of *Sparse t-Matching*:

$$\max \sum x_e \qquad\qquad \text{subject to}$$

$$\sum_{e \mid v \in e} x_e \leq t \qquad\qquad \forall v \in V \qquad (4)$$

$$X(S) \leq t \cdot \left\lfloor \frac{|S|}{2} \right\rfloor \qquad\qquad \forall S \subseteq V,\ |S| \text{ is odd} \qquad (5)$$

$$0 \leq x_e \leq 1 \qquad\qquad \forall e \in E \qquad (6)$$

Since every feasible solution for *Max edge t-coloring* is also feasible for *Sparse t-Matching*, the *Sparse t-Matching* LP is a relaxation also for *Max edge t-coloring*. We will show that fractional solutions of *Sparse t-Matching* LP can be rounded to give a solutions of *Max edge t-coloring*, of value at least a $1 - \left(1 - \frac{1}{t}\right)^t$ fraction of the value of the LP. This is the same approximation ratio that the greedy approach yields. It uses some ideas and intuition that will be useful later.

In order to solve the linear program using the ellipsoid method, we need an oracle for finding violated constraints of the LP. Dividing all the constraints by $t$ and multiplying the objective function by $t$ we get an equivalent LP. This linear program is similar (though not identical) to the *matching LP* (see Theorem 4), and can be solved using the same techniques (using the separation oracle given by Padberg and Rao at [21]).

## 5.1   Rounding the Linear Program Solution

The vector $\frac{1}{t}\boldsymbol{x}$ is valid for the matching LP and therefore according to Theorem 4 is a convex combination of a set of matchings. We will now describe how to find such a convex combination. Let $\mathcal{M}$ denote the set of all matchings, and define a variable $\lambda_M$ for each $M \in \mathcal{M}$. Solve the following LP:

$$\forall e \in E \qquad \sum_{M|e\in M} \lambda_M = \frac{x_e}{t}$$

$$\sum_{M\in\mathcal{M}} \lambda_M = 1$$

$$\forall M \in \mathcal{M} \qquad \lambda_M \geq 0$$

This LP has an exponential number of *variables* and a polynomial number of *constraints*. We sketch how to solve it: the dual of this LP can be solved using the ellipsoid method. By doing so we identify a polynomial number of constraints that are tight with respect to the solution. Moving back to the primal results in an LP that has a polynomial number of variables thus can be solved efficiently. See [25] for more details. Now we sample $t$ matchings. Each sample is done such that matching $M$ has probability of $\lambda_M$ to be chosen. Only a *polynomial* number of matchings have positive probability, therefore the sampling algorithm is efficient. We output as a solution the union of those $t$ matchings (denote it by $S$).

## Analysis

**Lemma 1.** $\Pr\{\, e \in S\} \geq \left(1 - \left(1 - \frac{1}{t}\right)^t\right) \cdot x_e$

*Proof.* Each time a matching is sampled, edge $e$ enters $S$ with probability $\frac{x_e}{t}$. Each sample is independent from the rest. Therefore:    $\Pr\{e \notin S\} \leq \left(1 - \frac{x_e}{t}\right)^t$

$$\Rightarrow \Pr\{e \in S\} \geq \left(1 - \left(1 - \frac{x_e}{t}\right)^t\right) \geq \left(1 - \left(1 - \frac{1}{t}\right)^t\right) \cdot x_e.$$

where the last inequality follows from the concavity of $\left(1 - \left(1 - \frac{x_e}{t}\right)^t\right)$ in $[0, 1]$.

From linearity of expectation it follows that $E(|S|) \geq \left(1 - \left(1 - \frac{1}{t}\right)^t\right) \cdot x^*$. Since $x^*$ is an upper bound on the optimum integral solution we conclude that the algorithm yields (on expectation) a $\left(1 - \left(1 - \frac{1}{t}\right)^t\right)$ approximation.

## 5.2   The Case t=2

In an interesting special case we are to find the 2-edge colorable subgraph with a maximum number of edges. In this case the greedy algorithm and the rounding procedure described in Sect. 5.1 both yield an approximation ratio of 0.75. We present an algorithm that achieves an approximation ratio of $\frac{10}{13} \approx 0.77$. Denote by $OPT$ the size of the optimal subgraph. Set a variable $x_e$ for each edge. Solve the *Sparse t-Matching* LP presented previously with $t = 2$. Denote by $x^*$ it's optimal value. The idea of the algorithm is to choose between two different kinds of roundings. If a large portion of the weight is in 'heavy' variables, then a threshold rounding is used. Otherwise the random rounding of Sect. 5.1 is used.

## Algorithm for t=2

1. Solve the linear program presented above. Let $x^*$ denote it's value.
2. If $\sum\{x_e \mid x_e > \frac{2}{3}\} \geq \frac{10}{13}x^*$ then remove all edges $e$ with $x_e \leq \frac{2}{3}$, remove one edge from every odd cycle that remains, and output the remaining edges. Otherwise perform the random rounding of Sect. 5.1.

**Analysis:** We will use the term 'light' edge for an edge whose weight is at most $\frac{2}{3}$, otherwise the edge is considered 'heavy'. Denote by $\alpha$ the fraction of weight (from $x^*$) of all the heavy edges; i.e. $\alpha \triangleq \frac{\sum\{x_e \mid x_e > \frac{2}{3}\}}{x^*}$.

**Lemma 2.** *There exists a 2-edge-colorable subgraph with at least $\alpha x^*$ edges.*

*Proof.* Let $H$ denote the subgraph induced by all the heavy edges. Constraint (4) implies that the maximum degree of $H$ is 2. Therefore $H$ consists of paths and cycles. We find a two edge colorable subgraph, by omitting one edge from each odd cycle. It remains to show that the solution built has at least $\alpha x^*$ edges. Let $C$ be a component of $H$. If $C$ is a path or an even length cycle, then the number of edges $C$ has contributed to the solution is $|C|$ while $\sum_{e \in C} x_e \leq |C|$. If $C$ is an odd length cycle, then the number of edges $C$ has contributed to the solution is $|C| - 1$, while constraint (5) implies that $\sum_{e \in C} x_e \leq |C| - 1$. We conclude that the cardinality of the solution is at least $\sum_{e \in H} x_e \geq \alpha x^*$.

**Lemma 3.** *There exists a 2-edge-colorable subgraph with at least $\left(\frac{5}{6} - \frac{\alpha}{12}\right) x^*$ edges.*

*Proof.* Perform a random rounding as in Sect. 5.1. The expected contribution of an edge $e$ to the solution is $(1 - (1 - \frac{x_e}{2})^2) = x_e(1 - \frac{x_e}{4})$. As before let $H$ be the subgraph induced by the heavy edges, and let $\overline{H}$ denote the subgraph induced by the light edges. The expected contribution of $H$ to the solution is at least $\frac{3}{4}(\alpha \cdot x^*)$, and that of $\overline{H}$ is at least $(1 - \frac{2}{3} \cdot \frac{1}{4})(1 - \alpha)x^*$. Linearity of expectation implies that the expected cardinality of the solution is at least $(\frac{5}{6} - \frac{\alpha}{12})x^*$.

**Corollary 1.** *The break even point between the two lemmas is $\alpha = \frac{10}{13}$, therefore we have a $\frac{10}{13}$ approximation for the problem.*

The LP approach as presented here cannot yield an approximation ratio better than 0.9, since the Petersen graph has an integrality ratio of 0.9. The optimal integer solution for the Petersen graph has 9 edges. However, assigning each edge a value of $\frac{2}{3}$ gives a feasible solution with value of 10. David Hartvigsen

[12] designed an algorithm that given a *simple* graph, finds a maximum triangle free subgraph with degree bounded by 2. This algorithm can be used to yield a $\frac{4}{5}$ approximation for simple graphs in the following way: apply the algorithm and delete from every odd cycle one edge. A 2-matching without odd cycles can be decomposed into two disjoint matchings. Since the output of the algorithm is triangle free, at most $\frac{1}{5}$ of the edges are deleted.

## 6  The *Sparse t-Matching* Scheme

Conjecture 1 states that the chromatic index of a multigraph is very close to the fractional chromatic index. This motivates us to try to deal with the *Max Sparse t-Matching* problem (Problem 2) as an intermediate problem, using an approximation for it to approximate *Max edge t-coloring*. In case $t = 2$ *Max Sparse t-Matching* is equivalent to *Max edge t-coloring* and therefore is known to be NP-Hard and does not admit a PTAS. In case $t \geq 3$ it is unknown whether the problem is hard, yet we believe it is. An example for an integrality gap of the *Sparse t-Matching* LP can be found at the appendix of [9].

The general algorithmic scheme we will use in this section is as follows:

1. Find a large subgraph with degree and odd density bounded by $t + o(t)$. Let $H$ be the subgraph returned.
2. Use an edge coloring algorithm to color $H$.
3. Output the $t$ largest color classes.

Note that in stage 2 the edge coloring algorithm is used as a black box. We will analyze our algorithm with respect to a general edge coloring algorithm, and then plug in the performance of the known algorithms.

### 6.1  A Few Simple Cases

In the two following examples we simplify step 1. The simplified step 1 is to find the maximum (in edges) subgraph with maximum degree bounded by $t$. This can be done by invoking Edmond's $b-$matching algorithm with $b_v = t$ for all $v \in V$.

If a multigraph $G$ has a multiplicity of $d$, then *Max edge t-coloring* can be approximated by a polynomial time algorithm up to a factor of $1 - \frac{d}{t+d}$. To see this let $OPT$ denote the maximum number of edges that can be colored by $t$ colors and follow the algorithmic scheme presented above. Clearly $|E(H)| \geq OPT$. Vizing's Theorem states that $H$ can be colored using $t + d$ colors. Therefore in step 3, we discard at most a $\frac{d}{t+d}$ fraction of the edges, which yields the approximation ratio.

For bipartite multigraphs, it is possible to solve *Max edge t-coloring* in polynomial time. It is well known that for bipartite multigraphs $\chi'(G) = \Delta(G)$ (see for example [5]), therefore replacing step 2 of the previous algorithm, by an algorithm that colors a bipartite graph $H$ with $\Delta(H)$ colors (for example [3]) yields an exact solution.

# 7    Approximating the *Sparse t-Matching* Problem

We present an approximation algorithm for the *Sparse t-Matching* problem. Our approximation however will not be in the usual sense. Given a multigraph $G$ and a parameter $t$, let $OPT$ denote the number of edges in the optimal subgraph. Our algorithm will output a subgraph $H$ such that $|E(H)| \geq OPT$, albeit $\Delta(H) \leq t+1$ and $\rho(H) \leq t + \sqrt{t+1} + 2$ (rather than $\rho(h), \Delta(H) \leq t$). The first stage of the algorithm for approximating the *Sparse t-Matching* problem is to solve the *Sparse t-Matching* LP from Sect. 5.1. This can be done in polynomial time and gives a fractional solution. The second stage is to round the fractional solution returned by the LP and is explained in Sect. 7.1. The integrality ratio of the *Sparse t-Matching* LP is at least $1 - \Omega(\frac{1}{t})$; details can be found at [9] Appendix B.

## 7.1    Rounding the Fractional Solution

The rounding technique consists of three stages. We hold a subset of the solution that increasingly grows as we proceed through the stages. Let $H$ be the set of edges that represents the solution. Initially $H = \phi$. In each stage of the algorithm edges are added to $H$, until finally $H$ is returned as the rounded solution. For each node $v$ define the variable $d_v$. As we proceed through the algorithm, $d_v$ will hold the number of edges adjacent to $v$ that can be added to $H$. Initially $d_v = t$. Every edge $(u, v)$ added to $H$ decreases $d_u$ and $d_v$ by one.

*Stage 1:* First we get rid of multiple edges. Let $(u, v)$ be a pair of vertices that are connected by multiple edges. If two parallel edges have positive fractional value, shift values between them until one of the edges either reaches the value of 0 or the value of 1. If an edge reached the value of 1 then add it to $H$ and decrease $d_u, d_v$ by one. Continue until the edges which have a positive fractional value form a *simple* graph. Denote this graph by $G'$. Clearly The edges of $H$ satisfy constraints (4), (5), (6).

*Stage 2:* We concentrate on $G'$ and get rid of even length cycles:

1. Find a cycle (not necessarily simple) in $G'$ with an even number of edges, and mark them *even*, *odd* alternately. An algorithm to do so appears in appendix C of [9].
2. Shift weights from the even edges to the odd edges until one of the two events happens:
   - An odd edge reached the weight of 1. In this case discard the edge from $G'$, add it to $H$ and modify the $d$-values of it's vertices.
   - An even edge reached the weight of 0. In this case discard it from $G'$
3. If there are even length cycles left in the graph then return to step one.

When this stage is done, it is evident that the edges of $H$ satisfy constraints (4),(6) though not necessarily constraint (5). The proof of lemma 4 can be found in appendix C of [9].

**Lemma 4.** *Let $G$ be a graph without even cycles. $G$ has the following properties:*

1. *All odd cycles are simple cycles.*
2. *All odd cycles are disjoint in vertices.*

From claim 4 we deduce that the graph at the end of stage 2 looks like a tree in which each node might represent an odd cycle. We use this restricted topology to complete the rounding.

*Stage 3:* Now we round the values of the edges in the 'tree' from the leaves up. A leaf of the remaining graph may be a node or an odd cycle. First we deal with the case of an odd cycle. Assume that the cycle $C = (v_1, v_2, \ldots, v_l)$ is an odd length cycle which forms a leaf of the 'tree'. Let the edge $w = (u, v_1)$ connect $C$ to the rest of the graph. Until this stage all that was done is value shifting, and the $d$-values were always updated correctly, so it holds that $d_{v_i} \geq 1$ for every node in the cycle. There are two cases:

**Case 1:** There exists a node $v_i \in C$ for which $d_{v_i} = 1$. Add to $H$ the edges of the cycle and discard the edge $w$ from the graph (Fig. 1). Since $d_{v_i} = 1$, it follows that the total value of edges in the cycle is bounded by $l - 1$. If we add the value of the edge $w$ the total value is bounded by $l$. $l$ edges of the cycle were entered to $H$ so no value was lost.

**Case 2:** For every node $v \in C$ we have $d_v \geq 2$. Shift values from other edges that contain $u$ into $w$, add the cycle *and* $w$ into $H$, and decrease $d_u$ by 1. The value shift did not change the degree of $u$, though the edges of $u$ may reach the value of 0 and therefore be discarded from the graph (Fig. 2). Since all the $l + 1$ edges were taken, no value was lost in this case.

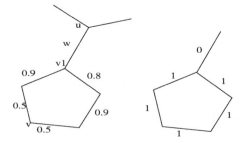

**Fig. 1.** *Case 1*, vertex $v$ has $d_v = 1$, so the cycle is taken and $w$ is discarded.

The case that a leaf of the tree is a node is simple and can be solved by a value shift downwards towards the node, i.e. the same as case 2. We iterate through stage 3 traversing the tree bottom up, until all edges are either discarded from the graph or added to $H$. Finally output $H$.

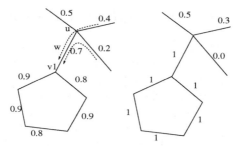

**Fig. 2.** *Case 2, $\forall v\ d_v \geq 2$. The cycle and $w$ are added to $H$ and $d_u$ is decreased.*

## 7.2   Analysis of the *Sparse t-Matching Algorithm*

Let $OPT$ denote the maximum size of a subgraph with degree and odd density bounded by $t$. From the description of the rounding it is clear that $|H| \geq OPT$. The only place in the rounding that the degree of a vertex $v$ might exceed $t$, is in stage 3 when $v$ belongs to a cycle. Note that when $d_v = 1$, at most two edges containing $v$ were added, and when $d_v \geq 2$ at most 3 edges were added. We conclude that $\Delta(H) \leq t + 1$. We show that the odd density of $H$ does not exceed $t$ by much.

**Theorem 7.** $\rho(H) \leq t + \sqrt{t + 1} + 2$

*Proof.* Let $S \subseteq V$ such that $|S| = 2k + 1$ for some integer $k$. Each $v \in S$ has degree at most $t + 1$ in $H$.
The number of edges of $S$ is at most $\frac{(2k+1)\cdot(t+1)}{2}$ therefore:

$$\rho_H(S) \leq \left\lceil \frac{(2k+1)(t+1)}{2k} \right\rceil \leq t + \left\lceil \frac{t+1}{2k} \right\rceil + 1 \leq t + \frac{t+1}{2k} + 2.$$

This bound is good for large sets as it tends to $t + 2$ when $k$ tends to infinity. We will give another bound which is good for small sets:
Let $x$ denote the fractional solution. Constraint (5) implies that $X(S) \leq k \cdot t$. Some mass was left for the manipulations of stage 2, thus at the beginning of stage 2 the number of edges in $S$ that were added to $H$ is at most $k \cdot t - 1$. Since at stage 2 we deal with a simple graph, the number of edges added to $S$ in stage 2 is at most $\binom{2k+1}{2} = (2k + 1)k$. Therefore:

$$\rho_H(S) \leq \left\lceil \frac{(2k+1)k + k \cdot t - 1}{k} \right\rceil \leq t + 2k + 1.$$

The break even point between the two bounds is when $k = \frac{1}{4} + \frac{1}{2}\sqrt{t + \frac{5}{4}}$. Setting $k$ to this value yields the desired result.

We remark here that the loss of 1 in $\Delta$ which is caused by the rounding is unavoidable, if the rounding procedure returns a solution with value at least as large as the value of the fractional solution (see appendix B in [9]).

# 8    Analysis of the *Sparse t-Matching* Algorithm

In Stage 2 of the algorithmic scheme presented in Sect. 6 we use an edge coloring algorithm as a black box. Naturally the quality of the result depends heavily on the performance of the algorithm used. Assume we have an efficient algorithm $A$ for the chromatic index problem, which edge colors a multigraph $G$ with $max\{\alpha\Delta(G) + \beta , \rho(G)\}$ colors. The algorithm $A$ can be used to color the subgraph $H$ found by the *Sparse t-Matching* algorithm with $max\{\alpha \cdot (t + 1) + \beta , t + \sqrt{t + 1} + 2\}$ colors. Let $L$ denote this maximum. Our algorithm yields then a $\frac{t}{L}$ approximation. If $t$ is large enough then the maximum would be achieved by $\alpha \cdot (t + 1) + \beta$. Taking the largest $t$ color classes results in a $(\frac{t}{\alpha(t+1)+\beta})$-approximation for the *Max edge t-coloring* problem. This value tends to $\alpha^{-1}$ as $t \to \infty$.

Currently the best approximation algorithm for the chromatic index problem is due to [18] where $\alpha = 1.1$. If conjecture 1 is true (with a suitable efficient algorithm), then there exists an efficient edge coloring algorithm for which $\alpha = 1$. This means that the larger $t$ is, the better is the approximation for the *Max edge t-coloring*. This phenomena combined with Kahn's result [15] that $\frac{\chi'(G)}{\chi'^*(G)} \to 1$ as $\chi'^*(G) \to \infty$ implies the following:

**Corollary 2.** *The value of the Sparse t-Matching LP is an estimation to the optimal value of Max edge t-coloring, with an approximation ratio that tends to 1 as $t \to \infty$.*

## Acknowledgments

This work was supported in part by a grant of the Israeli Ministry of Industry and Commerce through the consortium on Large Scale Rural Telephony. Some of the details that are omitted from this manuscript can be found in the M.Sc thesis of Eran Ofek [20] and in the M.Sc thesis of Ehud Wieder [25], both done at the Weizmann Institute of Science.

# References

1. L. D. Andersen: On edge-colourings of graphs. Math. Scand. 40:161-175, 1977.
2. Alberto Caprara and Romeo Rizzi: Improving a Family of Approximation Algorithms to Edge Color Multigraphs. Information Processing Letters, 68:11-15, 1998.
3. R. Cole and K. Ost and S. Schirra: Edge-coloring Bipartite Multigraphs in O(E log D) Time. Combinatorica, 21(1), 2001.
4. Gerard Cornuejols and William Pulleyblank: A matching Problem with Side Conditions. Discrete Mathematics, 29:135–159,1980.
5. Reinhard Diestel: Graph Theory. Springer 1996.
6. J. Edmonds: Maximum matching and a polyhedron with 0, 1 vertices. Journal of Research National Bureau of Standards, 69B:125–130, 1965.
7. J. Edmonds: Paths, trees, and flowers. Canadian Journal of Mathematics, 17:449–467, 1965.

8. Lene M. Favrholdt and Morten N. Nielsen: On-Line Edge-Coloring with a Fixed Number of Colors. Foundations of Software Technology and Theoretical Computer Science 20, 2000.
9. U. Feige and E. Ofek and U. Wieder: Approximating maximum edge coloring in multigraphs. Technical report, Weizmann Institute, 2002.
10. M.K. Goldberg: On multigraphs of almost maximal choratic class (in Russion. Diskret. Analiz, 1973.
11. M. Grotschel and L. Lovasz and A. Schrijver: Geometric algorithms and combinatorial optimization. Springer-Verlag, Berlin, 1988.
12. D. Hartvigsen: Extensions of Matching Theory. Carnegie-Mellon University, 1984.
13. Dorit Hochbaum: Approximation Algorithms For NP-Hard Problems. PWS Publishing Company, Boston, 1997.
14. Ian Holyer: The NP-completeness of edge-coloring. SIAM Journal on Computing, 10(4):718–720, 1981.
15. J. Kahn: Asymptotics of the Chromatic Index for Multigraphs. Journal of Combinatorial Theory, Series B, 68, 1996.
16. Daniel Leven and Zvi Galil: NP Completeness of Finding the Chromatic Index of Regular Graphs. Journal of Algorithms, 4(1):35–44, 1983.
17. L. Lovasz and Plummer :Matching Theory, Annals of Discrete Mathematics 29 North-Holland, 1986.
18. Takao Nishizeki and Kenichi Kashiwagi: On the 1.1 Edge-Coloring of Multigraphs. SIAM Journal on Discrete Mathematics 3:391–410, 1990.
19. T. Nishizeki and X. Zhou: Edge-Coloring and $f$-coloring for various classes of graphs. Journal of Graph Algorithms and Applications, 3, 1999.
20. Eran Ofek: Maximum edge coloring with a bounded number of colors. Weizmann Institute of Science, Rehovot, Israel, November 2001.
21. M. Padberg and M. R. Rao: Odd minimum cut-sets and $b$-matchings. Mathematics of Operations Research, 1982, 7:67–80.
22. Christos H. Papadimitriou and Mihalis Yannakakis: Optimization, Approximation, and Complexity Classes. Journal of Computer and System Sciences, 43(3):425–440, 1991.
23. P. Seymour: Some unsolved problems on one-factorizations of graphs. Graph Theory and Related Topics, Academic Press, 367-368, 1979.
24. V. G. Vizing: On an estimate of the chromatic class of a p-graph (in Russian). Diskret. Analiz 3:23–30, 1964.
25. Ehud Wieder: Offline satellite resource allocation or Maximum edge coloring with a fixed number of colors. Weizmann Institute of Science, Rehovot, Israel, November 2001.

# Approximating the Complement
# of the Maximum Compatible Subset
# of Leaves of $k$ Trees

Ganeshkumar Ganapathy and Tandy Warnow

Department of Computer Sciences, University of Texas, Austin, TX 78712
{gsgk,tandy}@cs.utexas.edu

**Abstract.** We address a combinatorial problem which arises in computational phylogenetics. In this problem we are given a set of unrooted
(not necessarily binary) trees each leaf-labelled by the same set $S$, and we
wish to remove a minimum number of leaves so that the resultant trees
share a common refinement (i.e. they are "compatible". If we assume the
input trees are all binary, then this is simply the Maximum Agreement
Subtree problem (MAST), for which much is already known. However, if
the input trees need not be binary, then the problem is much more computationally intensive: it is NP-hard for just two trees, and solvable in
polynomial time for any number $k$ of trees when all trees have bounded
degree. In this paper we present an $O(k^2 n^2)$ 4-approximation algorithm
and an $O(k^2 n^3)$ 3-approximation algorithm for the general case of this
problem.

## 1 Introduction

Let $\mathcal{T}$ be a set of unrooted, and not necessarily binary trees, each bijectively
leaf-labelled by the same set $S$ of objects. A tree $T$ on the set $S$ of leaves is said
to *refine* another tree $T'$ on the same set if $T'$ can be obtained by contracting
some selection of edges of $T$. If the trees share a common refinement, then the
set is said to be "compatible". In this case, the minimal common refinement is
unique, and can be computed in $O(nk)$ time, where $k = |\mathcal{T}|$ and $|S| = n$ [6,12].
Consider the following optimization problem:

**Maximum Compatible Subset Problem**

**Input:** Given $\mathcal{T}$, a collection of trees, each leaf-labelled by the same set $S$,
**Output:** Find a maximum cardinality $A \subseteq S$ so that the set $\{T|A : T \in \mathcal{T}\}$ (where $T|A$ denotes the tree obtained by restricting $T$ to the leaves
labelled by elements of $A$, and suppressing degree two nodes) has a common
refinement.

This problem is motivated by applications to computational phylogenetics (and
we describe the application in Section 5), and was initially posed by Hamel and

K. Jansen et al. (Eds.): APPROX 2002, LNCS 2462, pp. 122–134, 2002.
© Springer-Verlag Berlin Heidelberg 2002

Steel in [7]. It was proven NP-hard for six or more trees in [7], and later shown also NP-hard for two trees when at least one tree is of unbounded degree in [8]. It is also known to be solvable for any number of trees when all trees are of bounded degree [8].

We study the approximability of the complement of this problem, whereby we seek a minimum sized subset of the leaves so that restricted to the remaining leaves, all the trees are compatible. We call this the *Complement of the Maximum Compatible Subset Problem*, or *CMCS*. We will show that we can transform the input to the CMCS problem into an instance of the well-known Hitting Set problem, so that we can obtain a 4-approximation algorithm to the problem. The explicit calculation of this transformation yields a polynomial time algorithm which has high degree. We then show that we can avoid this explicit calculation, and obtain a 4-approximation algorithm in $O(n^2k^2)$ time, where $n = |S|$ and $k = |\mathcal{T}|$. We also present a slower $O(n^3k^2)$ algorithm using similar techniques which has a guaranteed 3-approximation ratio.

The rest of the paper is organized as follows. In Section **??** we present the basic definitions and lemmas, and we outline a brute force 4-approximation algorithm which is computationally expensive. In Section 3 we describe our $O(k^2n^2)$ algorithm which also achieves the 4-approximation ratio, and its proof of correctness and running time analysis. We present the 3-approximation algorithm for the same problem which runs in $O(k^2n^3)$ in Section 4. In Section 5 we describe the context in which this problem arises in computational biology, and we discuss the uses of these algorithms for this application.

## 2   Basics

In this section we present the basic definitions and lemmas that motivate a brute force algorithm that forms the basis for our $O(k^2n^2)$ 4-approximation algorithm for the CMCS problem.

**Definition 1.** *(Tree compatibility): A set of trees $\mathcal{T} = \{T_1, T_2, \ldots, T_k\}$ on the set of leaves $S$ is said to be compatible if there is a tree that is a common refinement of trees in $\mathcal{T}$.*

**Definition 2.** *(Restriction of a tree to a set of leaves): The restriction of a tree $T$ to a set $X$ of leaves, denoted $T|X$, is the tree obtained by deleting the leaves in $X$ from $T$ and then suppressing all internal nodes of degree 2.*

In the CMCS problem, we seek a minimum cardinality subset $X$ of $S$ such that $\{T|(S - X) : T \in \mathcal{T}\}$ is a compatible set of trees.

Much of the theory we develop rests upon the concept of compatibility of bipartitions, which we now define.

**Definition 3.** *(The set $C(T)$, the bipartition encoding of $T$): Removing an edge $e$ from a leaf-labelled tree $T$ induces a bipartition $\pi_e$ on its set $S$ of leaves. We denote by $C(T)$ the set $\{\pi_e : e \in E(T)\}$ and note that it uniquely identifies $T$, up to degree 2 nodes.*

**Definition 4.** *(Bipartition (or edge) compatibility): A set of bipartitions B is said to be* compatible *if and only if* $B \subseteq C(T)$ *for some tree T. We will call a pair of edges* $(e, e')$ *compatible if the corresponding bipartitions induced by these edges are compatible.*

We now state the relationship between tree compatibility and bipartition compatibility.

**Theorem 1.** *A set* $\mathcal{T}$ *of trees is compatible if and only if the set* $\bigcup_{T \in \mathcal{T}} C(T)$ *is compatible.*

Hence, if we wish to determine if a set of trees is compatible, we need only determine whether the set of bipartitions of all the trees in the set is compatible. This can be determined in $O(kn)$ time [6,12], where $\mathcal{T}$ has $k$ trees, each on the same $n$ leaves.

Our 4-approximation algorithm operates by eliminating taxa from the set of trees which cause the set of bipartitions to be incompatible. We therefore rely upon the following theory.

**Lemma 1 (From Buneman [2]).** *A set of bipartitions is compatible iff any two bipartitions in the set are are pairwise compatible. Furthermore, two bipartitions* $A = A_1 : A_2$ *and* $B = B_1 : B_2$ *are compatible iff at least one of the four sets* $A_1 \cap B_1$, $A_1 \cap B_2$, $A_2 \cap B_1$ *and* $A_2 \cap B_2$ *is empty.*

**Definition 5.** *(Quartet Tree): A* quartet tree *is an unrooted tree on four leaves that does not have any internal node of degree two. We denote the quartet tree induced by a tree T on a set q of four leaves by* $T|q$. *We will call q the* basis *of the quartet tree* $T|q$. *Two quartet trees on the same leaf set are said to be* incompatible *if both are binary trees and they differ.*

**Corollary 1.** *A set* $\mathcal{T}$ *of trees is incompatible if and only if there exists a quartet q of leaves and a pair of trees* $T_1$ *and* $T_2$ *in* $\mathcal{T}$ *such that* $T_1|q$ *and* $T_2|q$ *are incompatible quartet trees.*

*Proof.* If the set is incompatible, then there is a pair of bipartitions $A = A_1 : A_2$ and $B = B_1 : B_2$ (coming from trees $T$ and $T'$, respectively) and leaves $a, b, c, d$ such that $a \in A_1 \cap B_1, b \in A_1 \cap B_2, c \in A_2 \cap B_1$, and $d \in A_2 \cap B_2$. Hence, $T$ induces $ab|cd$ and $T'$ induces $ac|bd$, so that $T|\{a,b,c,d\}$ and $T'|\{a,b,c,d\}$ are incompatible. The converse is trivial.

We now introduce the Hitting Set problem.

**Definition 6.** *(Hitting Set:) Let* $\mathcal{X} \subseteq 2^A$ *be a collection of subsets of a ground set A. A* hitting set *for* $\mathcal{X}$ *is a subset* $V \subset A$ *such that for all* $X \in \mathcal{X}$ $\exists a \in V$ *such that* $a \in X$.

The best known version of the Hitting Set problem is the *Vertex Cover* problem: given a graph $G = (V, E)$, find a minimum subset $A \subset V$ so that every edge in $E$ has at least one of its endpoints in $A$. This problem is NP-Hard but can be 2-approximated in a simple algorithm which greedily accumulates a *maximal* matching $E_0$ within $E$ (so that no two edges within $E_0$ share an endpoint). Then the vertex set $A = \{v : \exists w \in V : (v, w) \in E_0\}$ is a vertex cover for $G$ which is at most twice the size of an optimal vertex cover. Thus, the NP-hard Vertex Cover problem can be 2-approximated in polynomial time.

More generally, the Hitting Set problem can be $p$-approximated in polynomial time using the same greedy technique, where $p = max\{|S| : S \in \mathcal{X}\}$.

Our algorithms are based upon a connection between the Hitting Set problem and the CMCS problem. We begin by defining the set $\mathcal{Q}_{inc}$:

**Definition 7.** *($\mathcal{Q}_{inc}$): Let $\mathcal{T}$ be a set of trees leaf-labelled by the same set $S$, and let $\mathcal{Q}_{inc}$ denote the set of all quartets on which the trees in $\mathcal{T}$ are incompatible.*

The following lemma explains the connection between the Hitting Set problem and the CMCS problems.

**Lemma 2.** *Let $\mathcal{T}$ be a set of trees on a set of leaves $S$. Then, $H$ is a hitting set for $\mathcal{Q}_{inc}$ if and only if $S - H$ is a compatible set for $\mathcal{T}$.*

*Proof.* Let $H$ be a hitting set for $\mathcal{Q}_{inc}$. Hence, for every $q \in \mathcal{Q}_{inc}$, there is at least one $s \in H$ such that $s \in q$. Now consider any pair of trees $T_1$ and $T_2$, when restricted to $S - H$. By Corollary 1 they are compatible, since they have no quartets on which they are incompatible. Hence the set of trees $\mathcal{T}$ restricted to $S - H$ is compatible. For the converse, if $X \subseteq S$ is such that all trees in $\mathcal{T}$ are compatible when restricted to $S - X$, then there are no incompatible quartets left; hence $X$ is a hitting set for $\mathcal{Q}_{inc}$.

Thus, for us $\mathcal{Q}_{inc}$ is the set $\mathcal{X}$ in the definition of the Hitting Set problem. Hence, as argued before, we can 4-approximate the CMCS problem in polynomial time:

**Theorem 2.** *Let $Q$ be a maximal collection of pairwise disjoint quartets drawn from $\mathcal{Q}_{inc}$. Let $S_Q = \bigcup_{q \in Q} q$. Then, $\mathcal{T}$ is compatible on $S - S_Q$, and $|S_Q| \leq 4|H^{opt}|$, where $H^{opt}$ is an optimal solution to the CMCS problem on input $\mathcal{T}$.*

(The proof follows from the previous lemma.)

As in all Hitting Set problems, this suggests a straightforward algorithm to obtain a 4-approximation for the CMCS problem. That is, compute $\mathcal{Q}_{inc}$, and then find a maximal collection of pairwise disjoint quartets within this set:

APPROX-HITTING-SET($\mathcal{Q}_{inc}$)

$H \leftarrow \emptyset$
$\mathcal{P} \leftarrow \mathcal{Q}_{inc}$
while $\mathcal{P} \neq \emptyset$
      pick an arbitrary set $q \in \mathcal{P}$
      $H \leftarrow H \cup q$
      remove every $X \in \mathcal{P}$ such that $q \cap X \neq \emptyset$
return $H$

The above procedure would give us a hitting set for $\mathcal{Q}_{inc}$ that is at most four times the size of an optimal hitting set. However, the above outlined procedure depends on explicitly enumerating all the incompatible quartets, and this is expensive. (There are $O(n^4)$ such quartets, and determining if a set of $k$ trees is incompatible on a given quartet takes $O(nk)$ time.) This brute force approach for finding a 4-approximation to the hitting set problem is therefore not practical, as it uses $O(kn^5)$ time.

## 3   The Efficient 4-Approximation Algorithm

### 3.1   Outline of the Algorithm

Note that the input to the CMCS problem is a set of $k$ unrooted trees, which *implicitly* codes for the set $\mathcal{Q}_{inc}$. The approach we use for obtaining a faster 4-approximation algorithm takes advantage of this implicit representation of $\mathcal{Q}_{inc}$, by never explicitly calculating $\mathcal{Q}_{inc}$.

The key to an efficient algorithm is efficient identification of incompatible quartets. In our algorithm we do not generate all the incompatible quartets - rather, we identify incompatible quartets through (on the fly) identification of incompatible *bipartitions*. That is, we show that we can efficiently identify a pair of incompatible bipartitions, and from that pair of bipartitions we can then efficiently obtain an incompatible quartet, so that in $O(k^2 n^2)$ time we have obtained a hitting set of size at most four times the optimal hitting set. This is the basic idea of our algorithm. Here we describe this basic idea just on two trees, noting that the algorithm for $k$ trees operates by repeatedly examining pairs of trees, finding incompatible quartets, and deleting each of the leaves in the incompatible quartets from all the trees in the input.

**MODIFIED-APPROX-HITTING-SET**$(T_1, T_2)$

```
        /* T₁ and T₂ are two trees on the set of leaves S */
1       H ← ∅; S' ← S
2       Mark all edges in T₁ as BAD
3       let e be the first BAD edge
4       while there are BAD edges
5             if e is compatible with all edges in T₂
6                   mark e as GOOD
7                   e ← next BAD edge, if there is one
8       else
9                   find an edge e' in T₂ that e and e' are incompatible
10                  let q be an incompatible basis as in Corollary 1
11                  S' ← S' − q
12                  H ← H ∪ q
13                  remove leaves in Q from T₁ and T₂
                    (but do not restrict T₁ and T₂ to S')
14      return H
```

It can be seen easily that MODIFIED-APPROX-HITTING-SET is correct. Note that since the while loop in line 4 terminates only when there are no BAD edges, all the edges remaining in $T_1$ at the end are compatible with the edges in $T_2$. Hence, by Lemma 1 $T_1|(S - H)$ is compatible with $T_2|(S - H)$. Hence $S - H$ is indeed a compatible set for $\{T_1, T_2\}$, and thus, $H$ is a hitting set for the collection of all incompatible bases for $\{T_1, T_2\}$. Moreover, since $H$ is the union of *disjoint* bases (each of which is of cardinality four), $|H| \leq 4|H^{opt}|$, where $S - H^{opt}$ is an MCS for $\{T_1, T_2\}$.

Note that the while loop terminates after $O(n)$ iterations, since in each iteration it eliminates four leaves or marks an edge as GOOD and there are $O(n)$ edges in $T_1$. Hence, if we ensure that each iteration can be completed in $O(n)$ time, we would have an $O(n^2)$ time algorithm. In the next section we will describe how this running time can be achieved.

## 3.2   Achieving $O(n^2)$ Running Time for the CMCS of Two Trees

From the outline of the algorithm in the previous section, it can be observed that the key to achieving $O(n^2)$ time bound is identifying an edge $e \in E(T_1)$ in $O(n)$ time such that $e$ is either compatible with all the edges in $T_2$ or is incompatible with some edge in $T_2$. We will now show how the above objective can be achieved.

We modify and extend MODIFIED-APPROX-HITTING-SET thus:

1. Pick a leaf arbitrarily, and root both the trees on this leaf. Let the rooted trees be $T_1'$ and $T_2'$.
2. For each node in $T_1'$ maintain a sorted list of all leaves below it. We denote the set of leaves below node $v$ as $L_v$. We maintain the sorted set as an indexed (doubly) linked list, to facilitate fast deletions. Such a data structure suffices since we will only delete from the list and never insert anything.
3. Let $(v, w)$ be an edge in $T_1'$ with $w$ below $v$. For each node $x$ in $T_2'$, compute two quantities $n_x$ and $n_x'$ defined as follows: $n_x = |L_x \cap L_w|$ and $n_x' = |L_x \cap (S' - L_w)|$. Here, $S'$ is the *current* set of leaves in $T_1'$ and $T_2'$.

The following lemma allows us to determine if an edge in $T_1$ is compatible with all other edges in $T_2$ or not.

**Lemma 3.** *Let $(v, w)$ be an edge in $T_1$ and let $(y, x)$ be an edge in $T_2$. Without loss of generality assume that $v$ is the parent of $w$ in $T_1'$ and that $y$ is the parent of $x$ in $T_2'$. Then $(v, w)$ is incompatible with $(y, x)$ iff $0 < n_x < |L_w|$ and $0 < n_x' < |S' - L_w|$.*

*Proof.* The proof follows from the following four observations.

1. $n_x > 0$ iff there exists a leaf $l \in L_x$ in $T_2'$ that is in $L_w$ in $T_1'$, in other words if and only if $L_x \cap L_w \neq \emptyset$.
2. $n_x < |L_w|$ iff there exists a leaf $l \in L_w$ that is *not* in $L_x$. In other words iff $L_w \cap (S' - L_x) \neq \emptyset$.

3. $n'_x > 0$ iff there is a leaf that is not in $L_w$ but is in $L_x$. In other words iff $(S' - L_w) \cap L_x \neq \emptyset$.
4. $n'_x < |S' - L_w|$ iff there is a leaf that is neither in $L_x$ nor in $L_w$. In other words iff $(S' - L_x) \cap (S' - L_w) \neq \emptyset$.

We can compute $n_x$ and $n'_x$ for all nodes $x$ in $T'_2$ in $O(n)$ time by doing the computations bottom up. We then inspect all edges in $T'_2$ to see if they are compatible with the edge $(v, w)$ (which is in $T'_1$). This can again be accomplished in $O(n)$ time since there are at most $O(n)$ edges. Hence, we can identify if an edge $(v, w)$ is GOOD or not in $O(n)$ time.

In case we determine it to be GOOD, we do no further processing. In case we find that the edge is incompatible with an edge $(y, x)$ we have to do some further processing.

1. We compute a basis $Q$ by computing $L_w \cap L_x$, $L_w \cap (S' - L_x)$, $(S' - L_w) \cap L_x$ and $(S' - L_w) \cap (S' - L_x)$ and picking one element from each set. Each of the four intersection computations can be carried out in $O(n)$ time since we maintain the sets sorted.
2. We remove the leaves in $Q$ from $T_1$ and $T_2$ (but we do not eliminate nodes of degree two that may be created due to the removal of $Q$) and hence we have to update the sorted set of leaves maintained in each node. A deletion from this set can be carried out in $O(1)$ time since the set is indexed and doubly linked. Hence, the time taken for the removal of $Q$ from the two trees is in $O(n)$.

The above discussion shows that both the *if* and *else* clause in the *while* loop in our algorithm can be accomplished in $O(n)$ time. We also noted that the loop terminates in $O(n)$ iterations. Hence the loop takes $O(n^2)$ time. The rest of the algorithm takes $O(n)$ time. Thus, the entire algorithm can be made to run in $O(n^2)$ time. The extension to $k$ trees is straightforward, as we just greedily compute incompatible quartet trees, and delete the leaves from each of the trees in the input. Hence we have the following theorem.

**Theorem 3.** *Given a set $\mathcal{T}$ of $k$ unrooted trees on a set of leaves $S$, let $S - H^{opt}$ be a maximum compatible set for $\mathcal{T}$. We can compute a compatible set $S - H$ such that $|H| \leq 4|H^{opt}|$ in $O(k^2 n^2)$ time.*

## 4    A Slower Algorithm with a Better Approximation Ratio

In this section we describe how to modify the previously described algorithm to achieve a 3-approximation ratio, albeit using more time. The algorithm operates by approximating the solution to the rooted version of the CMCS problem, but must apply it to $n$ subproblems, where the $i^{th}$ subproblem considers all the trees in $\mathcal{T}$ rooted at leaf $i$. By taking the minimum of all the solutions obtained, we can compute the overall solution with the desired approximation bound. (Later we give an example of two unrooted trees which shows why it does not suffice to just look at a single rooting.) Although we can show that each individual rooted problem can be solved again in $O(k^2n^2)$ problem, because we look at $n$ subproblems this approach is $O(n)$ slower than the 4-approximation algorithm we showed earlier. Hence, this is a slower algorithm, but it achieves a better guaranteed approximation ratio than our first algorithm.

We begin by making some definitions.

**Definition 8.** *(Triplet Tree): A* triplet tree *is a rooted tree on three leaves that does not have any internal node of degree two. We use the notation $T|q$ to denote the triplet tree induced by a rooted tree $T$ on a set $q = \{a, b, c\}$ of three leaves. The star triplet tree (where all three leaves are children of the root) is denoted by $(a, b, c)$, whereas $((a, b), c)$ (or its equivalent form, $(c, (a, b))$) denotes the triplet tree in which $a$ and $b$ are siblings, and have a least common ancestor below the root. We will call $q$ the* basis *of the triplet tree $T|q$.*

**Definition 9.** *(Triplet Compatibility): Two triplet trees, each on the same leaf set, are said to be* compatible *if at least one of them is a star or they are the same. Otherwise they are said to be* incompatible.

As in unrooted trees, we can talk about the subtree of a rooted tree induced by a set of leaves. Here, however, we do not suppress a node of degree two if it is the root. Corresponding to the set $Q_{inc}$ of incompatible quartet trees, we can define the set $Trip_{inc}$ of incompatible triplet trees:

**Definition 10.** *($Trip_{inc}$): Given a set $\mathcal{T}$ of rooted trees, each leaf-labelled by the same set $S$ of leaves, we define the set $Trip_{inc} = \{\{a, b, c\} \subseteq S : \exists \{A, B\} \subseteq \mathcal{T} : A|\{a, b, c\}$ and $B|\{a, b, c\}$ induce incompatible triplet trees}.*

Now let $A$ be an unrooted tree on the set $S$ of leaves $S = \{1, 2, \ldots, n\}$, and let $i \in S$ be arbitrary.

**Definition 11.** *($A^i$): $A^i$ refers to the rooted tree on the leaf set $S - \{i\}$, obtained by rooting $A$ at the leaf $i$, and deleting $i$ and its incident edge.*

As before, we let $L_v$ represent the leaves below the node $v$ in a tree (whose identity will be known in context).

We now show through the following lemma that the incompatibility question for unrooted trees can be rephrased in terms of incompatibility of induced triplets, an idea crucial to improving the approximation ratio. We state the lemma in terms of two trees, but the result obviously extends to any number of trees.

**Lemma 4.** *Let $A$ and $B$ be two trees on leaf set $S$, and let $i \in S$ be a fixed leaf. Then $A^i$ and $B^i$ induce no incompatible triplet trees if and only if the unrooted trees $A$ and $B$ are compatible.*

*Proof.* First we prove that if the rooted pair of trees $A^i$ and $B^i$ have a pair of incompatible triplets, then $A$ and $B$ are incompatible. Suppose that the set of three leaves $\{a, b, c\}$ is an incompatible triplet, so that $A^i$ induces the triplet tree $(a, (b, c))$ and $B^i$ induces the triplet tree $((a, b), c)$. Then we can see that on $\{a, b, c, i\}$, tree $A$ induces the split $\{a, i\}|\{b, c\}$ whereas tree $B$ induces the split $\{a, b\}|\{c, i\}$. Hence, $A$ and $B$ are incompatible.

Suppose now that $A$ and $B$ are incompatible trees. Hence, there is a set $q = \{a, b, c, d\}$ on which the two trees induce two different splits. Recall that $i$ is an arbitrary fixed leaf. We will show no matter how $i$ is chosen, we can select three leaves from $q$ so that $A^i$ and $B^i$ induce incompatible triplet trees for that subset. Let $A$ induce $\{a, b\}|\{c, d\}$ and $B'$ induce $\{a, c\}|\{b, d\}$ on $q$. We first address the case where $i \in \{a, b, c, d\}$. In this case, without loss of generality assume $i = d$. Then $A^i \ (= A^d)$ induces $((a, b), c)$ on leaf set $a, b, c$, whereas $B^i \ (= B^d)$ induces $((a, c), b)$. Hence, $A^i$ and $B^i$ induce incompatible triplet trees. Now we consider the case where $i \notin \{a, b, c, d\}$. Consider the restriction of the rooted trees $A^i$ and $B^i$ to the leaves $a, b, c, d$. In each of these rooted trees there is at least one sibling pair of leaves. Since $A$ induces the quartet $ab|cd$, the only possible sibling pairs in $A^i|\{a, b, c, d\}$ are $a, b$ and $c, d$. Assume, without loss of generality, that $a, b$ are siblings in $A^i|\{a, b, c, d\}$. Similarly, the only possible sibling pairs in $B^i|\{a, b, c, d\}$ are $a, c$ and $b, d$. Again, without loss of generality, assume $a, c$ are siblings in $B^i|\{a, b, c, d\}$. Then on the set $a, b, c$ of leaves the rooted trees $A^i$ and $B^i$ differ: $A^i|\{a, b, c\} = ((a, b), c)$ whereas $B^i|\{a, b, c\} = ((a, c), b)$. Hence, $a, b, c$ induces incompatible quartet trees in $A^i$ and $B^i$.

We now state the **Maximum Compatible Subset of Rooted Trees (MCSR)** problem:

- **Input:** Set $\mathcal{T}$ of rooted trees, each leaf-labelled by the same set $L$ of leaves.
- **Output:** Subset $L' \subseteq L$ so that the trees in $\mathcal{T}$ restricted to the leaves in $L'$ do not induce any incompatible triplet trees.

The next observation follows directly from Lemma 4, and we state it without proof:

**Observation 1.** *Let $\mathcal{T}$ be a set of unrooted trees leaf-labelled by the same set $S$, and assume that $|S| \geq 1$. Let $i \in S$ be arbitrary. For every subset $S_0 \subset S - \{i\}$, if $S_0$ is a feasible solution to the MCSR problem on set $\{T^i : T \in \mathcal{T}\}$ of trees, then $S_0 \cup \{i\}$ is a feasible solution to the MCS problem on $\mathcal{T}$.*

This suggests the following general strategy for the MCS problem:

- For each $i \in S$, let $L(i)$ denote the MCSR of the set of rooted trees $\{T^i : T \in \mathcal{T}\}$.
- Let $i^*$ be selected so that $|L(i^*)|$ is maximum among $\{|L(i)| : 1 \leq i \leq n = |S|\}$. Return $L(i^*) \cup \{i^*\}$.

*Comments:* (1) Observation 1 implies that this algorithm exactly solves the MCS problem, but that it requires $n$ iterations of an exact solution to the MCSR problem. However, a little arithmetic and Observation 1 also imply that an $r$-approximation algorithm for the CMCSR problem (that is, for the complement of the MCSR problem) would yield an $r$-approximation algorithm for the CMCS problem. Hence, although CMCSR is also NP-hard, if we can approximate CM-CSR we can use that algorithm to approximate CMCS, and with the same guaranteed performance ratio. This is the approach we will take.

(2) Note also that we cannot simply pick a single leaf and root the trees at this leaf, and use the solution obtained for the resultant rooted trees. This approach can be arbitrarily bad. Consider the two caterpillar trees $T$ and $T'$ on leaf set $1, 2, 3, \ldots, 100, x$, as shown in Figure 2. These two trees clearly have a very large compatible subset: namely, the set $1, 2, \ldots, 100$ of leaves. However, if we root $T$ and $T'$ at leaf $x$, then they have very small compatible subsets as rooted trees. Similarly, arbitrary rootings of the trees $T$ and $T'$ will greatly affect the size of the compatible subset of the leaves that is obtained.

*Approximating the CMCSR problem:* We now show how we can 3-approximate the complement of the MCSR problem. The technique is essentially the same as for the complement of the MCS problem, but rather than 4-approximating the Hitting Set for $Q_{inc}$ we 3-approximate the Hitting Set for $Trip_{inc}$. Here, too, we can use the same techniques we used in the unrooted version (the CMCS problem) in order to get an $O(n^2)$ algorithm for the CMCSR of two trees. Since we do this for each possible leaf at which to root each of the trees, this gives us an $O(n^3)$ 3-approximation algorithm for the CMCSR problem on 2 trees, and hence, an $O(k^2 n^3)$ 3-approximation algorithm for $k$ trees.

We now show how we do this. Let $A$ and $B$ be two unrooted trees on leaf set $L$, and root each at leaf $i$ thus producing $A^i$ and $B^i$. Let $L' = L - \{i\}$. As before, let $L_x$ denote the leaves below node $x$. The only variant is how we calculate incompatible triplets "on the fly". As before, we look for "bad" edges. Suppose we find a bad edge $(v, w)$ in $A^i$ that is incompatible with edge $(x, y)$ in $B^i$. As before, we identify one leaf from each of the four sets $L_w \cap L_y$, $L_w \cap (L' - L_y)$ and $(L' - L_w) \cap L_y$ and $(L' - L_y) \cap (L' - L_w)$. Let these leaves be $a$, $b$, $c$ and $d$ respectively. Earlier, we eliminated all the four leaves, whereas now we eliminate only $a$, $b$ and $c$ from $A^i$ and $B^i$ leaving $d$ untouched.

Observe that the set $\{a, b, c\}$ can not be part of any compatible set for the rooted trees $A^i$ and $B^i$ since when restricted to $\{a, b, c\}$, $A^i$ induces $((a, b), c)$ and $B^i$ induces $(a, (b, c))$. Hence, at least one leaf from $\{a, b, c\}$ has to be eliminated. Moreover observe that $\{a, b, c, i\}$ forms an incompatible quartet for the unrooted trees $A$ and $B$. Hence, all incompatible induced triplets can be identified in the above manner. Hence we have the following theorem:

**Theorem 4.** *Given a set $\mathcal{T}$ of $k$ unrooted trees on a set of leaves $L$, let $L - H^{opt}(\mathcal{T})$ be a maximum compatible set for the trees. We can compute a compatible set $L - H$ such that $|H| \leq 3|H^{opt}|$ in $O(k^2 n^3)$ time.*

# 5   Application in Computational Biology

A "phylogeny" (or evolutionary tree) represents the evolutionary history of a set of species $S$ by a rooted (typically binary) tree in which the leaves are labelled with elements from $S$, and with internal nodes unlabelled.

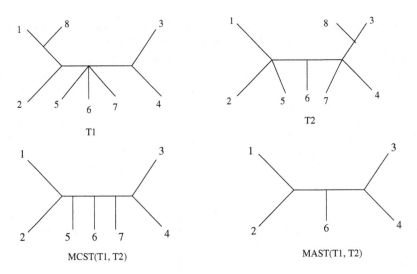

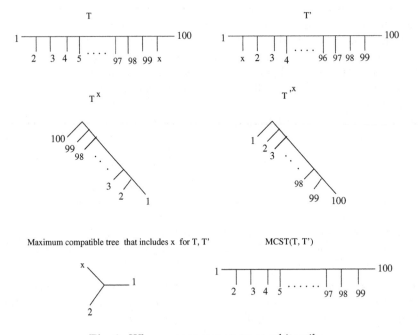

**Fig. 1.** We calculate the maximum agreement subtree (MAST) and the maximum compatible subtree (MCST) of two trees, and show they are different.

**Fig. 2.** Why we cannot root trees arbitrarily.

A standard approach for inferring phylogenies is to attempt to solve a hard optimization problem (such as Maximum Parsimony or Maximum Likelihood), and this can result in hundreds or thousands of equally good trees obtained. A consensus tree is then used to represent the entire set of trees. However, standard consensus methods, such as the majority consensus and strict consensus, can return highly unresolved trees. For example, the strict consensus tree is that tree which is the most resolved common contraction of all the trees in the input set. This consensus tree is highly unresolved when one species is sufficiently distantly related to the other species, so that its location in the tree is not well localized. Such a species is called a "rogue taxon", and is known to cause problems in phylogenetic analysis. The majority consensus tree is similarly poorly affected. (See [11] for a discussion of phylogenetic inference in practice.)

The identification, and elimination, of rogue taxa from phylogenetic analyses has led to alternative consensus approaches which reduce the set of species (by eliminating a small set of species) so that the resultant trees have a well resolved consensus tree. One such approach, originally posed in [5], is the "Maximum Agreement Subtree Problem", or MAST: this consensus method eliminates the smallest subset of the species so that all the trees agree on the remaining species. (The largest common subtree in all the trees is the maximum agreement subtree, or MAST, of the input.) The MAST problem is NP-hard for three or more trees [1], but polynomial time for two trees [4,9] or for any number of trees if at least one tree has bounded degree [3]. Software for the MAST problem exists in the PAUP [10] software package, but is not very useful since the size of the MAST is often not large enough to be interesting (David Swofford, personal communication). The problem seems to be that requiring complete agreement between the trees, even when restricted to a smaller set of taxa, is too strong a requirement.

A potential explanation for this phenomenon lies in the difficulty in correctly resolving "short edges" in the true tree. In this case, optimal solutions for problems such as Maximum Parsimony or Maximum Likelihood may differ significantly in terms of how they resolve around these short edges. This results in having the MAST of the set of trees being very small. On the other hand, the identification and contraction of short edges in each of the ML or MP trees could result in a collection of trees that, while not binary, have a common refinement on some large subset of leaves. This suggests the following approach:

- First, contract all short enough edges, and
- Second, identify and delete a minimum set of leaves so that after these leaves are deleted and degree two nodes suppressed, the resultant trees share a common refinement (i.e. are compatible). If a consensus tree is desired, the minimum common refinement (i.e. the maximum compatibility subtree, or MCST) of the resultant trees is returned.

The first step is straightforward (though establishing how short is short enough needs to be studied), but the second step is essentially the CMCS problem that we studied in this paper.

(Note that there is a relationship between the number of leaves in the MAST (maximum agreement subtree) of a set of trees and its MCST (maximum compatibility subtree). By definition, the number of leaves in a MCST is at least as big as that of the MAST, and this can often be significantly larger. See Figure

1 for an example of two trees and their MAST and MCST, for which there is a difference in sizes. Thus, the MCST problem is a better model for consensus trees, as it retains more information and is less affected by noise in the input.)

Thus, our approximation algorithm for the CMCS problem allows us to achieve two goals: obtain meaningful and informative consensus trees and identify rogue taxa. Because our algorithms are approximations, they are only useful when the trees do share a large common subset of the taxa on which they are compatible. This is an assumption which seems to be believed by many systematic biologists, but one which we can now test through the use of these algorithms. Thus, these algorithms can serve an additional purpose: enabling us to validate (or prove incorrect) the assumptions of systematic biologists about the level of phylogenetic signal in their datasets, and the performance of their methods for estimating phylogenetic trees.

## Acknowledgments

This research was supported by a Fellowship in Science and Engineering from the David and Lucile Packard Foundation to Warnow, and by NSF grant EIA 01-21680 to Warnow.

## References

1. A. Amir and D. Keselman. Maximum agreement subtrees in a set of evolutionary evolutionary trees: Metrics and efficient algorithms. *SIAM Journal of Computing*, 26(6):1656–1669, 1997. A preliminary version of this paper appeared in FOCS '94.
2. P. Buneman. The recovery of trees from measures of dissimilarity. *Mathematics in the Archaelogical and historical Sciences*, pages 387–395, 1971.
3. M. F. Colton, T.M. Przytycka, and M. Thorup. On the agreement of many trees. *Information Processing Letters*, 55:297–301, 1995.
4. M. F. Colton and M. Thorup. Fast comparison of evolutionary trees. In *Proc. of the 5th Annual ACM-SIAM Symposium on Discrete Algorithms*, pages 481–488, 1994.
5. C.R. Finden and A.D. Gordon. Obtaining common pruned trees. *Journal of Classification*, 2:255–276, 1985.
6. D. Gusfield. Efficient algorithms for inferring evolutionary trees. *Networks*, 21:19–28, 1991.
7. A. Hamel and M. A. Steel. Finding a maximum compatible tree is NP-hard for sequences and trees. *Applied Mathematics Letters*, 9(2):55–60, 1996.
8. J. Hein, T. Jiang, L. Wang, and K. Zhang. On the complexity of comparing evolutionary trees. *Discrete and Applied Mathematics*, 71:153–169, 1996.
9. M. Steel and T. Warnow. Kaikoura tree theorems: computing the maximum agreement subtree. *Information Processing Letters*, 48:77–82, 1993.
10. D. Swofford. PAUP*: Phylogenetic analysis using parsimony (and other methods), version 4.0. 1996.
11. D. L. Swofford, G. J. Olsen, P. J. Waddell, and D. M. Hillis. Phylogenetic inference. In D. M. Hillis, B. K. Mable, and C. Moritz, editors, *Molecular Systematics*, pages 407–514. Sinauer Assoc., 1996.
12. T. Warnow. Tree compatibility and inferring evolutionary history. *Journal of Algorithms*, 16:388–407, 1994.

# A 27/26-Approximation Algorithm for the Chromatic Sum Coloring of Bipartite Graphs

Krzysztof Giaro, Robert Janczewski, Marek Kubale, and Michał Małafiejski

Foundations of Informatics Department,
Gdańsk University of Technology, Poland
kubale@eti.pg.gda.pl

**Abstract.** We consider the CHROMATIC SUM PROBLEM on bipartite graphs which appears to be much harder than the classical CHROMATIC NUMBER PROBLEM. We prove that the CHROMATIC SUM PROBLEM is NP-complete on planar bipartite graphs with $\Delta \leq 5$, but polynomial on bipartite graphs with $\Delta \leq 3$, for which we construct an $O(n^2)$-time algorithm. Hence, we tighten the borderline of intractability for this problem on bipartite graphs with bounded degree, namely: the case $\Delta = 3$ is easy, $\Delta = 5$ is hard. Moreover, we construct a 27/26-approximation algorithm for this problem thus improving the best known approximation ratio of 10/9.

## 1 Introduction

Let $G = (V, E)$ be a simple graph with vertex set $V = V(G)$ and edge set $E = E(G)$. By $n$ and $m$ we denote the number of vertices and the number of edges of $G$, respectively. By $\Delta(G)$ we denote the maximum degree over all vertices of graph $G$. If $W \subset V(G)$ is a nonempty set then by $G[W]$ we denote a subgraph of $G$ induced by $W$. The chromatic sum is defined as follows [6].

**Definition 1.** *By the **chromatic sum** of graph $G$ we mean $\sum(G) = \min_c \sum(G, c)$, where $\sum(G, c) = \sum_{v \in V(G)} c(v)$ and $c : V(G) \to \mathbb{N}$ is a proper vertex coloring of $G$, i.e. $c(v) \neq c(w)$ whenever $\{v, w\} \in E(G)$. A coloring $c$ of $G$ is said to be a **best coloring** if $\sum(G, c) = \sum(G)$.*

The problem of verifying the inequality $\sum(G) \leq k$ for a graph $G$ and arbitrary positive integer $k$ is known as the CHROMATIC SUM PROBLEM. This differs from the SUM COLORING PROBLEM, which requires a best coloring $c$ in addition. The notion of chromatic sum was first introduced in [6], where the authors showed that the CHROMATIC SUM PROBLEM is NP-complete on arbitrary graphs. Another complexity result comes from [11], where NP-completeness has been proved for interval graphs. In [2] the authors have shown that there exists $\varepsilon > 0$, such that there is no $(1 + \varepsilon)$-ratio approximation algorithm for the SUM COLORING PROBLEM on bipartite graphs, unless P=NP. In [7] the author proved

K. Jansen et al. (Eds.): APPROX 2002, LNCS 2462, pp. 135–145, 2002.
© Springer-Verlag Berlin Heidelberg 2002

the NP-completeness of the CHROMATIC SUM PROBLEM on cubic planar graphs. Moreover, in [7] the CHROMATIC SUM PROBLEM on $r$-regular graphs was proved to be NP-complete for any $r \geq 3$. The 2-approximation algorithm for interval graphs have been shown in [9]. In [1] the authors showed $(\Delta+2)/3$-approximation algorithm for graphs with bounded degree and the 2-approximation algorithm for line graphs.

In this paper we deal with the CHROMATIC SUM PROBLEM on bipartite graphs. We establish a borderline of intractability for this problem on bipartite graphs with bounded degree, namely the case $\Delta = 3$ is easy, $\Delta = 5$ is hard. We construct an $O(n^2)$-time algorithm for bipartite graphs with $\Delta \leq 3$, and a 27/26-approximation algorithm for the SUM COLORING PROBLEM on any bipartite graph. This improves the previously best known 10/9-approximation algorithm for this problem [2].

## 1.1  NP-Completeness Results
## on Bipartite Planar Graphs with $\Delta \leq 5$

In this extended abstract we omit the proofs of NP-completeness.

**Theorem 1 ([8]).** *The* CHROMATIC SUM PROBLEM *is NP-complete on planar bipartite graphs with* $\Delta \leq 5$.

**Corollary 1 ([8]).** *The* CHROMATIC SUM PROBLEM *is NP-complete on planar bipartite graphs with* $\Delta \leq 5$, *even when restricted to graphs for which there exists a best 3-coloring.*

## 2   Exact and Approximation Algorithms

In this section we introduce an idea of 3-pseudocolorings of bipartite graphs and construct an algorithm for finding the best pseudocoloring for any bipartite graph in $O(mn)$ time.

**Definition 2.** *By a **pseudocoloring** (3-**pseudocoloring**) of bipartite graph $G$ we mean any mapping $q : V(G) \rightarrow \{1,2,3\}$ satisfying conditions: every set $C_i := q^{-1}(\{i\})$ is an independent set in graph $G$ for $i = 1$ and $i = 2$. Analogously to Definition 1, by the **pseudochromatic sum** of a bipartite graph $G$ we mean $\sum_{qs}(G) := \min_q \sum(G,q)$, where $\sum(G,q) := \sum_{v \in V(G)} q(v)$ and $q$ is a pseudocoloring of $G$. A pseudocoloring $q$ of graph $G$ is said to be a **best pseudocoloring** if $\sum(G,q) = \sum_{qs}(G)$.*

For any bipartite graph $G$ we have an obvious

**Proposition 1.** $\sum_{qs}(G) \leq \sum(G) \leq 3n/2$.

We get at once

**Proposition 2.** *For any best pseudocoloring $q$ of subcubic (i.e. $\Delta(G) \leq 3$) bipartite graph $G$ we have $\Delta(G[C_2 \cup C_3]) \leq 2$.*

Before we show the algorithm, we need some well-known notation of minimum cuts in weighted digraphs (e.g. see [10]). Let $D = (V, A)$ be any digraph without loops and multiple edges, and let $w$ be a vector of positive weights (including $\infty$) on the edges of $D$. For any two different vertices $s, t \in V(D)$ by the $s - t$ cut (or simply cut) we mean a partition $(S, T)$ of the set $V(D)$ such that $s \in S$, $t \in T$, $S \cap T = \emptyset$ and $S \cup T = V(D)$. By the capacity of the cut $(S, T)$ we mean $f(S, T) := \sum_{e \in A(D) \cap (S \times T)} w(e)$. By the minimum cut $f_o(D, w, s, t)$ we mean the $s - t$ cut of weighted digraph $D$ which minimizes $f(S, T)$.

**Theorem 2.** *There exists an algorithm for finding the best pseudocoloring of any bipartite graph in $O(mn)$ time.*

*Proof.* Let $G = (V_1 \cup V_2, E)$ be a bipartite graph. We construct the digraph $D$ with weights $w$ such that a minimum $s - t$ cut $(P, Q)$ for some vertices $s \in P$ and $t \in Q$ is equal to $f_o(D, w, s, t) = f(P, Q) = \sum_{qs}(G) - n(G)$.

Let $G^* = (V_1^* \cup V_2^*, E^*)$ be the isomorphic copy of $G$ such that $V(G) \cap V(G^*) = \emptyset$. By $v^*$ we denote an image of vertex $v \in V(G)$ under isomorphism $h : V(G) \to V(G^*)$, i.e. $h(v) = v^*$ $(h^{-1}(v^*) = v)$, analogously $h(V_i) = V_i^*$. The directed graph $D$ with weights $w$ shown in Figure 1 is formally defined as follows:

$$V(D) = V(G^*) \cup V(G) \cup \{s\} \cup \{t\} \qquad (1)$$
$$A(D) = A_{1,2} \cup A_{2,1} \cup A_{s,1} \cup A_{2,t} \cup A_{1,1} \cup A_{2,2}$$
$$w(e) = \begin{cases} 1 & \text{if } e \in A_{s,1} \cup A_{2,t} \cup A_{1,1} \cup A_{2,2} \\ \infty & \text{if } e \in A_{1,2} \cup A_{2,1} \end{cases}$$

where

$$A_{1,2} = \{(v_1, v_2) : v_1 \in V_1 \wedge v_2 \in V_2 \wedge \{v_1, v_2\} \in E(G)\}$$
$$A_{2,1} = \{(v_2, v_1) : v_1 \in V_1^* \wedge v_2 \in V_2^* \wedge \{v_1, v_2\} \in E(G^*)\}$$
$$A_{s,1} = \{s\} \times V_1$$
$$A_{2,t} = V_2 \times \{t\}$$
$$A_{1,1} = \{(v_1^*, v_1) : v_1 \in V_1\}$$
$$A_{2,2} = \{(v_2, v_2^*) : v_2 \in V_2\}$$

Let $(P, Q)$ be the minimum $s - t$ cut in $D$. We introduce auxiliary notations (see Figure 1) for $i = 1, 2$:

$$P_i = V_i \cap P, \ P_i^* = V_i^* \cap P \qquad (2)$$
$$Q_i = V_i \cap Q, \ Q_i^* = V_i^* \cap Q,$$

moreover, using the isomorphism $h$ we define $P_{1,Q}^* = h(Q_1) \cap P$, $Q_{2,P}^* = h(P_2) \cap Q$. Because $f(P, Q) \leq \sum_{e \in A_{s,1}} w(e) = |V_1| < \infty$, from the infinity of weights of edge sets $A_{1,2}$ and $A_{2,1}$ we get

$$A(D) \cap ((P_2^* \times Q_1^*) \cup (P_1 \times Q_2)) = \emptyset. \qquad (3)$$

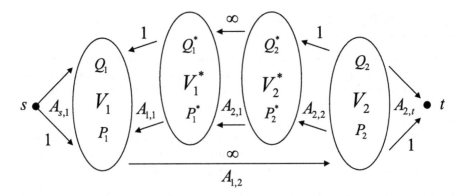

**Fig. 1.** The directed graph $D$ with specified sets of vertices, edges and its weights.

So, from definition of capacity we obtain $f(P,Q) = |Q_1| + |P_2| + |P^*_{1,Q}| + |Q^*_{2,P}|$. Moreover, if $h(P_1) \cap Q^*_1 \neq \emptyset$ then we can change the $(P,Q)$ partitioning by moving these vertices from $Q$ to $P$. Observe, that this operation cannot increase the cut capacity and can be done in linear time. Analogously, if $h(Q_2) \cap P^*_2 \neq \emptyset$ the we can move these vertices from $P$ to $Q$. Therefore in the following we assume that $Q^*_1 \subseteq h(Q_1)$ and $h(Q_2) \subseteq Q^*_2$. So, we have

$$
\begin{aligned}
f(P,Q) &= |Q_1| + |P_2| + |P^*_{1,Q}| + |Q^*_{2,P}| \qquad\qquad (4)\\
&= |Q_1| + |P_2| + |h(Q_1)| - |h(Q_1) \cap Q| + |h(P_2)| - |h(P_2) \cap P|\\
&= 2 \cdot |Q_1| + 2 \cdot |P_2| - |Q^*_1| - |P^*_2|.
\end{aligned}
$$

Now, we shall show the connection between the constructed minimum cut $(P,Q)$ and some pseudocoloring of $G$. We prove the following claims:

**Claim 1.** $C_1 := P_1 \cup Q_2$ and $C_2 := h^{-1}(Q^*_1 \cup P^*_2)$ are independent sets in $G$.

**Claim 2.** Defining $C_3 := V(G) \backslash (C_1 \cup C_2)$ we get the pseudocoloring $q$ defined as follows: $q^{-1}(\{i\}) = C_i$ with $\sum(G,q) = f(P,Q) + n(G)$.

**Claim 3.** Pseudocoloring $q$ is the best one, i.e. $\sum(G,q) = \sum_{qs}(G)$.

By (3) $C_1 = P_1 \cup Q_2$ is an independent set in $G$ and $Q^*_1 \cup P^*_2$ is an independent set in $G^*$ and because $h^{-1}$ is an isomorphism so $C_2$ is an independent set in $G$. Claim 1 is proved. Then $q$ is a pseudocoloring of $G$, hence by (4) we get Claim 2:

$$
\begin{aligned}
\sum(G,q) &= |C_1| + 2 \cdot |C_2| + 3 \cdot |C_3| = 3 \cdot n(G) - 2 \cdot |C_1| - |C_2|\\
&= n(G) + 2 \cdot (n(G) - |C_1|) - |C_2| = n(G) + f(P,Q).
\end{aligned}
$$

Now, observe that for any pseudocoloring $p$ of $G$ the following partition $(S^p, T^p)$:

$$
S^p = \{s\} \cup (C'_1 \cap V_1) \cup h(C'_1 \cap V_1) \cup (V_2 \backslash C'_1) \cup h(V_2 \cap C'_2) \cup h(V_1 \cap C'_3)
$$
$$
T^p = \{t\} \cup (C'_1 \cap V_2) \cup h(C'_1 \cap V_2) \cup (V_1 \backslash C'_1) \cup h(V_1 \cap C'_2) \cup h(V_2 \cap C'_3)
$$

is an $s - t$ cut of capacity $f(S^p, T^p) = \sum(G, p) - n(G)$, where $C_i' := p^{-1}(\{i\})$. Because $(P, Q)$ is the minimum cut in $D$ we get that $q$ is the best pseudocoloring of $G$, so we have proved Claim 3.

We can construct the minimum cut $(P, Q)$ in $O(mn)$ time using the Ford-Fulkerson algorithm (see [10]), hence we have constructed the best pseudocoloring $q$ of graph $G$ in polynomial time.

As the first consequence of Theorem 2 we get an $O(n^2)$-time algorithm for solving the SUM COLORING PROBLEM on subcubic bipartite graphs.

**Theorem 3.** *The* SUM COLORING PROBLEM *on subcubic bipartite graphs can be solved in* $O(n^2)$ *time.*

*Proof.* Let $G$ be any subcubic bipartite graph. Because $m = O(n)$, so by Theorem 2 we can construct in time $O(n^2)$ the best pseudocoloring $q$ such that every $C_i := q^{-1}(\{i\})$ is an independent set in $G$ for $i = 1, 2$. By Proposition 2 we conclude that the subgraph of $G$ induced by $C_2 \cup C_3$ is of degree at most 2. Because $q$ is the best pseudocoloring of $G$, we can easily recolor graph $G[C_2 \cup C_3]$ with colors $2, 3$ and get a proper coloring $c$ of graph $G$ using only 3 colors with the same sum of colors. From Proposition 1 it follows $\sum(G, c) = \sum(G, q) = \sum_{qs}(G) \leq \sum(G)$, hence $c$ is the best coloring of $G$.

In [1] the authors proposed a 9/8-approximation algorithm, which has been improved in [2].

**Theorem 4 ([2]).** *There exists a 10/9-approximation algorithm for the* SUM COLORING PROBLEM *on bipartite graphs.*

Now, we improve on this result by using the pseudocoloring algorithm given in the proof of Theorem 2.

**Theorem 5.** *There exists a 27/26-approximation algorithm for the* SUM COLORING PROBLEM *on bipartite graphs of complexity* $O(mn)$.

*Proof.* Let $G = (V_1 \cup V_2, E)$ be any bipartite graph with $m$ edges, $n$ vertices and assume that $|V_1| \geq |V_2|$. By Theorem 2 we can construct the best pseudocoloring $q$ in $O(mn)$ time. Let us denote, $C_i := q^{-1}(\{i\})$ and $a_i := |C_i|$ for $i = 1, 2, 3$. Proposition 1 implies

$$\sum(G, q) = a_1 + 2a_2 + 3a_3 = 2n - a_1 + a_3 \leq \sum(G). \tag{5}$$

Now, consider three algorithms $A_1$, $A_2$ and $A_3$ for coloring a bipartite graph $G$. By $A_1$ we mean an algorithm that colors $V_1$ with color 1 and $V_2$ with color 2. It is easy to see that

$$S(A_1) \leq 3n/2, \tag{6}$$

where by $S(A_i)$ we denote the sum of colors used by algorithm $A_i$ for $i = 1, 2, 3$. The algorithm $A_2$ colors all the vertices from $C_1$ with color 1 and colors graph $G[C_2 \cup C_3]$ analogously to $A_1$ with colors 2 and 3. It is easy to see that

$$S(A_2) \leq a_1 + 5(a_2 + a_3)/2 = 5n/2 - 3a_1/2. \tag{7}$$

Finally, let $A_3$ be an algorithm that colors $C_1$ with 1, $C_2$ with 2 and colors graph $G[C_3]$ similarly to $A_1$ with colors 3 and 4, hence we get

$$S(A_3) \le a_1 + 2a_2 + 7a_3/2 = 2n - a_1 + 3a_3/2. \tag{8}$$

Now, let $A$ be an algorithm that colors graph $G$ using $A_1$, $A_2$, $A_3$ and chooses the solution with minimum sum of colors. Using 6, 7, 8 and 5 we get

$$26S(A) \le 2S(A_1) + 6S(A_2) + 18S(A_3)$$
$$\le 54n - 27a_1 + 27a_3 = 27(2n - a_1 + a_3) \le 27 \sum(G).$$

In contrast to the general case, where the CHROMATIC SUM PROBLEM on $r$-regular graphs is NP-complete [7], the CHROMATIC SUM PROBLEM on bipartite regular graphs appears to be polynomially solvable. In fact, we get an exact formula for the chromatic sum.

**Theorem 6.** *The chromatic sum of a connected bipartite regular graph is equal to $3n/2$ for any $n > 1$. Moreover, any coloring $c$ using more than two colors has a greater sum.*

*Proof.* Consider an arbitrary feasible coloring $c$ of $k$-regular graph with $n$ vertices. Then

$$k \sum_{v \in V} c(v) = \sum_{\{v,u\} \in E} (c(v) + c(u)) \ge \sum_{\{v,u\} \in E} 3 = 3|E| = 3kn/2,$$

hence $\sum(G) \ge 3n/2$. The lower bound is attained for bipartite regular graphs by coloring with 1 all vertices in one part of the bipartition, and by coloring with 2 all vertices in the other part.

## 3    Conclusions

The results given in the previuos section tighten the borderline between P and NP-completeness for the CHROMATIC SUM PROBLEM on low-degree bipartite graphs, namely: graphs with $\Delta \le 3$ are easy instances and those with $\Delta \le 5$ are

**Table 1.** Complexity classification for the chromatic sum problem on graphs with bounded degree.

| Problem: CSP or SCP on graphs | Complexity | Reference |
|---|---|---|
| $\Delta \le 2$ | $O(n)$ | [7] |
| regular bipartite | $O(n)$ | Thm. 6 |
| planar cubic graphs | NPC | [7] |
| $k$-regular ($k \ge 3$) | NPC | [7] |
| bipartite subcubic ($\Delta \le 3$) | $O(n^2)$ | Thm. 3 |
| bipartite with $\Delta \le 5$ | NPC | Thm. 7 |

hard. A still open question is the complexity of the problem on bipartite graphs with $\Delta = 4$. The authors conjecture that this problem is polynomially solvable, but this case claim seems to be very hard to prove.

The proposed approximation algorithm produces a coloring that is less than 4% worse than the value of optimal solution. In [2] the authors show that there exists an $\varepsilon > 0$, such that there is no $(1 + \varepsilon)$-ratio approximation algorithm (unless P=NP). We still don't know how far is 1/26 from this $\varepsilon$. Table 1 summarizes the complexity results proved for graphs with small degree.

# References

1. Bar-Noy A., Bellare M., Halldórsson M.M., Shachnai H., Tamir T.: On chromatic sums and distributed resource allocation. Information and Computation **140** (1998) 183-202
2. Bar-Noy A., Kortsarz G.: Minimum color sum of bipartite graphs. Journal of Algorithms **28** (1998) 339-365
3. Dyer M.E., Frieze A.M.: Planar $3DM$ is NP-complete. Journal of Algorithms **7** (1986) 174-184
4. Garey M.R., Johnson D.S.: Computers and Intractability: A Guide to the Theory of NP-Completeness. W.H. Freeman (1979)
5. Giaro K., Kubale M.: Edge-chromatic sum of trees and bounded cyclicity graphs. Inf. Proc. Letters **75** (2000) 65-69
6. Kubicka E., Schwenk A.J.: An introduction to chromatic sums. Proceedings of ACM Computer Science Conference (1989) 39-45
7. Małafiejski M.: The complexity of the chromatic sum problem on planar graphs and regular graphs. The First Cologne Twente Workshop on Graphs and Combinatorial Optimization, Cologne (2001), Electronic Notes in Discrete Mathematics **8** (2001)
8. Małafiejski M.: Scheduling Multiprocessor Tasks to Minimize Mean Flow Time (in Polish), Ph.D. Thesis, Gdańsk University of Technology (2002)
9. Nicolso S., Sarrafzadeh M., Song X.: On the sum coloring problem on interval graphs. Algorithmica **23** (1999) 109-126
10. Papadimitriou C.H., Steiglitz K.: Combinatorial Optimization: Algorithms and Complexity. Prentice-Hall, New Jersey (1982)
11. Szkaliczki T.: Routing with minimum wire length in the dogleg-free Manhattan model is NP-complete. SIAM Journal on Computing **29** (1999) 274-287

# 4   Appendix

The reduction uses the restriction of the classical NP-complete problem $3DM$ [4], namely a planar $3DM$ problem introduced and proved in [3].

**Definition 3.** $3DM_p$: *let* $W, X, Y$ *be three disjoint sets satisfying* $|W| = |X| = |Y| = q$ *and let* $M$ *be any subset of* $W \times X \times Y$. *For every* $a \in W \cup X \cup Y$ *we define* $\#a := |\{(w, x, y) \in M : w = a \vee y = a \vee x = a\}|$ *which is equal to 2 or 3. Moreover, a bipartite graph* $G = (W \cup X \cup Y \cup M, \{\{a, m\} : a \in m, a \in W \cup X \cup Y, m \in M\})$ *is planar, where* $a \in m$ *means that* $a$ *is one of the coordinates of vector* $m$. *The question that we state is as follows: is there a subset* $M' \subseteq M$

*satisfying $|M'| = q$, such that every two elements $m_1, m_2 \in M'$, $m_1 \neq m_2$, differ on each coordinate?*

The following easy observation holds for any best colorings of graph $G$.

**Proposition 3.** *Given a graph $G$ and a decomposition of $G$ into vertex disjoint subgraphs $G_1, ..., G_k$ such that $\bigcup_{i=1}^{k} V(G_i) = V(G)$ and $\bigcup_{i=1}^{k} E(G_i) \subset E(G)$ implies $\sum(G) \geq \sum_{i=1}^{k} \sum(G_i)$. Moreover, if $c_i$ is a best coloring of $G_i$ for all $i = 1, ..., k$ and all these colorings form a coloring $c$ of $G$, then $c$ is a best coloring of $G$ and $\sum(G) = \sum(G, c) = \sum_{i=1}^{k} \sum(G_i, c_{|V(G_i)})$.*

**Theorem 7.** *The* CHROMATIC SUM PROBLEM *is NP-complete on planar bipartite graphs with $\Delta \leq 5$.*

*Proof.* We show a polynomial reduction from problem $3DM_p$ to the CHROMATIC SUM PROBLEM on planar bipartite graphs with degree bounded by 5. This reduction is a modification of NP-completeness proof of the CSP for subcubic planar graphs showed in [7]. Let $W, X, Y, q, M$ be given as in Definition 3 and let $x_i$ be the number of elements $a \in W \cup X \cup Y$ such that $\#a = i$ ($i = 2$ or $i = 3$).
  We define a graph $G$ as follows

$$V(G) = \{v_m : m \in M\} \cup \bigcup_{a \in W \cup X \cup Y} V(A_{\#a}^a) \tag{9}$$

$$E(G) = \{\{a_m, v_m\}, \{b_m, v_m\}, \{c_m, v_m\} : m = (a, b, c) \in M\} \cup \bigcup_{a \in W \cup X \cup Y} E(A_{\#a}^a),$$

where $a \in W \cup X \cup Y$ and $A_2^a$ ($\#a = 2$) or $A_3^a$ ($\#a = 3$) are bipartite graphs with the desired properties of the best colorings.
  First, we construct an auxiliary bipartite graph $B$ with non-symmetry property of every best coloring. Consider the bipartite graph $B$ with $\Delta = 5$ shown in Figure 2.

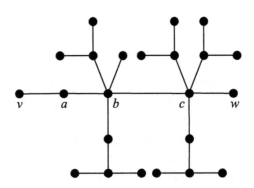

**Fig. 2.** The auxiliary graph $B$ with the chromatic sum 33.

We will show the following property: for every best coloring $c$ of $B$ vertex $v$ is colored with 2 and $w$ is colored with 1. Moreover, if we color the pair of vertices $(v, w)$ with a pair of colors $(1, 2)$ we can extend this partial colorig to the coloring of graph $B$ in such a way that the sum of colors exceeds the chromatic sum of $B$ exactly by 1.

By $T_b$ we denote a connected subgraph of $B \backslash \{a, c\}$ including vertex $b$, analogously by $T_c$ we mean that tree of $B \backslash \{b, w\}$ including vertex $c$. Let $T_v = B[\{v, a\}]$ and $T_w = B[\{w\}]$. For a given graph $G$ and a vertex $v' \in V(G)$ let $bc(G, v') = \{k \in \mathbb{N} : c(v') = k \wedge c \text{ any best coloring of } G\}$ be a list of colors. Analogously, for any $v', w' \in V(G)$ let $bc(G, (v', w')) = \{(k, l) : c(v') = k \wedge c(w') = l \wedge c \text{ any best coloring of } G\}$. Analyzing all the best colorings of the defined trees we get $bc(T_v, v) = \{1, 2\}$, $bc(T_v, a) = \{1, 2\}$, $bc(T_w, w) = \{1\}$, $bc(T_b, b) = \{2, 3\}$ and $bc(T_c, c) = \{1, 3\}$. Moreover $\sum(T_v) = 3$, $\sum(T_w) = 1$, $\sum(T_b) = 13$ and $\sum(T_c) = 16$, hence by Proposition 3 we obtain $\sum(G) = 33$ and there is only one coloring $c_p$ of the path $B[\{v, a, b, c, w\}]$ that can be extended to best coloring of the whole graph $B$, namely $c_p$ colors the vertices $v, a, b, c, w$ with colors $2, 1, 2, 3, 1$, respectively. Now, color vertex $v$ with 1 and $w$ with 2. Coloring the vertex $a$ with 2, $b$ with 3 and $c$ with 1 we can extend this pre-coloring to the whole graph $B$ with the sum of colors equal to 34.

We construct a graph $A_2^a$ shown in Figure 3 for a given element $a \in W \cup X \cup Y$ occuring only in $x, y \in M$.

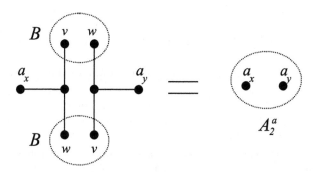

**Fig. 3.** Graph $A_2^a$ with the chromatic sum 73.

Notice, that graph $A_2^a$ is bipartite with $\Delta = 5$ and $bc(A_2^a, a_x) = \{1, 2\}$, $bc(A_2^a, a_y) = \{1, 2\}$. By Proposition 3 we have $\sum(A_2^a) \geq 2 \cdot 33 + 6 = 72$, from the properties of $B$ it follows $\sum(A_2^a) > 72$. On the other hand, one can easily construct colorings of $A_2^a$ with the chromatic sum equal to 73. Considering all possibilities of coloring of the vertices $a_x$ and $a_y$ we get at once $bc(A_2^a, (a_x, a_y)) = \{(1, 2), (2, 1)\}$.

At last, for a given element $a \in W \cup X \cup Y$ occuring only in $x, y, z \in M$ we construct a graph $A_3^a$ shown in Figure 4.

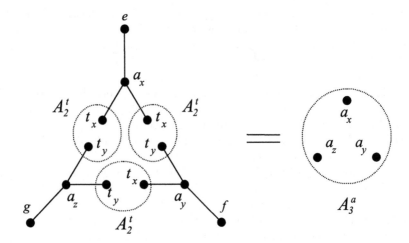

**Fig. 4.** Bipartite graph $A_3^a$ with the chromatic sum 229.

Notice, that graph $A_2^t$ is just an auxiliary graph. First, let us note that graph $A_3^a$ is bipartite with $\Delta \leq 5$ and by Proposition 3 we have $\sum(A_3^a) \geq 3 \cdot 73 + 3 \cdot 3 = 228$, but it is impossible to extend best colorings of all $A_2^t$-graphs to the whole graph $A_3^a$, so $\sum(A_3^a) > 228$. On the other hand, one can easily construct a coloring with the sum equal to 229. Let $c$ be any best coloring of $A_3^a$, i.e. $\sum(A_3^a, c) = 229$. There are only two possibilities:

(1) $c(\{a_x, a_y, a_z\}) = \{1, 2, 3\}$ and the coloring $c$ restricted to any $A_2^t$-graph is the best coloring, or

(2) $c(\{a_x, a_y, a_z\}) = \{1, 2\}$ and only one $A_2^t$-graph is colored with sum greater than 73.

In both cases $\{1, 2\} \subset c(\{a_x, a_y, a_z\})$. Moreover, coloring any vertex from set $\{a_x, a_y, a_z\}$ with 1 and the others with 2 we can extend this pre-coloring to the best coloring of $A_3^a$.

Now we are able to show that there exists a proper solution $M'$ to $3DM_p$ if and only if there exists a coloring $c$ satisfying $\sum(G, c) \leq k$, where $k = 73 \cdot x_2 + 229 \cdot x_3 + 2 \cdot q + (|M| - q)$. Let us notice that the graph defined in (9) is bipartite with $\Delta(G) \leq 5$ and by Definition 3 it is planar.

Now, suppose that $M'$ is a proper solution of $3DM_p$. We define a coloring $c$ as folows: $c(v_m) = 2$ if $m \in M'$ and $c(v_m) = 1$ if $m \in M \backslash M'$. For any $a \in W \cup X \cup Y$ we color the graphs $A_{\#a}^a$ with 3 colors such that $c(a_m) = 1$, whenever $m \in M'$ and $c(a_m) = 2$, if $m \notin M'$. Based on the properties of graphs $A_{\#a}^a$ we can extend the coloring $c$ to the whole graph $G$ so that $\sum(G, c) \leq k$.

Conversely, suppose that $c$ is a coloring of the graph $G$ satisfying $\sum(G, c) \leq k$. Now let $\sum_M := \sum_{m \in M} c(v_m) > |M| + q$. We conclude that

$$\sum_{a \in W \cup X \cup Y} \sum(A_{\#a}^a) = \sum(G, c) - \sum_M < 73 \cdot x_2 + 229 \cdot x_3,$$

which is impossible. Thus suppose that exactly $p < q$ vertices among all $|M|$ vertices $v_m$ are colored with a color different from 1. Hence at most $3 \cdot p$ graphs $A^a_{\#a}$ have neighbors in set $\{v_m : m \in M \wedge c(v_m) \geq 2\}$ and at least $3 \cdot (q - p)$ graphs $A^a_{\#a}$ are colored with $2, 3, \ldots$. This gives

$$\sum (G, c) = \sum_{a \in W \cup X \cup Y} \sum (A^a_{\#a}) + \sum_M \geq$$
$$\geq 73 \cdot x_2 + 229 \cdot x_3 + 3 \cdot (q - p) + |M| + p > k$$

which is impossible. Hence $\sum_M = |M| + q$ and exactly $q$ vertices $v_m$ are colored with 2, we get the desired equality $\sum (G, c) = k$. Thus we have constructed the solution $M' = \{m \in M : c(v_m) = 2\}$ in polynomial time.

Note that simply replacing graphs $A^a_2$ by edges $\{a_x, a_y\}$ and similarly $A^a_3$ by triangles we can prove NP-completeness for planar subcubic graphs [7].

# Facility Location and the Geometric Minimum-Diameter Spanning Tree[*]

Joachim Gudmundsson[1], Herman Haverkort[2],
Sang-Min Park[3], Chan-Su Shin[4], and Alexander Wolff[5]

[1] Dept. of Comp. Science, Utrecht University, The Netherlands.
joachim@cs.uu.nl
[2] Dept. of Comp. Science, Utrecht University, The Netherlands.
herman@cs.uu.nl
[3] Dept. of Comp. Science, KAIST, Korea.
smpark@jupiter.kaist.ac.kr
[4] School of Electr. and Inform. Engineering,
Hankuk University of Foreign Studies, Korea.
cssin@hufs.ac.kr
[5] Institut für Mathematik und Informatik, Universität Greifswald, Germany.
awolff@uni-greifswald.de

**Abstract.** Let $P$ be a set of $n$ points in the plane. The geometric minimum-diameter spanning tree (MDST) of $P$ is a tree that spans $P$ and minimizes the Euclidian length of the longest path. It is known that there is always a mono- or a dipolar MDST, i.e. a MDST with one or two nodes of degree greater 1, respectively. The more difficult dipolar case can so far only be solved in slightly subcubic time.

This paper has two aims. First, we present a solution to a new data structure for facility location, the minimum-sum dipolar spanning tree (MSST), that mediates between the minimum-diameter dipolar spanning tree and the discrete two-center problem (2CP) in the following sense: find two centers $p$ and $q$ in $P$ that minimize the sum of their distance plus the distance of any other point (client) to the closer center. This is of interest if the two centers do not only serve their customers (as in the case of the 2CP), but frequently have to exchange goods or personnel between themselves. We show that this problem can be solved in $O(n^2 \log n)$ time and that it yields a factor-4/3 approximation of the MDST.

Second, we give two fast approximation schemes for the MDST. One uses a grid and takes $O^*(E^{6-1/3} + n)$ time, where $E = 1/\varepsilon$ and the $O^*$-notation hides terms of type $O(\log^{O(1)} E)$. The other uses the well-separated pair decomposition and takes $O(nE^3 + En \log n)$ time. A combination of the two approaches runs in $O^*(E^5 + n)$ time. Both schemes can also be applied to MSST and 2CP.

---

[*] J. Gudmundsson was supported by the Swedish Foundation for International Co-operation in Research and Higher Education, H. Haverkort by the Netherlands' Organization for Scientific Research, C.-S. Shin by grant No. R05-2002-000-00780-0 from the Basic Research Program of the Korea Science and Engineering Foundation (KOSEF), and A. Wolff by the KOSEF program Brain Korea 21.

K. Jansen et al. (Eds.): APPROX 2002, LNCS 2462, pp. 146–160, 2002.
© Springer-Verlag Berlin Heidelberg 2002

# 1    Introduction

The MDST can be seen as a network without cycles that minimizes the maximum travel time between any two sites connected by the network. This is of importance e.g. in communication systems where the maximum delay in delivering a message is to be minimized. Ho et al. showed that there always is a mono- or a dipolar MDST [10]. In the same paper they also gave an $O(n \log n)$-time algorithm for the monopolar and an $O(n^3)$-time algorithm for the dipolar case. The latter time bound was recently improved by Chan [6] to $\tilde{O}(n^{3-c_d})$, where $c_d = 1/((d+1)(\lceil d/2 \rceil + 1))$ is a constant that depends on the dimension $d$ of the point set and the $\tilde{O}$-notation hides factors that are $o(n^\varepsilon)$ for any fixed $\varepsilon > 0$. In the planar case $c_d = 1/6$. Chan speeds up the exhaustive-search algorithm of Ho et al. by using new semi-dynamic data structures.

Note that in the dipolar case the objective is to find the two poles $x, y \in P$ of the tree such that the function $r_x + |xy| + r_y$ is minimized, where $|xy|$ is the Euclidean distance of $x$ and $y$, and $r_x$ and $r_y$ are the radii of two disks centered at $x$ and $y$ whose union covers $P$. On the other hand the *discrete k-center problem* is to determine $k$ points in $P$ such that the union of $k$ congruent disks centered at the $k$ points covers $P$ and the radius of the disks is minimized. This is a typical facility location problem: there are $n$ supermarkets and in $k$ of them a regional director must be placed such that the maximum director-supermarket distance is minimized. This problem is NP-hard provided that $k$ is part of the input [8]. Thus, the main research on this problem has focused on small $k$, especially on $k = 1, 2$. For $k = 1$, the problem can be solved in $O(n \log n)$ time using the farthest-point Voronoi diagram of $P$. For $k = 2$, the problem becomes considerably harder. Using the notation from above, the discrete two-center problem consists of finding two centers $x, y \in P$ such that the function $\max\{r_x, r_y\}$ is minimized. Agarwal et al. [2] gave the first subquadratic-time algorithm for this problem. It runs in $O(n^{4/3} \log^5 n)$ time.

In this paper we are interested in (a) a new facility location problem that mediates between the minimum-diameter dipolar spanning tree (MDdST) and the two-center problem and (b) fast approximations of the computationally expensive MDdST. As for our first aim we observe the following. Whereas the dipolar MDdST minimizes $|xy| + (r_x + r_y)$, the discrete two-center problem is to minimize $\max\{r_x, r_y\}$, which means that the distance between the two centers is not considered at all. If, however, the two centers need to communicate with each other for cooperation, then their distance should be considered as well—not only the radius of the two disks. Therefore our aim is to find two centers $x$ and $y$ that minimize $|xy| + \max\{r_x, r_y\}$, which is a compromise between the two previous objective functions. We will refer to this problem as the *discrete minimum-sum two-center problem* and call the resulting graph the *minimum-sum dipolar spanning tree* (MSST). As it turns out, our algorithm for the MSST also constitutes a compromise, namely in terms of runtime between the subcubic-time MDdST-algorithm and the superlinear-time 2CP-algorithm. More specifically, in Section 2 we will describe an algorithm that solves the discrete minimum-sum two-center problem in the plane in $O(n^2 \log n)$ time using

$O(n^2)$ space. For dimension $d < 5$ a variant of our algorithm is faster than the more general $\tilde{O}(n^{3-c_d})$-time MDST-algorithm of Chan [6] that can easily be modified to compute the MSST instead.

In Section 3 we turn to our second aim, approximations for the MDST. To our knowledge nothing has been published on that field so far.[1] We combine a slight modification of the MSST with the minimum-diameter monopolar spanning tree (MDmST). We identify two parameters that depend on the MDdST and help to express a very tight estimation of how well the two trees approximate it. It turns out that at least one of them is a factor-4/3 approximation of the MDST.

Finally, in Section 4 we show that there are even approximating schemes for the MDdST. More precisely, given a set $P$ of $n$ points and some $\varepsilon > 0$ we show how to compute a dipolar tree whose diameter is at most $(1 + \varepsilon)$ times as long as the diameter of a MDdST. Our first approximation scheme uses a grid of $O(E) \times O(E)$ square cells (where $E = 1/\varepsilon$) and runs Chan's exact algorithm [6] on one representative point per cell. The same idea has been used before [3,?] to approximate the diameter of a point set, i.e. the longest distance between any pair of the given points. Our scheme takes $O^*(E^6 + n)$ time, where the $O^*$-notation hides terms of type $O(\log^{O(1)} E)$.

Our second approximation scheme is based on the well-separated pair decomposition [4] of $P$ and takes $O(E^3 n + En \log n)$ time. Well-separated pair decompositions make it possible to consider only a linear number of pairs of points on the search for the two poles of an approximate MDdST. If we run our second scheme on the $O(E^2)$ representative points in the grid mentioned above, we get a new scheme with a running time of $O^*(E^5 + n)$. Both schemes can also be applied to the MSST and the 2CP.

We will refer to the diameter $d_P$ of the MDST of $P$ as the *tree diameter* of $P$. We assume that $P$ contains at least four points.

## 2    The Minimum-Sum Dipolar Spanning Tree

It is simple to give an $O(n^3)$-time algorithm for computing the MSST. Just go through all $O(n^2)$ pairs $\{p, q\}$ of input points and compute in linear time the point $m_{pq}$ whose distance to the current pair is maximum. In order to give a faster algorithm for computing the MSST, we need a few definitions. Let $h_{pq}$ be the open halfplane that contains $p$ and is delimited by the perpendicular bisector $b_{pq}$ of $p$ and $q$. Note that $h_{pq}$, $h_{qp}$, and $b_{pq}$ partition the plane. Let $\mathcal{T}_{pq}$ be the tree with dipole $\{p, q\}$ where all other points are connected to the closer pole. (Points on $b_{pq}$ can be connected to either.) Clearly the tree $\mathcal{T}_{pq}$ that minimizes $|pq| + \min\{|pm_{pq}|, |qm_{pq}|\}$ is a MSST.

The first important idea of our algorithm is to split the problem of computing all points of type $m_{pq}$ into two halves. Instead of computing the point $m_{pq}$ farthest from the pair $\{p, q\}$, we compute for each ordered pair $(p, q)$ the

---

[1] Very recently we were informed of [11] where the authors give an approximation scheme for the MDST that runs in $O(\varepsilon^{-3} + n)$ time using $O(n)$ space.

point $f_{pq} \in P \setminus h_{qp}$ that is farthest from $p$ ignoring the halfplane of $q$, see Figure 1. We call $f_{pq}$ the $q$-*farthest point from* $p$. Now we want to find the tree $T_{pq}$ that minimizes $|pq| + \max\{|pf_{pq}|, |qf_{qp}|\}$. This strategy enables us to reuse information.

Our algorithm consists of two phases. In phase I we go through all points $p$ in $P$. The central (and time-critical) part of our algorithm is the computation of $f_{pq}$ for all $q \in P \setminus \{p\}$. In phase II we then use the above form of our target function to determine the MSST.

The second important observation that helped us to reduce the running time of the central part of our algorithm is the following. Let $p$ be fixed. Instead of going through all $q \in P \setminus \{p\}$ and computing $f_{pq}$ we characterize all $q$ for which the $q$-farthest point $f_{pq}$ of $p$ is identical:

**Lemma 1.** *If $x \in P$ is the farthest point from $p \in P$, then $x$ is the $q$-farthest point from $p$ if and only if $q \notin D(x,p)$, where $D(a,b)$ (for points $a \neq b$) is the open disk that is centered at $a$ and whose boundary contains $b$.*

*Proof.* Since $x$ is farthest from $p$, $x$ is $q$-farthest from $p$ if and only if $x \in h_{pq}$. This is the case iff the angle $\alpha = \angle pcx$ in the midpoint $c$ of $pq$ is at most 90 degrees, see Figure 2. Due to the Theorem of Thales this is equivalent to $c \notin D(m,p)$, where $m$ is the midpoint of $px$. Finally this is equivalent to $q \notin D(x,p)$, since $d(p,q) = 2d(p,c)$ and $D(x,p)$ is the result of scaling $D(m,p)$ relative to $p$ by a factor of 2. □

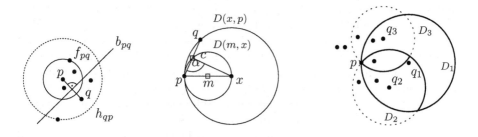

**Fig. 1.**   The   $q$-farthest point $f_{pq}$ from $p$ is farthest from $p$ among all points closer to $p$ than to $q$.

**Fig. 2.** If $x$ is farthest from $p$ then $x$ is $q$-farthest from $p$ iff $q \notin D(x,p)$.

**Fig. 3.** Labeling points with their $q$-farthest point.

Using the above characterization we can label all points $q \in P \setminus \{p\}$ with the $q$-farthest point $f_{pq}$ as follows. We first sort $P$ in order of "decreasing" (i.e. non-increasing) distance from $p$. Let $q_1, q_2, \ldots, q_n = p$ be the resulting order. Label all points in the complement of $D(q_1, p)$ with $q_1$. Then label all *unlabeled* points in the complement of $D(q_2, p)$ with $q_2$. Continue this process until all points are labeled. Figure 3 visualizes the first three steps of this process. In that figure the areas shaded light, medium, and dark correspond to the areas in which all points are labeled with $q_1$, $q_2$, and $q_3$, respectively.

It remains to show how all points $q \in P \setminus \{p\}$ can be labeled with $f_{pq}$ efficiently. One approach would be to use dynamic circular range searching, which is equivalent to halfspace range searching in $\mathbb{R}^3$ [1]. The necessary data structure can be build in $O(n^{1+\varepsilon})$ time and space. After each query with a disk $D(q_i, p)$ all points that are *not* returned must be deleted. The total time for querying and updating is also $O(n^{1+\varepsilon})$. This would yield an $O(n^{2+\varepsilon})$-time algorithm. We will show that we can do better in the plane. However, it is not clear how our results can be generalized to higher dimensions. For dimensions $d \in \{3, 4\}$ computing the MSST with range searching takes $O(n^{2.5+\varepsilon})$ time [1] and thus is still faster than Chan's $\tilde{O}(n^{3-c_d})$-time algorithm [6], where $c_d = 1/((d+1)(\lceil d/2 \rceil + 1))$.

**Lemma 2.** *Given a set $P$ of $n$ points in the plane and given $n$ disks $D_1, \ldots, D_n$ that all touch a point $p$, there is a data structure that allows to determine in $O(\log^2 n)$ time for each point $q \in P$ the smallest integer $i$ such that $q \in D_1 \cap \ldots \cap D_{i-1}$ and $q \notin D_i$ if such an integer exists. The data structure needs $O(n \log n)$ preprocessing time and space.*

*Proof.* To simplify the presentation we assume $n = 2^k$. We build a complete binary tree $\mathcal{B}$ with $k$ levels over $n$ leaves with labels $1, \ldots, n$, see Figure 4. Each inner node $v$ with left child $l$ and right child $r$ is labeled by a set of consecutive integers $\{a(v), \ldots, b(v)\} \subset \{1, \ldots, n\}$ that is recursively defined by $a(v) = a(l)$ and $b(v) = a(r) - 1$. For each leaf $w$ we set $a(w) = b(w) = \text{label}(w)$. Note that the root is labeled $\{1, \ldots, n/2\}$. In Figure 4 $[a, b]$ is shorthand for $\{a, \ldots, b\}$.

A query with a point $q \in P$ consists of following a path from the root to a leaf whose label $i$ is the index of the $q$-furthest point from $p$, in other words $p_i = f_{pq}$. In each inner node $v$ the path of a query with point $q$ is determined by testing whether $q \in D_{a(v)} \cap \ldots \cap D_{b(v)}$. If yes the next node of the query path is the right child, otherwise the left child. (Why such a query in deed gives the desired answer can be proven by induction over $k$.)

A query time of $O(\log^2 n)$ can be achieved by storing in each internal node $v$ a decomposition of $D_{a(v)} \cap \ldots \cap D_{b(v)}$ into at most $b(v) - a(v) + 2$ vertical strips. The strips are bounded by all verticals through the endpoints of the arcs that for the boundary of $D_{a(v)} \cap \ldots \cap D_{b(v)}$. In this decomposition $q$ can then be located in $O(\log n)$ time, and this has to be done $O(\log n)$ times on the way from the root to a leaf.

In order to construct the tree $\mathcal{B}$ we first build a tree $\mathcal{B}'$. The tree $\mathcal{B}'$ is also a binary tree over $\{1, \ldots, n\}$, but in $\mathcal{B}'$ each internal node $v$ is labeled with the set of the labels of all leaves in the subtree rooted at $v$. The tree $\mathcal{B}'$ can be built in a bottom-up fashion since we can construct $D_1 \cap \ldots \cap D_{2m}$ from $D_1 \cap \ldots \cap D_m$ and $D_{m+1} \cap \ldots \cap D_{2m}$ by a merge-sort style procedure in $O(m)$ time. Note that each disk contributes at most one piece to the boundary of the intersection. Hence the construction of $\mathcal{B}'$ takes $O(n \log n)$ time (and space) in total.

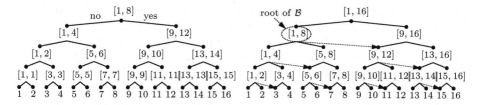

**Fig. 4.** The binary search tree $\mathcal{B}$. A label $[a, b]$ means that the test $q \in D_a \cap D_{a+1} \cap \ldots \cap D_b$ is performed for a query point $q$.

**Fig. 5.** The auxiliary tree $\mathcal{B}'$ for the construction of $\mathcal{B}$ (see dotted arrows).

From $\mathcal{B}'$ we obtain $\mathcal{B}$ in three steps. First we make the left child of the root of $\mathcal{B}'$ the root of $\mathcal{B}$. Second for each node $v$ we change the pointer from the right child of $v$ to the left child of the sister of $v$, see the dotted arrows in Figure 5. Third, we make each leaf node $w$ with label $i$ in $\mathcal{B}'$ an internal node with label $\{i\}$ in $\mathcal{B}$ and add to $w$ a left child with label $i$ and a right child with label $i+1$ as new leaves. The vertical decomposition of each internal node $v$ can then be computed in time linear in the size of the complexity of $D_{a(v)} \cap \ldots \cap D_{b(v)}$. Thus the construction of $\mathcal{B}$ takes $O(n \log n)$ time and space. $\square$

The time complexity for querying the data structure can be reduced from $O(\log^2 n)$ to $O(\log n)$ by applying fractional-cascading techniques [7]. The time we need to reorganize the tree $\mathcal{B}$ to support fractional cascading is proportional to its space consumption, which is $O(n \log n)$ since there are $O(n)$ items on each level.

**Theorem 1.** *There is an algorithm that computes a* MSST *in* $O(n^2 \log n)$ *time using quadratic space.*

*Proof.* With the procedure described before Lemma 2 we can compute, for each $p \in P$, all points $f_{pq}$ that are $q$-farthest from $p$ in $O(n \log n)$ time using the data structure of Lemma 2 and fractional cascading. Thus we can compute all points of type $f_{pq}$ in $O(n^2 \log n)$ time. Since we can only determine the MSST *after* computing all points of type $f_{pq}$ we must store them explicitly, which requires quadratic space. $\square$

## 3   Approximating the Minimum-Diameter Spanning Tree

We first make the trivial observation that the diameter of *any* monopolar tree on $P$ is at most twice as long as the tree diameter $d_P$ of $P$. We use the following notation. Let $\mathcal{T}_{\mathrm{di}}$ be a fixed MDdST and $\mathcal{T}_{\mathrm{mono}}$ a fixed MDmST of $P$. The tree $\mathcal{T}_{\mathrm{di}}$ has minimum diameter among those trees with vertex set $P$ in which all but two nodes—the poles—have degree 1. The tree $\mathcal{T}_{\mathrm{mono}}$ is a minimum-diameter star with vertex set $P$. Let $x$ and $y$ be the poles of $\mathcal{T}_{\mathrm{di}}$, and let $\delta = |xy|$ be their distance. Finally let $r_x$ ($r_y$) be the length of the longest edge in $\mathcal{T}_{\mathrm{di}}$ incident to $x$ ($y$) without taking into account the edge $xy$. We assume $r_x \geq r_y$.

In order to get a good approximation of the MDST, we slightly modify the algorithm for the MSST described in Section 2. After computing the $O(n^2)$ points of type $f_{pq}$, we go through all pairs $\{p,q\}$ and consider the tree $\mathcal{T}_{pq}$ with dipole $\{p,q\}$ in which each point is connected to its closer dipole. In Section 2 we were searching for a tree of type $\mathcal{T}_{pq}$ that minimizes $|pq| + \max\{|f_{pq}p|, |qf_{qp}|\}$. Now we go through all trees $\mathcal{T}_{pq}$ to find the tree $\mathcal{T}_{\text{bisect}}$ with minimum *diameter*, i.e. the tree that minimizes $|pq| + |f_{pq}p| + |qf_{qp}|$. Note that the only edge in $\mathcal{T}_{pq}$ that crosses the perpendicular bisector of $pq$ is the edge $pq$ itself. This is of course not necessarily true for the MDdST $\mathcal{T}_{\text{di}}$. We will show the following:

**Lemma 3.** *Given a set $P$ of $n$ points in the plane there is a tree with the following two properties: it can be computed in $O(n^2 \log n)$ time using $O(n^2)$ storage, and its diameter is at most $4/3 \cdot d_P$.*

*Proof.* Due to Theorem 1 it suffices to show the approximation factor. We will first compute upper bounds for the approximation factors of $\mathcal{T}_{\text{bisect}}$ and $\mathcal{T}_{\text{mono}}$ and then analyze where the minimum of the two takes its maximum.

For the analysis of $\mathcal{T}_{\text{bisect}}$ consider the tree $\mathcal{T}_{xy}$ whose poles are those of $\mathcal{T}_{\text{di}}$. The diameter of $\mathcal{T}_{xy}$ is an upper bound for that of $\mathcal{T}_{\text{bisect}}$. Let $r'_x$ $(r'_y)$ be the length of the longest edge of $\mathcal{T}_{xy}$ incident to $x$ $(y)$ without taking into account the edge $xy$. Note that $r'_x = |xf_{xy}|$ and $r'_y = |yf_{yx}|$.

Now we compare the diameter of $\mathcal{T}_{xy}$ with that of $\mathcal{T}_{\text{di}}$. We have $\max\{r'_x, r'_y\} \leq r_x$. This is due to our assumption $r_x \geq r_y$ and to the fact that $f_{xy}$ and $f_{yx}$ have at most distance $r_x$ from both $x$ and $y$. This observation yields $\text{diam } \mathcal{T}_{xy} = r'_x + \delta + r'_y \leq 2\max\{r'_x, r'_y\} + \delta \leq 2r_x + \delta$. Now we define two constants $\alpha$ and $\beta$ that only depend on $\mathcal{T}_{\text{di}}$. Let $\alpha = \delta/(r_x + r_y)$ and $\beta = r_x/r_y$. Note that $\alpha > 0$ and $\beta \geq 1$. Introducing $\alpha$ and $\beta$ yields

$$\frac{\text{diam } \mathcal{T}_{\text{bisect}}}{\text{diam } \mathcal{T}_{\text{di}}} \leq \frac{\text{diam } \mathcal{T}_{xy}}{\text{diam } \mathcal{T}_{\text{di}}} \leq \frac{2r_x + \delta}{r_x + \delta + r_y} = \frac{\alpha(1+\beta) + 2\beta}{(1+\alpha)(1+\beta)} =: f_{\text{bisect}}(\alpha, \beta),$$

since $2r_x = 2\beta(r_x + r_y)/(1+\beta)$ and $\delta = \alpha(r_x + r_y)$. The function $f_{\text{bisect}}(\alpha, \beta)$ is an upper bound for the approximation factor that $\mathcal{T}_{\text{bisect}}$ achieves.

Now we apply our $\alpha$-$\beta$-analysis to $\mathcal{T}_{\text{mono}}$. The stability lemma $r_x < \delta + r_y$ [10] implies that all points in $P$ are contained in the disk $D_{x,\delta+r_y}$ of radius $\delta + r_y$ centered at $x$. Due to that, the diameter of a monopolar tree $\mathcal{T}$ that spans $P$ and is rooted at $x$ is at most twice the radius of the disk. We know that $\text{diam } \mathcal{T}_{\text{mono}} \leq \text{diam } \mathcal{T}$ since $\mathcal{T}_{\text{mono}}$ is the MDmST of $P$. Thus

$$\text{diam } \mathcal{T}_{\text{mono}} \leq 2(\delta + r_y) = 2\alpha(r_x + r_y) + \frac{2}{1+\beta}(r_x + r_y),$$

since $\delta = \alpha(r_x + r_y)$ and $1 + \beta = (r_x + r_y)/r_y$. Using $\text{diam } \mathcal{T}_{\text{di}} = (1+\alpha)(r_x + r_y)$ yields

$$\frac{\text{diam } \mathcal{T}_{\text{mono}}}{\text{diam } \mathcal{T}_{\text{di}}} \leq \frac{2\alpha(1+\beta) + 2}{(1+\alpha)(1+\beta)} =: f_{\text{mono}}(\alpha, \beta),$$

and the function $f_{\text{mono}}(\alpha, \beta)$ is an upper bound of $\mathcal{T}_{\text{mono}}$'s approximation factor.

In order to compute the maximum of the minimum of the two bounds we first analyze where $f_{\text{bisect}} \leq f_{\text{mono}}$. This is always the case if $\alpha \geq 2$ but also if $\alpha < 2$ and $\beta \leq g_{\text{equal}}(\alpha) := \frac{\alpha+2}{2-\alpha}$. Since neither $f_{\text{bisect}}$ nor $f_{\text{mono}}$ have any local or global maxima in the interior of the $(\alpha, \beta)$-range we are interested in, we must consider their boundary values.

1. For $\beta \equiv 1$ the tree $\mathcal{T}_{\text{bisect}}$ is optimal since $f_{\text{bisect}}(\alpha, 1) \equiv 1$.
2. Note that the stability lemma $r_x \leq \delta + r_y$ is equivalent to $\beta \leq g_{\text{stab}}(\alpha) := \frac{\alpha+1}{1-\alpha}$. Along $g_{\text{stab}}$ the tree $\mathcal{T}_{\text{mono}}$ is optimal since there $f_{\text{mono}} \equiv 1$.
3. Along $g_{\text{equal}}$ both functions equal $(3\alpha+2)/(2\alpha+2)$. This expression increases monotonically from 1 towards 4/3 when $\alpha$ goes from 0 towards 2.

The partial derivatives show that $f_{\text{mono}}$ increases while $f_{\text{bisect}}$ decreases monotonically when $\alpha$ goes to infinity. Thus the maximum of $\min(f_{\text{mono}}, f_{\text{bisect}})$ is indeed attained at $g_{\text{equal}}$.                                                                    □

# 4   Approximation Schemes for the MDST

In this section we give some fast approximation schemes for the MDST, i.e. factor-$(1+\varepsilon)$ approximation algorithms. The first scheme uses a grid, the second and third the well-separated pair decomposition, and the forth is a combination of the first and the third method. The reason for this multitude of approaches is that we want to take into account the way the running time depends not only on $n$, the size of the point set, but also on $\varepsilon$, the approximation factor.

Chan [5] uses the following notation. Let $E = 1/\varepsilon$ and let the $O^*$-notation be a variant of the $O$-notation that hides terms of type $O(\log^{O(1)} E)$. (Such terms come into play e.g. when the use of the floor function is replaced by binary search with precision $\varepsilon$.) Then a *linear-time approximation scheme (LTAS) of order c* is a scheme with a running time of the form $O^*(E^c n)$ for some constant $c$. A *strong LTAS of order c* has a running time of $O^*(E^c + n)$. Our best scheme for approximating the MDST is a strong LTAS of order 5.

## 4.1   A Grid-Based Approximation Scheme

The idea of our first scheme is based on a grid which has been used before e.g. to approximate the diameter of a point set [3,?], i.e. the longest distance between any pair of the given points. We lay a grid of $O(E) \times O(E)$ cells over $P$, choose an arbitrary representative point for each cell and use the exact algorithm of Ho et al. [10] to compute the MDST $\mathcal{T}_R$ of the set $R$ of all representative points. By connecting the remaining points in $P \setminus R$ to the pole adjacent to their representatives, we get a dipolar tree $\mathcal{T}_\varepsilon$ whose diameter is at most $(1+\varepsilon)$ times the tree diameter $d_P$ of $P$.

The details are as follows. Let $M = \max_{p,q \in P}\{|x(p)x(q)|, |y(p)y(q)|\}$ be the edge length of the smallest enclosing square of $P$ and let $l = \varepsilon M/(10\sqrt{2})$ be the edge length of the square grid cells. Clearly $M \leq d_P$. Since each path in $T_\varepsilon$ is at most by two edges of length $l\sqrt{2}$ longer than the corresponding path in

$T_R$ we have diam $T_\varepsilon \leq$ diam $T_R + 2l\sqrt{2} \leq$ diam $T_R + \varepsilon d_P/5$. To see that diam $T_\varepsilon \leq (1 + \varepsilon) d_P$ it remains to prove:

**Lemma 4.** diam $\mathcal{T}_R \leq (1 + 4\varepsilon/5) d_P$.

*Proof.* Let $\mathcal{T}_P$ be a MDST of $P$ that is either mono- or dipolar. Such a tree always exists according to [10].

**Case I:** $\mathcal{T}_P$ is monopolar. Let $x \in P$ be the pole of $\mathcal{T}_P$ and let $\rho_p \in R$ be the representative point of $p \in P$. Due to the definition of $T_R$ we have

$$\text{diam } \mathcal{T}_R \leq \min_{x' \in R} \max_{s \neq t \in R} |sx'| + |x't| \leq \max_{s \neq t \in R} |s\rho_x| + |\rho_x t|.$$

(The first two terms are equal if there is a monopolar MDST of $R$, the last two terms are equal if there is a MDmST of $R$ with pole $\rho_x$.) By triangle inequality

$$\text{diam } \mathcal{T}_R \leq \max_{s \neq t \in R} |sx| + |x\rho_x| + |\rho_x x| + |xt|,$$

i.e. we maximize the length of the polygonal chain $(s, x, \rho_x, x, t)$ over all $s \neq t \in R$. By appending edges to points $a$ and $b \in P$ in the grid cells of $s$ and $t$, respectively, the length of the longest chain does not decrease, even if we now maximize over all $a, b \in P$ with $a \neq b$.

$$\text{diam } \mathcal{T}_R \leq \max_{a \neq b \in P} |a\rho_a| + |\rho_a x| + 2|x\rho_x| + |x\rho_b| + |\rho_b b|.$$

Using $|a\rho_a|, |x\rho_x|, |\rho_b b| \leq l\sqrt{2}$ and the inequalities $|\rho_a x| \leq |\rho_a a| + |ax|$ and $|x\rho_b| \leq |xb| + |b\rho_b|$ yields diam $\mathcal{T}_R \leq 6l\sqrt{2} + \max_{a \neq b \in P} |ax| + |xb| = (1 + 3\varepsilon/5)d_P$.

**Case II:** $\mathcal{T}_P$ is dipolar. The analysis is very similar to case I, except the chains consist of more pieces. This yields diam $\mathcal{T}_R \leq 8l\sqrt{2} + \text{diam } \mathcal{T}_P = (1 + 4\varepsilon/5) d_P$. □

**Theorem 2.** *A spanning tree $\mathcal{T}_P$ of $P$ with diam $\mathcal{T}_P \leq (1 + 1/E) \cdot d_P$ can be computed in $O^*(E^{6-1/3} + n)$ time using $O^*(E^2 + n)$ space.*

*Proof.* In order to determine the grid cell of each point in $P$ without the floor function, we do binary search—once on an $x$- and once on a $y$-interval of size $M$ until we have reached a precision of $l$, i.e. we need $O(\log E)$ steps for each point. Using Chan's algorithm [6] to compute $T_R$ takes $\tilde{O}(|R|^{3-1/6})$ time and $\tilde{O}(|R|)$ space, where $|R| = O(E^2)$. □

## 4.2  The Well-Separated Pair Decomposition

Our second scheme uses the well-separated pair decomposition of Callahan and Kosaraju [4]. We briefly review this decomposition below.

**Definition 1.** *Let $\tau > 0$ be a real number, and let $A$ and $B$ be two finite sets of points in $\mathbb{R}^d$. We say that $A$ and $B$ are well-separated w.r.t. $\tau$, if there are two disjoint d-dimensional balls $C_A$ and $C_B$ both of radius $r$ such that $A \subset C_A$, $B \subset C_B$, and the distance between $C_A$ and $C_B$ is at least equal to $\tau r$.*

The parameter $\tau$ will be referred to as the *separation constant*. The following lemma follows easily from Definition 1.

**Lemma 5.** *Let $A$ and $B$ be two finite sets of points that are well-separated w.r.t. $\tau$, let $x$ and $p$ be points of $A$, and let $y$ and $q$ be points of $B$. Then (i) $|xy| \le (1 + 2/\tau) \cdot |xq|$, (ii) $|xy| \le (1 + 4/\tau) \cdot |pq|$, (iii) $|px| \le (2/\tau) \cdot |pq|$, and (iv) the angle between the line segments $pq$ and $py$ is at most $\arcsin(2/\tau)$.*

**Definition 2.** *Let $P$ be a set of $n$ points in $\mathbb{R}^d$, and $\tau > 0$ a real number. A well-separated pair decomposition (WSPD) for $P$ (w.r.t. $\tau$) is a sequence of pairs of non-empty subsets of $P$, $(A_1, B_1), (A_2, B_2), \ldots, (A_\ell, B_\ell)$, such that*

1. *$A_i$ and $B_i$ are well-separated w.r.t. $\tau$ for $i = 1, 2, \ldots, \ell$, and*
2. *for any two distinct points $p$ and $q$ of $P$, there is exactly one pair $(A_i, B_i)$ in the sequence such that (i) $p \in A_i$ and $q \in B_i$, or (ii) $q \in A_i$ and $p \in B_i$,*

The integer $\ell$ is called the *size* of the WSPD. Callahan and Kosaraju show that a WSPD of size $\ell = O(\tau^2 n)$ can be computed using $O(n \log n + \tau^2 n)$ time and space.

### 4.3   A Straight-Forward Approximation Scheme

The approximation algorithm consists of two subalgorithms: the first algorithm computes a MDmST and the second computes an approximation of the MDdST. We always output the one with smaller diameter. According to [10] there exists a MDST that is either a monopolar or a dipolar tree. The MDmST can be computed in time $O(n \log n)$, hence we will focus on the problem of computing a MDdST. Let $d_{\min}$ be the diameter of a MDdST and let $\mathcal{S}_{pq}$ denote a spanning tree with dipole $\{p, q\}$ whose diameter is minimum among all such trees. For any dipolar spanning tree $\mathcal{T}$ with dipole $\{u, v\}$ let $r_u(\mathcal{T})$ ($r_v(\mathcal{T})$) be the length of the longest edge of $\mathcal{T}$ incident to $u$ ($v$) without taking into account the edge $uv$. When it is clear which tree $\mathcal{T}$ we refer to, we will use $r_u$ and $r_v$.

**Observation 1.** *Let $(A_1, B_1), \ldots, (A_\ell, B_\ell)$ be a WSPD of $P$ w.r.t. $\tau$, and let $p$ and $q$ be any two points in $P$. Then there is a pair $(A_i, B_i)$ such that for every point $u \in A_i$ and every point $v \in B_i$ the inequality $\mathrm{diam}\,\mathcal{S}_{uv} \le (1+8/\tau)\cdot\mathrm{diam}\,\mathcal{S}_{pq}$ holds.*

*Proof.* According to Definition 2 there is a pair $(A_i, B_i)$ in the WSPD such that $p \in A_i$ and $q \in B_i$. If $u$ is any point in $A_i$ and $v$ is any point in $B_i$, then let $\mathcal{T}$ be the tree with poles $u$ and $v$ where $u$ is connected to $v$, $p$ and each neighbor of $p$ in $\mathcal{S}_{pq}$ except $q$ is connected to $u$, and $q$ and each neighbor of $q$ in $\mathcal{S}_{pq}$ except

$p$ is connected to $v$. By Lemma 5(ii) $|uv| \leq (1+4/\tau)|pq|$ and by Lemma 5(iii) $r_u \leq |up| + r_p \leq 2|pq|/\tau + r_p$. Since diam $\mathcal{T} = r_u + |uv| + r_v$ we have

$$\text{diam } \mathcal{T} \leq \left(r_p + 2\frac{|pq|}{\tau}\right) + \left(|pq| + 4\frac{|pq|}{\tau}\right) + \left(r_q + 2\frac{|pq|}{\tau}\right) < \left(1 + \frac{8}{\tau}\right) \cdot \text{diam } \mathcal{S}_{pq}.$$

The lemma follows due to the minimality of $\mathcal{S}_{uv}$.    □

A first algorithm is now obvious. For each of the $O(\tau^2 n)$ pairs $(A_i, B_i)$ in a WSPD of $P$ w.r.t. $\tau = 8E$ pick *any* point $p \in A_i$ and *any* point $q \in B_i$, sort $P$ according to distance from $p$, and compute $\mathcal{S}_{pq}$ in linear time by checking every possible radius of a disk centered at $p$ as in [10].

**Lemma 6.** *A dipolar tree $\mathcal{T}$ with diam $\mathcal{T} \leq (1+1/E) \cdot d_{\min}$ can be computed in $O(E^2 n^2 \log n)$ time using $O(E^2 n + n \log n)$ space.*

### 4.4    A Fast Approximation Scheme

Now we describe a more involved algorithm. It is faster than the previous algorithm for $n = \Omega(E)$. The correctness proof is in the full version [9, Section 4.5].

**Theorem 3.** *A dipolar tree $\mathcal{T}$ with diam $\mathcal{T} \leq (1+1/E) \cdot d_{\min}$ can be computed in $O(E^3 n + En \log n)$ time using $O(E^2 n + n \log n)$ space.*

The idea of the algorithm is again to check a linear number of pairs of points, using the WSPD, but to speed up the computation of the disks around the two poles. Note that we need to find a close approximation of the diameters of the disks to be able to guarantee a $(1+\varepsilon)$-approximation of the MDdST. Obviously we cannot afford to try all possible disks for all possible pairs of poles. Instead of checking the disks we will show in the analysis that it suffices to check a constant number of partitions of the points among the poles. The partition of points is done by cuts that are orthogonal to the line through the poles. We cannot afford to do this for each possible pair. Instead we select a constant number of orientations and use a constant number of orthogonal cuts for each orientation. For each cut we calculate for each point in $P$ the approximate distance to the farthest point on each side of the cut. Below we give a more detailed description of the algorithm. For its pseudocode refer to Algorithm 1.

**Phase 1: Initializing.** Choose an auxiliary positive constant $\kappa < \min\{0.9\varepsilon, 1/2\}$. As will be clear later, this parameter can be used to fine-tune which part of the algorithm contributes how much to the uncertainty and to the running time. In phase 3 the choice of the separation constant $\tau$ will depend on the value of $\kappa$ and $\varepsilon$.

**Definition 3.** *A set of points $P$ is said to be $l$-ordered if the points are ordered with respect to their orthogonal projection onto the line $l$.*

Let $l_i$ be the line with angle $\frac{i\pi}{\gamma}$ to the horizontal line, where $\gamma = \lceil 4/\kappa \rceil$. This implies that for an arbitrary line $l$ there exists a line $l_i$ such that $\angle l_i l \leq \frac{\pi}{2\gamma}$.

**Algorithm 1.** Approx-MDdST$(P, \varepsilon)$

**Ensure:** diam $\mathcal{T} \le (1 + \varepsilon) \, d_{\min}$
**Phase 1:** initializing
  1: choose $\kappa \in (0, \min\{0.9\varepsilon, 1/2\})$; set $\gamma \leftarrow \lceil 4/\kappa \rceil$
  2: **for** $i \leftarrow 1$ **to** $\gamma$ **do**
  3:    $l_i \leftarrow$ line with angle $i\frac{\pi}{\gamma}$ to the horizontal
  4:    $F_i \leftarrow l_i$-ordering of $P$
  5: **end for** $i$
  6: **for** $i \leftarrow$ **to** $\gamma$ **do**
  7:    rotate $P$ and $l_i$ such that $l_i$ is horizontal
  8:    let $p_1, \ldots, p_n$ be the points in $F_i$ from left to right
  9:    $d_i \leftarrow |p_1.x - p_n.x|$
 10:    **for** $j \leftarrow 1$ **to** $\gamma$ **do**
 11:       $b_{ij} \leftarrow$ point on $l_i$ at dist. $j\frac{d_i}{\gamma+1}$ to the right of $p_1$
 12:       **for** $k \leftarrow 1$ **to** $\gamma$ **do**
 13:          $L'_{ijk} \leftarrow l_k$-ordered subset of $F_k$ to the left of $b_{ij}$
 14:          $R'_{ijk} \leftarrow l_k$-ordered subset of $F_k$ to the right of $b_{ij}$
 15:       **end for** $k$
 16:    **end for** $j$
 17: **end for** $i$
**Phase 2:** computing approximate farthest neighbors
 18: **for** $i \leftarrow 1$ **to** $\gamma$ **do**
 19:    **for** $j \leftarrow 1$ **to** $\gamma$ **do**
 20:       **for** $k \leftarrow 1$ **to** $n$ **do**
 21:          $N(p_k, i, j, L) \leftarrow p_k$ {dummy}
 22:          **for** $l \leftarrow 1$ **to** $\gamma$ **do**
 23:             $p_{\min} \leftarrow$ first point in $L'_{ijl}$
 24:             $p_{\max} \leftarrow$ last point in $L'_{ijl}$
 25:             $N(p_k, i, j, L) \leftarrow$ the point in $\{p_{\min}, p_{\max}, N(p_k, i, j, L)\}$ furthest from $p_k$
 26:          **end for** $l$
 27:       **end for** $k$
 28:       repeat lines 20–27 with $R$ instead of $L$
 29:    **end for** $j$
 30: **end for** $i$
**Phase 3:** testing pole candidates
 31: $\tau = 8(\frac{1+\varepsilon}{(1+\varepsilon-(1+\kappa)(1+\kappa/24)} - 1)$
 32: build WSPD for $P$ with separation constant $\tau$
 33: $d \leftarrow \infty$ {smallest diameter so far}
 34: **for** each pair $(A, B)$ in WSPD **do**
 35:    choose any two points $u \in A$ and $v \in B$
 36:    find $l_i$ with the smallest angle to the line through $u$ and $v$
 37:    $D \leftarrow \infty$ {approximate diameter of tree with poles $u$ and $v$, ignoring $|uv|$}
 38:    **for** $j \leftarrow 1$ **to** $\gamma$ **do**
 39:       $D \leftarrow \min\{D, |N(u, i, j, L)u| + |vN(v, i, j, R)|, |N(u, i, j, R)u| + |vN(v, i, j, L)|\}$
 40:    **end for** $j$
 41:    **if** $D + |uv| < d$ **then** $u' \leftarrow u$; $v' \leftarrow v$; $d \leftarrow D + |uv|$ **end if**
 42: **end for** $(A, B)$
 43: compute $\mathcal{T} \leftarrow \mathcal{S}_{u'v'}$
 44: **return** $\mathcal{T}$

For each $i$, $1 \leq i \leq \gamma$, sort the input points with respect to the $l_i$-ordering. We obtain $\gamma$ sorted lists $F = \{F_1, \ldots, F_\gamma\}$. Each point $p$ in $F_i$ has a pointer to itself in $F_{(i \bmod \gamma)+1}$. The time to construct these lists is $O(\gamma n \log n)$.

For each $l_i$, rotate $P$ and $l_i$ such that $l_i$ is horizontal and consider the orthogonal projection of the points in $P$ onto $l_i$. For simplicity we denote the points in $P$ from left to right on $l_i$ by $p_1, \ldots, p_n$. Let $d_i$ denote the horizontal distance between $p_1$ and $p_n$. Let $b_{ij}$, $1 \leq j \leq \gamma$, be the point on $l_i$ at distance $\frac{j d_i}{\gamma+1}$ to the right of $p_1$. Let $L_{ij}$ and $R_{ij}$ be the set of points to the left and to the right of $b_{ij}$ respectively.

For each point $b_{ij}$ on $l_i$ we construct $\gamma$ pairs of lists, denoted $L'_{ijk}$ and $R'_{ijk}$, where $1 \leq k \leq \gamma$. A list $L'_{ijk}$ ($R'_{ijk}$) contains the set of points in $L_{ij}$ ($R_{ij}$) sorted according to the $l_k$-ordering. Such a list can be constructed in linear time since the ordering is given by the list $F_k$. (Actually it is not necessary to store the lists $L'_{ijk}$ and $R'_{ijk}$: we only need to store the first and the last point in each list.) Hence the total time complexity needed to construct the lists is $O(\gamma^3 n + \gamma n \log n)$, see lines 1–17 in Algorithm 1. These lists will help us to compute an approximate farthest neighbor in $L_{ij}$ and $R_{ij}$ for each point $p \in P$ in time $O(\gamma)$, as we describe below.

**Phase 2: Computing approximate farthest neighbors.** Compute, for each point $p$, an approximate farthest neighbor in $L_{ij}$ and an approximate farthest neighbor in $R_{ij}$, denoted $N(p, i, j, L)$ and $N(p, i, j, R)$ respectively. This can be done in time $O(\gamma)$ by using the lists $L'_{ijk}$ and $R'_{ijk}$: just compute the distance between $p$ and the first respectively the last point in each list. There are $\gamma$ lists for each pair $(i, j)$ and given that at most two entries in each list have to be checked, an approximate farthest neighbor can be computed in time $O(\gamma)$. Hence the total time complexity of this phase is $O(\gamma^3 n)$, as there are $O(\gamma^2 n)$ triples of type $(p, i, j)$. The error we make by using approximate farthest neighbors is small:

**Observation 2.** *If $p$ is any point in $P$, $p_L$ the point in $L_{ij}$ farthest from $p$ and $p_R$ the point in $R_{ij}$ farthest from $p$, then    (a)  $|pp_L| \leq (1+\kappa/24) \cdot |pN(p, i, j, L)|$ and    (b)  $|pp_R| \leq (1+\kappa/24) \cdot |pN(p, i, j, R)|$.*

*Proof.* Due to symmetry it suffices to check (a). If the algorithm did not select $p_L$ as farthest neighbor it holds that for each of the $l_i$-orderings there is a point further from $p$ than $p_L$. Hence $p_L$ must lie within a symmetric $2\gamma$-gon whose edges are at distance $|pN(p, i, j, L)|$ from $p$. This implies that $|pN(p, i, j, L)| \geq |pp_L| \cos(\pi/(2\gamma)) \geq |pp_L|/(1+\kappa/24)$, using some basic calculus and $\kappa \leq 1/2$.  □

**Phase 3: Testing pole candidates.** Compute the WSPD for $P$ with separation constant $\tau$. To be able to guarantee a $(1+\varepsilon)$-approximation algorithm the value of $\tau$ will depend on $\varepsilon$ and $\kappa$ as follows:

$$\tau = 8 \left( \frac{1+\varepsilon}{1+\varepsilon - (1+\kappa)(1+\kappa/24)} - 1 \right).$$

Note that the above formula implies that there is a trade-off between the values $\tau$ and $\kappa$, which can be used to fine-tune which part of the algorithm contributes how much to the uncertainty and to the running time. Setting for instance $\kappa$ to $0.9\varepsilon$ yields for $\varepsilon$ small $16/\varepsilon + 15 < \tau/8 < 32/\varepsilon + 31$, i.e. $\tau = \Theta(1/\varepsilon)$. For each pair $(A, B)$ in the decomposition we select two arbitrary points $u \in A$ and $v \in B$. Let $l_{(u,v)}$ be the line through $u$ and $v$. Find the line $l_i$ that minimizes the angle between $l_i$ and $l_{(u,v)}$. That is, the line $l_i$ is a close approximation of the direction of the line through $u$ and $v$. From above we have that $l_i$ is divided into $\gamma + 1$ intervals of length $d_i/(\gamma + 1)$. For each $j$, $1 \le j \le \gamma$, compute $\min(|N(u, i, j, L)u| + |vN(v, i, j, R)|, |N(u, i, j, R)u| + |vN(v, i, j, L)|)$. The smallest of these $O(\gamma)$ values is saved, and is a close approximation of diam $\mathcal{S}_{uv} - |uv|$, see [9, Section 4.5].

The number of pairs in the WSPD is $O(\tau^2 n)$, which implies that the total running time of the central loop of this phase (lines 41–51 in Algorithm 1) is $O(\gamma \cdot \tau^2 n)$. Building the WSPD and computing $\mathcal{S}_{u'v'}$ takes an extra $O(\tau^2 n + n \log n)$ time. Thus the whole algorithm runs in $O(\gamma^3 n + \gamma \tau^2 n + \gamma n \log n)$ time and uses $O(n \log n + \gamma^2 n + \tau^2 n)$ space. Setting $\kappa = 0.9\varepsilon$ yields $\gamma = O(E)$ and $\tau = O(E)$ and thus the time and space complexities we claimed.

## 4.5 Putting Things Together

Combining grid- and WSPD-based approach yields a strong LTAS of order 5:

**Theorem 4.** *A spanning tree $\mathcal{T}$ of $P$ with diam $\mathcal{T} \le (1 + 1/E) d_P$ can be computed in $O^*(E^5 + n)$ time using $O(E^4 + n)$ space.*

*Proof.* Applying Algorithm 1 to the set $R \subseteq P$ of the $O(E^2)$ representative points takes $O(E^3|R| + E|R| \log |R|)$ time using $O(E^2|R| + |R| \log |R|)$ space according to Theorem 3. Connecting the points in $P \setminus R$ to the poles adjacent to their representative points yields a $(1 + \varepsilon)$-approximation of the MDdST of $P$ within the claimed time and space bounds as in Section 4.1. The difference is that now the grid cells must be slightly smaller in order to compensate for the fact that we now approximate the MDdST of $R$ rather than compute it exactly. A $(1 + \varepsilon)$-approximation of the MDmST of $P$ can be computed via the grid and an exact algorithm of Ho et al. [10] in $O^*(E^2 + n)$ time using $O(E^2 + n)$ space. The tree with smaller diameter is a $(1+\varepsilon)$-approximation of the MDST of $P$.  □

# Conclusions

On the one hand we have presented a new planar facility location problem, the discrete minimum-sum two-center problem that mediates between the discrete two-center problem and the minimum-diameter dipolar spanning tree. We have shown that there is an algorithm that computes the corresponding MSST in $O(n^2 \log n)$ time and that a variant of this tree is a factor-4/3 approximation of the MDST. Is there a near quadratic-time algorithm for the MSST that uses $o(n^2)$ space?

On the other hand we have given a number of fast approximation schemes for the MDST. The fastest is a combination of a grid-based approach with an algorithm that uses the well-separated pair decomposition. It computes in $O^*(1/\varepsilon^5 + n)$ time a tree whose diameter is at most $(1 + \varepsilon)$ times that of a MDST. Such a scheme is called a strong linear-time approximation scheme of order 5. Are there schemes of lower order? Are there exact algorithms that are faster than Chan's [6]?

Our scheme also works for higher-dimensional point sets, but the running time increases exponentially with the dimension. Linear-time approximation schemes for the discrete two-center problem and the MSST can be constructed similarly.

## Acknowledgments

We thank an anonymous referee of an earlier version of this paper for suggesting Theorem 2. We also thank Pankaj K. Agarwal for pointing us to [6] and Timothy Chan for sending us an updated version of [6].

## References

1. P. K. Agarwal and J. Matoušek. Dynamic half-space range reporting and its applications. *Algorithmica*, 13:325–345, 1995.
2. P. K. Agarwal, M. Sharir, and E. Welzl. The discrete 2-center problem. *Discrete & Computational Geometry*, 20, 1998.
3. G. Barequet and S. Har-Peled. Efficiently approximating the minimum-volume bounding box of a point set in three dimensions. In *Proc. 10th Annual ACM-SIAM Symp. on Discr. Algorithms (SODA'99)*, pages 82–91, Baltimore, 1999.
4. P. B. Callahan and S. R. Kosaraju. A decomposition of multidimensional point sets with applications to $k$-nearest-neighbors and $n$-body potential fields. *Journal of the ACM*, 42(1):67–90, Jan. 1995.
5. T. M. Chan. Approximating the diameter, width, smallest enclosing cylinder, and minimum-width annulus. In *Proc. 16th Annual Symposium on Computational Geometry (SoCG'00)*, pages 300–309, New York, 12–14 June 2000. ACM Press.
6. T. M. Chan. Semi-online maintenance of geometric optima and measures. In *Proc. 13th Symp. on Discr. Algorithms (SODA'02)*, pages 474–483, 2002.
7. B. Chazelle and L. J. Guibas. Fractional cascading: I. A data structuring technique. *Algorithmica*, 1(3):133–162, 1986.
8. M. R. Garey and D. S. Johnson. *Computers and Intractability: A Guide to the Theory of NP-Completeness*. W. H. Freeman, New York, NY, 1979.
9. J. Gudmundsson, H. Haverkort, S.-M. Park, C.-S. Shin, and A. Wolff. Approximating the geometric minimum-diameter spanning tree. Technical Report 4/2002, Institut für Mathematik und Informatik, Universität Greifswald, Mar. 2002. See www.uni-greifswald.de/~wwwmathe/preprints/shadow/wolff02_4.rdf.html.
10. J.-M. Ho, D. T. Lee, C.-H. Chang, and C. K. Wong. Minimum diameter spanning trees and related problems. *SIAM Journal on Computing*, 20(5):987–997, 1991.
11. M. J. Spriggs, J. M. Keil, S. Bespamyatnikh, M. Segal, and J. Snoeyink. Computing a $(1 + \epsilon)$-approximate geometric minimum-diameter spanning tree. Private communication, 2002.

# Improved Approximation Algorithms for the Partial Vertex Cover Problem

Eran Halperin[1,*] and Aravind Srinivasan[2,**]

[1] CS Division, Soda Hall, University of California Berkeley, CA 94720-1776
`eran@eecs.berkeley.edu`
[2] Department of Computer Science and Institute for Advanced Computer Studies,
University of Maryland, College Park, MD 20742
`srin@cs.umd.edu`

**Abstract.** The partial vertex cover problem is a generalization of the vertex cover problem: given an undirected graph $G = (V, E)$ and an integer $k$, we wish to choose a minimum number of vertices such that at least $k$ edges are covered. Just as for vertex cover, 2-approximation algorithms are known for this problem, and it is of interest to see if we can do better than this. The current-best approximation ratio for partial vertex cover, when parameterized by the maximum degree $d$ of $G$, is $(2 - \Theta(1/d))$. We improve on this by presenting a $(2 - \Theta(\frac{\ln \ln d}{\ln d}))$-approximation algorithm for partial vertex cover using semidefinite programming, matching the current-best bound for vertex cover. Our algorithm uses a new rounding technique, which involves a delicate probabilistic analysis.
**Key words and phrases:** Partial vertex cover, approximation algorithms, semidefinite programming, randomized rounding

## 1 Introduction

The vertex cover problem is basic in combinatorial optimization, and is as follows. Given an undirected graph $G = (V, E)$, we wish to choose a minimum number of vertices such that every edge is *covered* (has at least one end-point chosen). In recent years, *partial covering* versions of such covering problems have received quite a bit of attention. For instance, a partial vertex cover problem gets as input a graph $G$ and an integer $k$, and we aim to choose a minimum number of vertices such that at least $k$ edges are covered. Just as for the vertex cover problem, 2-approximation algorithms are known for this problem [2,7,1], and it is an interesting question if we can do better than a factor of 2. The current-best approximation ratio for partial vertex cover, when parameterized by the maximum degree $d$ of $G$, is $(2 - \Theta(1/d))$ [4,10]; such a bound is unavoidable if we use the (natural) linear programming (LP) relaxation considered in [4,10]. In this paper, we develop a $(2 - \Theta(\frac{\ln \ln d}{\ln d}))$-approximation algorithm for partial vertex

* Supported in part by NSF grants CCR-9820951 and CCR-0121555 and DARPA cooperative agreement F30602-00-2-0601.
** Supported in part by NSF Award CCR-0208005.

cover via semidefinite programming (SDP), matching the current-best bound for the standard "full coverage" version of vertex cover (wherein $k$ equals $|E|$) [5].

While the work of [5] on vertex cover also uses SDP, our work differs from it in some crucial ways. At a high level, just as in [5], the SDP outputs a unit vector $v_0$ and, for each vertex $i$, a unit vector $v_i$; ideally, each $v_i$ (for $i \in V$) would be of the form $\pm v_0$, naturally indicating whether to place $i$ in the vertex cover or not, but the SDP can of course output an arbitrary collection of unit vectors $v_i$. Nevertheless, as pointed out in [5], the condition that every edge must be covered in the vertex cover problem leads to some good valid constraints. For instance, the troublesome case for the vertex cover problem is when, for some edge $(i, j)$, the vectors $v_i$ and $v_j$ are roughly orthogonal to $v_0$. However, since every edge must be covered in vertex cover, this forces $v_i$ to be approximately $-v_j$, a fact which can then be exploited [5]. However, in our partial cover setting, an edge may only be partially covered; this leads to at least two different troublesome cases, which need to be handled simultaneously. Also, as opposed to current algorithms based on SDP, our rounding scheme does a "semidefinite rounding" and also interprets some of the variables as components of an LP and does an "LP-style" rounding on them. Finally, our algorithm succeeds with a small probability such as $\Theta(1/\log |V|)$ and hence, the analysis needs to be somewhat delicate; in particular, we need certain types of random variables to be very well concentrated around their means, for which we use recently-developed ideas from [10]. (The success probability of $\Omega(1/\log |V|)$ can then be boosted by repetition.)

The rest of this paper is organized as follows. We present the known LP relaxation and our SDP relaxation in Section 2. Some useful tools are then described in Section 3. Our rounding algorithm and its analysis are presented in Sections 4 and 5 respectively. Section 6 then shows some extensions of our main result, and concluding remarks are made in Section 7.

## 2 Relaxations of the Partial Vertex Cover Problem

Let $G = (V, E)$ be a graph with $|E| = m$ edges and $|V| = n$ vertices. Let $S \subseteq V$ be a subset of the vertices of $G$. We say that $S$ covers an edge $(u, v) \in E$ if $u \in S$ or $v \in S$. We say that $S$ is a $k$-cover of $G$ if $S$ covers at least $k$ edges. In the partial vertex cover problem, we are given a graph $G = (V, E)$ with $|E| = m$ and $|V| = n$, and a number $k \leq m$, and are interested in finding a minimum size $k$-cover of $G$.

Let $V = \{1, \ldots, n\}$, and consider the following LP relaxation of the partial vertex cover problem.

$$\text{Minimize} \sum_{i=1}^{n} x_i$$

$$
\begin{aligned}
0 \leq x_i \leq 1 &\quad,\quad i \in V \\
0 \leq z_{ij} \leq 1 &\quad,\quad (i,j) \in E \\
z_{ij} \leq x_i + x_j &\quad,\quad (i,j) \in E \\
\sum_{(i,j) \in E} z_{ij} \geq k
\end{aligned}
\tag{1}
$$

Relaxation (1) was used in [4] and in [10] to show a 2-approximation to the problem. In fact, these two papers show that using this LP relaxation, one can get a $2 - \Theta(1/d)$-approximation, where $d$ is the maximal degree in $G$. In [10], the rounding procedure takes linear time. The integrality ratio of relaxation (1) is $2 - \frac{1}{n}$: it is attained, e.g., for the complete graph with $n$ vertices, and by setting $k = |E|$. In the complete graph, the maximal degree is $d = n - 1$; therefore, in order to get an approximation better than $(2 - \Theta(1/d))$ for the problem, we need a different relaxation.

In [5] it is shown that using semidefinite programming one can achieve a $2 - \frac{(2-o(1)) \ln \ln d}{\ln d}$-approximation algorithm for the vertex cover problem; here, the "$o(1)$" term is a function of only $d$, and tends to 0 as $d$ increases. We extend these ideas by using the following semidefinite relaxation for the partial vertex cover problem. In the following, $\|v\|$ denotes the 2-norm of a vector $v$.

$$
\begin{aligned}
\text{Minimize} \sum_{i=1}^{n} \frac{1 + v_0 \cdot v_i}{2} \\
\sum_{(i,j) \in E} \frac{3 + v_0 \cdot v_j + v_0 \cdot v_i - v_i \cdot v_j}{4} \geq k \\
\frac{3 + v_0 \cdot v_j + v_0 \cdot v_i - v_i \cdot v_j}{4} \leq 1 &\quad,\quad (i,j) \in E \\
v_0 \cdot v_i + v_0 \cdot v_j + v_i \cdot v_j \geq -1 &\quad,\quad (i,j) \in E \\
v_i \in \mathbb{R}^n, \|v_i\| = 1 &\quad,\quad i \in (\{0\} \cup V)
\end{aligned}
\tag{2}
$$

We first show that this is indeed a relaxation. Let $S \subseteq V$ be a $k$-cover of $G$. Assign for each vertex $i \in V$ a unit vector $v_i$. If $i \in S$ we set $v_i = v_0$, and otherwise we set $v_i = -v_0$. It is easy to verify that in this case, for every covered edge $(i,j) \in E$ we have that $\frac{3 + v_0 \cdot v_j + v_0 \cdot v_i - v_i \cdot v_j}{4} = 1$, and since $S$ is a $k$-cover, all the constraints are satisfied. This relaxation can be solved (to within an exponentially small additive error) in polynomial time.

Note that relaxation (2) is an extension of relaxation (1), as one can assign

$$
x_i = \frac{1 + v_0 \cdot v_i}{2} \quad \text{and} \quad z_{ij} = \frac{3 + v_0 \cdot v_j + v_0 \cdot v_i - v_i \cdot v_j}{4}.
\tag{3}
$$

Thus, every solution of (2) corresponds to a solution of (1), while the other direction is not necessarily true. This implies that the semidefinite constraints are tighter then the linear constraints, and thus, there is hope to gain more by using the semidefinite programming relaxation.

# 3    Some Useful Ingredients

Before introducing the algorithm, we state some useful lemmas and notations. We will use some basic $n$-dimensional geometry, and some probability distributions presented in [10].

## 3.1    Properties of $n$-Dimensional Geometry

Recall that $\phi(x) = \frac{1}{\sqrt{2\pi}} e^{\frac{-x^2}{2}}$ is the density function of a standard normal random variable. Let $N(x)$ be the tail of the standard normal distribution, i.e., $N(x) = \int_x^\infty \phi(y)dy$. We will use the following bounds on $N(x)$ (see Lemma VII.2 of Feller [3]):

**Lemma 1.** *For every* $c \geq 1$, $\phi(c)(\frac{1}{c} - \frac{1}{c^3}) < N(c) < \frac{\phi(c)}{c}$.

A random vector $r = (r_1, \ldots, r_n)$ is said to have the $n$-*dimensional standard normal distribution* if the components $r_i$ are independent random variables, each one having the standard normal distribution. In this paper a random $n$-dimensional vector will always denote a vector chosen from the $n$-dimensional standard normal distribution. We will use the following (see Theorem IV.16.3 in [9]):

**Lemma 2.** *Let* $u$ *be any unit vector in* $\mathbb{R}^n$. *Let* $r$ *be a random* $n$-*dimensional vector. The projection of* $r$ *on* $u$, *given by the product* $r \cdot u$, *is distributed according to the standard normal distribution.*

The following lemma is proved in [8] (see also [6]). For completeness, we will give the proof here.

**Lemma 3.** *Given two unit vectors* $u, v \in \mathbb{R}^n$, *and a random* $n$-*dimensional vector* $r$,

$$\Pr(u \geq r, v \geq r) \leq N\left(\frac{2r}{\|u + v\|}\right).$$

*Proof.* By Lemma 2,

$$\Pr(u \geq r, v \geq r) \leq \Pr((u + v) \geq 2r) = \Pr\left(\frac{u + v}{\|u + v\|} \geq \frac{2}{\|u + v\|}r\right) = N\left(\frac{2r}{\|u + v\|}\right).$$

## 3.2    Certain Distributions on Level-Sets

One of the ingredients of our algorithm is the distribution on level-sets presented in [10]. Suppose we are given a sequence of $t$ numbers $P = (p_1, p_2, \ldots, p_t)$, such that $0 \leq p_i \leq 1$, and $p_1 + p_2 + \ldots p_t = \ell$ for some integer $\ell$. Then, the distribution $\mathcal{D}(t, P)$ is defined to be any distribution on $\{0, 1\}^t$, such that if $(X_1, \ldots, X_t) \in \{0, 1\}^t$ denotes a vector sampled from $\mathcal{D}(t, P)$, then the following properties hold:

**(A1)** $\Pr(X_i = 1) = p_i$ for every $i \le t$.

**(A2)** $\Pr(\sum_{i=1}^{t} X_i = \ell) = 1$.

**(A3)** For all $S \subseteq \{1, \ldots, t\}$,

$$\Pr\left(\bigwedge_{i \in S}(X_i = 0)\right) \le \prod_{i \in S}\Pr(X_i = 0); \quad \Pr\left(\bigwedge_{i \in S}(X_i = 1)\right) \le \prod_{i \in S}\Pr(X_i = 1).$$

It is shown in [10] that $\mathcal{D}(t, P)$ exists for any $P$ whose entries add up to an integer, and that we can generate a random sample from $\mathcal{D}(t, P)$ in $O(t)$ time.

In our applications, the condition that $p_1 + p_2 + \ldots p_t$ is an integer may not always hold. To take care of this, we define a distribution $\mathcal{D}'(t, P)$ as follows. Let $\psi \in [0, 1)$ be the smallest non-negative value such that $p_1 + p_2 + \ldots p_t + \psi$ is an integer; define $P' = (p_1, p_2, \ldots, p_t, \psi)$. Then, the distribution $\mathcal{D}'(t, P)$ is defined to be the projection of $\mathcal{D}(t + 1, P')$ on to its first $t$ components: i.e., to sample from $\mathcal{D}'(t, P)$, we sample from $\mathcal{D}(t + 1, P')$ and discard the last component of the sample. It is easy to check that (A1) and (A3) still hold for the sample from $\mathcal{D}'(t, P)$, and that instead of (A2), we get

**(A2')** $\Pr((\sum_{i=1}^{t} X_i) \in \{\lfloor \sum_{i=1}^{t} p_i \rfloor, \lceil \sum_{i=1}^{t} p_i \rceil\}) = 1$.

## 4   A $(2 - \Theta(\frac{\ln \ln d}{\ln d}))$-Approximation Algorithm

We first give an outline of the algorithms given in [4,10]. Their algorithm is based on LP relaxation (1). We first solve the LP, and then, for each vertex $i$ such that $x_i \ge \frac{1}{2-\epsilon}$, we round $x_i$ to 1; $\epsilon$ is a term of the form $\Theta(1/d)$. For each vertex such that $x_i \le \frac{1}{2-\epsilon}$, we round it to 1 with probability $x_i(2 - \epsilon)$. The latter is done independently for each vertex in [4] and using a certain distribution on level-sets in [10]. If at least $k$ edges are covered then we are done. Otherwise, we add vertices arbitrarily, each vertex covering at least one new edge, until the number of covered edges is at least $k$. The analysis needs to show that not many vertices are added in the last step. It turns out that the worst case for such an analysis is when $x_i = x_j = \frac{1}{2}$ for all $(i, j) \in E$. In this case, the LP "pays" one unit for covering the edge $(i, j)$ since $x_i + x_j = 1$, but there is still a small probability that the algorithm will not cover this edge.

Supplied with the ingredients described in the last section, we are now ready to describe our algorithm. The flow of the algorithm is given in Figure 1; please do not confuse the parameter $x$ (which will be $\Theta(\frac{\ln \ln d}{\ln d})$) with the variables $x_i$ defined in (3). Also, the distributions $\mathcal{D}'(\cdot, \cdot)$ are those defined in Section 3.2. The algorithm starts by solving the semidefinite relaxation. It then partitions the set of vertices into three sets: $S_1, S_2$, and $S_3$. $S_1$ is the set of vectors that are close to $v_0$, $S_3$ is the set of vectors that are close to $-v_0$, and $S_2$ is the set of vectors that are almost orthogonal to $v_0$. We then take care of each set separately.

We now give some intuition. We first add to the cover all the vertices in $S_1$. Clearly, since all vertices of $S_1$ contribute at least $\frac{1+x}{2}$ to the SDP, we get an approximation ratio of $2/(1 + x) \approx (2 - 2x)$ on these. For $S_2$, consider the following two extreme cases. The first case is an edge $(i, j)$ in $S_2 \times S_2$, such that $z_{ij} = 1$. In this case, since $v_i$ and $v_j$ are almost orthogonal to $v_0$, it can be shown that $v_i \cdot v_j \approx -1$. Therefore, in step 4, it is not likely that both $v_i$ and $v_j$ will be removed from the cover, and therefore, the edge will be covered with good probability; thus, we can handle the case that was difficult for the LP relaxation. The other extreme case is when $v_i = v_j$, and then $z_{ij} \approx 0.5$. If $v_i$ and $v_j$ are not removed from the cover in step 4 we are fine. Otherwise, by putting each of them back independently with probability $1 - \epsilon$, we get that the probability of covering the edge is much more than $z_{ij}$. Step 5 is taking care of these edges; we use a distribution on level-sets since we need to make sure that the total number of vertices added to the cover is small. In $S_3$ we take each vertex to be in the cover with probability $\sim (2 - 4x)x_i$, and since we use a level-sets distribution, we get that the total number of vertices taken there is $\sim (2 - 4x)$ times the contribution of $S_3$ to the semidefinite program. Moreover, property (A3) of level-sets distributions can be used to show that the number of edges covered is sufficient with a reasonable probability.

## 5    Analysis of the Algorithm

We first need some notation. Let $Z$ be the value of the semidefinite program. For $a, b \in \{1, 2, 3\}$, let $S_{ab}$ be the set of edges with one endpoint in $S_a$ and the other endpoint in $S_b$. For any set of edges $D$, let $Val(D) = \sum_{(i,j)\in D} z_{ij}$ be the total value of the set, and $Cov(D)$ be the number of edges in $D$ covered by the set $C$ in our algorithm. Let $C_1, C_2, C_3$ be the number of vertices chosen in $S_1, S_2, S_3$ respectively before step 7.

By step 7 of the algorithm we are sure that the algorithm produces a set $C$ which covers at least $k$ edges. In order to show that we get a good approximation ratio, we have to bound the size of $C$. We show that the number of vertices added in step 7 is relatively small with high probability, and that the total number of vertices added before step 7 is small enough. In order to show this, we first show that with high probability for each of the sets $S_{ij}$, $Cov(S_{ij})$ is not much smaller than $Val(S_{ij})$ before step 7. Thus, the total number of covered edges before step 7 is not much smaller than the sum of all the $z_{ij}$, which is at least $k$ as ensured by our relaxation.

Clearly, we cover all the edges with one endpoint in $S_1$. We therefore have to show that we cover many edges in $S_{22}, S_{23}$ and $S_{33}$.

For each $(i, j) \in S_{22}$, let $y_{ij} = 1 + v_i \cdot v_j$, $D_1 = \{(i, j) \mid y_{ij} > 4x\}$, $D_2 = S_{22} - D_1$, and $Y_k = \sum_{(i,j)\in D_k} y_{ij}$ for $k = 1, 2$. We first present a useful bound on the probability $P_{ij}$ that edge $(i, j) \in S_{22}$ is not covered at the end of step 5. By Lemma 3 and by property (A3) of Section 3.2, it can be verified that

$$P_{ij} \leq \epsilon^2 N \left( \frac{2c}{\|v_i + v_j\|} \right) = \epsilon^2 N \left( \frac{\sqrt{2}c}{\sqrt{y_{ij}}} \right). \tag{4}$$

We let $\mathbf{E}[R]$ denote the expectation of a random variable $R$. We start by showing that with high probability we will cover many edges of $D_1$.

---

**ALGORITHM part-cover**

**Input:** A graph $G = (V, E)$ with maximum degree $d$, $|E| = m$, and $|V| = n$; a number $k \leq m$ is also given.
**Output:** A subset $C \subseteq V$ which covers at least $k$ edges.

1. Solve relaxation (2); let the $x_i$ and $z_{ij}$ be as in (3).
2. Let $x$ be a value to be determined later; $x$ will be $\Theta(\frac{\ln \ln d}{\ln d})$. Let $S_1 = \{i \mid v_0 \cdot v_i \geq x\}$, $S_2 = \{i \mid -2x \leq v_0 \cdot v_i < x\}$, $S_3 = V \setminus (S_1 \cup S_2)$.
3. $C \leftarrow S_1 \cup S_2$.
4. Find a (hopefully large) subset $I$ of $S_2$, using the following procedure (see [8]). Let $c$ be such that $N(c) = \Theta(\frac{\ln \ln d}{\ln d})$; $c \approx \sqrt{2 \ln \ln d}$ will be chosen precisely later. Choose an $n$-dimensional random vector $r$, and let $I = \{i \in S_2 \mid v_i \cdot r \geq c\}$. Let $C = C \setminus I$.
5. Renumber the vertices so that $I = \{1, 2, \ldots, |I|\}$. Let $\epsilon$ be a parameter to be determined later. Draw a vector $(X_1, \ldots, X_{|I|})$ from the distribution $\mathcal{D}'(|I|, (1 - \epsilon, \ldots, 1 - \epsilon))$. For each vertex $i \in I$, add $i$ to $C$ if $X_i = 1$.
6. If all edges with at least one end-point in $S_3$ have already been covered, skip to the next step; otherwise proceed as follows. Renumber the vertices so that $S_3 = \{1, 2, \ldots, |S_3|\}$. For each vertex $i \in S_3$, let $p_i = \frac{1 + v_0 \cdot v_i}{1 + 2x} = \frac{2x_i}{1 + 2x}$. Let $(X_1, \ldots, X_{|S_3|})$ be a vector drawn from the distribution $\mathcal{D}'(|S_3|, (p_1, \ldots, p_{|S_3|}))$. Let $S = \{i \mid X_i = 1\}$. Repeat this step $\lceil 4|E|/x \rceil$ times independently, and finally choose the set $S$ which covers the maximum number of yet-uncovered edges with at least one endpoint in $S_3$. If this maximum number is zero, re-define $S$ to be the set containing any one vertex of $S_3$ which covers at least one yet-uncovered edge. (Such a vertex must exist, for otherwise we would have skipped this step.) Add $S$ to $C$.
7. If $C$ does not cover at least $k$ edges, arbitrarily add vertices to $C$, such that each vertex adds at least one new covered edge. When $C$ covers at least $k$ edges, stop and output $C$.

---

**Fig. 1.** Algorithm **part-cover** for finding a small vertex cover in a graph

**Lemma 4.** *If $c \geq \sqrt{2}$, then the probability that $Val(D_1) \geq Cov(D_1)$ is at most $4\epsilon^2 N(c)$.*

*Proof.* Let $(i, j) \in D_1$. We have

$$1 - z_{ij} = \frac{1 - v_0 \cdot v_i - v_0 \cdot v_j + v_i \cdot v_j}{4} \geq \frac{y_{ij} - 2x}{4} \geq \frac{y_{ij}}{8}.$$

By applying Lemma 1, one can verify that the function $(8\epsilon^2/y) \cdot N\left(\frac{\sqrt{2}c}{\sqrt{y}}\right)$ is an increasing function of $y$ whenever $0 \leq y \leq 2$ and $c \geq \sqrt{2}$. Therefore, by (4), the ratio $\frac{P_{ij}}{1 - z_{ij}}$ is bounded by $4\epsilon^2 N(c)$, and we get

$$\mathbf{E}[|D_1| - Cov(D_1)] \leq 4\epsilon^2 N(c)(|D_1| - Val(D_1)).$$

Hence,

$$\Pr(Cov(D_1) \leq Val(D_1)) = \Pr(|D_1| - Cov(D_1) \geq |D_1| - Val(D_1)) \leq 4\epsilon^2 N(c),$$

by Markov's inequality.

Next we consider the set $D_2$. We show that almost all the edges of $D_2$ are covered before step 7.

**Lemma 5.** *Let* $\gamma$ *be such that* $2N(c)^{\frac{1}{2x}} < \gamma$. *For every* $\beta > 0$, *the probability that* $Cov(D_2) \leq |D_2| - \beta|D_2|$ *after step 5 is at most* $\frac{\epsilon^2\gamma}{\beta}$.

*Proof.* By (4), we have

$$\mathbf{E}[|D_2| - Cov(D_2)] \leq \epsilon^2 \sum_{(i,j)\in D_2} N\left(\frac{\sqrt{2c}}{\sqrt{y_{ij}}}\right).$$

It is easy to see that $N\left(\frac{\sqrt{2c}}{\sqrt{y}}\right)$ is an increasing function of $y$, and therefore, since for every $(i,j) \in D_2$ $y_{ij} \leq 4x$, by Lemma 1 we get that

$$\mathbf{E}[|D_2| - Cov(D_2)] \leq \epsilon^2|D_2|N\left(\frac{c}{\sqrt{2x}}\right) \leq 2\epsilon^2|D_2|\sqrt{x}N(c)^{1/2x} \leq \gamma\epsilon^2|D_2|.$$

By Markov's inequality, the probability that $|D_2| - Cov(D_2) > \beta|D_2|$ is at most $\frac{\epsilon^2\gamma}{\beta}$.

We now consider the set $S_{23}$. For each $i \in S_2$, let $W_i$ be the indicator for the event that $i$ was chosen in step 4. As in Figure 1, let $X_i$ be the indicator for the event that $i$ was chosen in step 5. The following lemma shows that after step 5, the expected number of edges covered in $S_{23}$ is high.

**Lemma 6.** *Let* $W_i, X_i$ *be as defined a few lines above. Let* $A$ *be a random variable defined by*

$$A = \sum_{(i,j)\in S_{23}} (1 - W_i)(1 - X_i)\left(1 - \frac{2x_j}{1+x}\right).$$

*After step 5,* $\Pr(A \geq \frac{|S_{23}|-Val(S_{23})}{2}) \leq 12\epsilon N(c)$.

*Proof.* Let $(i,j) \in S_{23}$, where $i \in S_2, j \in S_3$. We first bound the expression $1 - z_{ij}$.

$$1 - z_{ij} = \frac{1 + v_i \cdot v_j - v_0 \cdot v_j - v_0 \cdot v_i}{4} \geq \frac{-v_0 \cdot v_i - v_0 \cdot v_j}{2} \geq \frac{-x+1-2x_j}{2}.$$

We also have that the expectation $\mathbf{E}[A]$ of $A$ equals $\sum_{(i,j)\in S_{23}} \epsilon N(c)(1 - \frac{2x_j}{1+x})$. Now, since $j \in S_3$, we have $2x_j \leq 1 - 2x$; for any such $j$, we can check that

$1 - \frac{2x_j}{1+x} \leq 3(1 - x - 2x_j)$. So,

$$\mathbf{E}[A] \leq \sum_{(i,j) \in S_{23}} 3\epsilon N(c)(-x + 1 - 2x_j)$$

$$\leq 6\epsilon N(c) \sum_{(i,j) \in S_{23}} (1 - z_{ij}) = 6\epsilon N(c)(|S_{23}| - Val(S_{23})).$$

Markov's inequality now completes the proof.

**Notation.** For $i = 1, 2, 3$, let $\Phi_i = \sum_{j \in S_i} x_j$.

Using Lemma 6, we next show that enough edges are covered from $S_{23}$ and $S_{33}$ in step 6.

**Lemma 7.** *There is an absolute constant $\delta > 0$ such that the following holds as long as $x \leq \delta$. Let $A$ be as in the statement of Lemma 6. Fix the random choices in the steps up to and including step 5, conditional on the bound $A \leq \frac{|S_{23}| - Val(S_{23})}{2}$. Then, with probability at least $1 - 1/e$, both of the following hold at the end of step 6:*

*(a) at most $(2 - 2x)\Phi_3 + 1$ elements of $S_3$ have been chosen; and*
*(b) $Cov(S_{23}) + Cov(S_{33}) \geq Val(S_{23}) + Val(S_{33})$.*

*This probability is only with respect to the random choices made in step 6.*

*Proof.* If the test made at the beginning of step 6 led to the rest of the step being skipped, then items (a) and (b) hold trivially; so suppose the rest of the step was not skipped.

Item (a) is easy. Recall that step 6 repeats a "level-sets" trial $\lceil 4|E|/x \rceil$ times. By (A2'), we know that in each trial, the number of vertices chosen from $S_3$ is at most $\Phi_3 \cdot \frac{2}{1+2x} + 1 \leq \Phi_3(2 - 2x) + 1$, with probability one. Also, if we do re-define $S$ to be an appropriate singleton set in step 6, item (a) still holds with probability one. Thus, we only need to focus on showing that item (b) holds with probability at least $1 - 1/e$, which we do now.

Note that $A$ is now deterministic. In each of the $\lceil 4|E|/x \rceil$ "level-sets" trials, $\mathbf{E}[|S_{23}| - Cov(S_{23})] = A$. Consider one such trial. By the assumption of the lemma,

$$\mathbf{E}[|S_{23}| - Cov(S_{23})] \leq \frac{|S_{23}| - Val(S_{23})}{2}. \tag{5}$$

Consider the set $S_{33}$. Since the vertices chosen in step 6 are drawn from a level-set distribution, for every two vertices $i, j \in S_3$, the probability that neither of them will be chosen in $S$ is at most $(1 - p_i)(1 - p_j)$; this follows from property (A3) of Section 3.2. Thus,

$$\mathbf{E}[Cov(S_{33})] \geq \sum_{(i,j) \in S_{33}} (1 - (1 - p_i)(1 - p_j))$$

$$= \sum_{(i,j) \in S_{33}} \left( \frac{2x_i + 2x_j}{1 + x} - \frac{4x_i x_j}{(1+x)^2} \right)$$

$$\geq \sum_{(i,j)\in S_{33}} ((x_i + x_j)(\frac{2}{1+x} - \frac{x_i + x_j}{(1+x)^2}))$$

$$\geq \sum_{(i,j)\in S_{33}} z_{ij} \cdot \frac{1 + (2+2)x}{(1+x)^2} \tag{6}$$

$$\geq (1+x)Val(S_{33}). \tag{7}$$

Inequality (6) follows from the fact that $z_{ij} \leq x_i + x_j$ (see the last constraint of the semidefinite program), and since $i, j \in S_3$ implies that $x_i + x_j \leq 1 - 2x$. Bound (7) holds for all $x$ small enough, since $1 + 4x \geq (1+x)^3$ holds for all small enough positive $x$.

The definitions of $S_2$ and $S_3$ imply that

$$Val(S_{23}) \leq \sum_{(i,j)\in S_{23}} (x_i + x_j) \leq \sum_{(i,j)\in S_{23}} (\frac{1+x}{2} + \frac{1-2x}{2}) = (1 - x/2) \cdot |S_{23}|.$$

Thus, bounds (5) and (7) can be used to show that $\mathbf{E}[R] \geq (1+x/4) \cdot a$, where $R = Cov(S_{23}) + Cov(S_{33})$ and $a = Val(S_{23}) + Val(S_{33})$ for notational convenience. We wish to show that at the end of step 6,

$$\Pr(R \geq a) \geq 1 - 1/e. \tag{8}$$

This certainly holds if $a \leq 1$, since we ensure in step 6 that at least one yet-uncovered edge is chosen; so, suppose $a > 1$. Note that $R \leq |E|$ always; so by applying Markov's inequality to the random variable $|E| - R$, we get that in each trial of step 6,

$$\Pr(R < a) \leq \frac{|E| - (1 + x/4) \cdot a}{|E| - a}$$

$$= 1 - (x/4) \cdot \frac{a}{|E| - a}$$

$$\leq 1 - (x/4) \cdot \frac{1}{|E|}$$

since $a > 1$. Thus, the probability that none of the $\lceil 4|E|/x \rceil$ trials succeeds in achieving "$R \geq a$" is at most

$$\left(1 - (x/4) \cdot \frac{1}{|E|}\right)^{\lceil 4|E|/x \rceil} \leq 1/e;$$

this proves (8).

We now prove the main theorem. In the theorems below, we content ourselves with showing that an algorithm produces a satisfactory solution with probability $\Omega(1/\log n)$; repeating the algorithm $a \log n$ times for a large enough constant $a$, will clearly lead to a satisfactory solution with high probability. Also, our results below have an extra "+2" term in the objective function. After Corollary 1, we discuss how to remove this extra term. We will let "$o(1)$" denote any function of $d$ alone that goes to zero as $d$ increases; different usages of such a term can correspond to different choices of such functions.

**Theorem 1.** *Let $G = (V, E)$ be a graph with maximal degree $d$. For sufficiently large $d$, with probability at least $\frac{1}{\ln d}$, algorithm **part-cover** will produce a $k$-cover of size at most $(2 - \Omega(\frac{\ln \ln d}{\ln d}))OPT + 2$, where $OPT$ is the size of a minimal $k$-cover.*

*Proof.* We now choose our parameters; the choice of $c$ is motivated by our aim to have $|N(c) - \frac{48 \ln \ln d}{\ln d}| = o(\frac{\ln \ln d}{\ln d})$. We choose:

$$c = \sqrt{2 \ln \left( \frac{\ln d}{96 \sqrt{\pi} (\ln \ln d)^{3/2}} \right)}; \quad x = \frac{\ln \ln d}{6 \ln d}; \text{ and } \epsilon = \frac{1}{48}. \tag{9}$$

One can verify that the following properties hold for $d$ large enough (to verify (10), we may use Lemma 1):

$$\left| N(c) - \frac{48 \ln \ln d}{\ln d} \right| = o(\frac{\ln \ln d}{\ln d}); \tag{10}$$

$$2N(c)^{1/2x} \le \frac{2}{d^{3-o(1)}}; \tag{11}$$

$$(1 - 1/e) \cdot (N(c)/2 - 4\epsilon^2 N(c) - \frac{2\epsilon^2}{d} - 12\epsilon N(c)) \ge \frac{1}{\ln d}. \tag{12}$$

Recall that $C_1, C_2, C_3$ are the sizes of $C$ intersecting $S_1, S_2$ and $S_3$ respectively. Since for each $i \in S_1$, $x_i \ge \frac{1+x}{2}$, we have that $C_1 \le \Phi_1 \cdot \frac{2}{1+x} = (2 - 2x + O(x^2))\Phi_1$. Next, the expectation of $C_2$ after step 4 is $|S_2|(1 - N(c))$. Therefore, by applying Markov's inequality, the probability that $C_2 \ge (1 - \frac{N(c)}{2})|S_2|$ after step 4 is at most $p_1 = \frac{1-N(c)}{1-N(c)/2} \le 1 - N(c)/2$. Thus, we can use property (A2') of the level-sets distribution to see that after step 5, $C_2 \le |S_2|(1 - \epsilon N(c)/2) + 1$ holds with probability at least $1 - p_1$.

It is easy to see that the precondition "$c \ge \sqrt{2}$" of Lemma 4 and the precondition "$x$ small enough" of Lemma 7 hold for all $d$ sufficiently large. Since $\sum_{i,j=1,2,3} Val(S_{ij}) \ge k$, then by (11) and Lemmas 4, 5, 6 and 7, the following holds just before executing step 7. The probability that: (i) the number of covered edges is at least $k - |D_2|/d^{2-o(1)}$ (for some specific choice of the "$o(1)$" term) *and* (ii) $C_2 \le |S_2|(1 - \epsilon N(c)/2) + 1$, is at least

$$p_2 = (1 - p_1 - 4\epsilon^2 N(c) - \frac{2\epsilon^2}{d} - 12\epsilon N(c)) \cdot (1 - 1/e).$$

Since the maximal degree of $G$ is $d$, we have $|D_2| \le |S_2|d$; thus, with probability at least $p_2$, we add in step 7 at most $\frac{|S_2|}{d^{1-o(1)}}$ vertices to $C_2$. Hence, the total number of vertices chosen after the algorithm terminates is at most

$$(2 - 2x + O(x^2))\Phi_1 + |S_2|(1 - \epsilon N(c)/2 + \frac{1}{d^{1-o(1)}}) + 1 + (2 - 2x)\Phi_3 + 1$$

with probability at least $p_2$. On the other hand, $\Phi_2 \ge |S_2|\frac{1-2x}{2}$. Thus, the algorithm finds a $k$-cover $C$, such that

$$|C| \le \max\{2 - 2x + O(x^2), 2 - (1 - o(1)) \cdot (\epsilon N(c) - 4x)\}Z + 2$$

$$\leq (2 - \frac{\ln \ln d}{(3 + o(1)) \ln d})OPT + 2$$

with probability at least $p_2$, for sufficiently large $d$; by (12), this probability is at least $\frac{1}{\ln d}$.

## 6  Extensions of the Analysis to Other Cases

The proof of Theorem 1 is based on the proper choice of $c, x$ and $\epsilon$. For graphs with bounded degree $d$, the choice of the parameters was given in the proof of the theorem. In this section we show how to choose the corresponding parameters in other graphs.

A $p$-claw in a graph is an *induced* star with $p + 1$ vertices (the center of the star and its $p$ neighbors). A $p$-claw-free graph is a graph which does not contain any $p$-claw. Theorem 2 is motivated by a similar result from [5].

**Theorem 2.** *Let $G = (V, E)$ be a $p$-claw-free graph. For a proper choice of $c, x$ and $\epsilon$, algorithm **part-cover** will produce a $k$-cover $C$ of size at most*

$$|C| \leq (2 - \Omega(\frac{\ln \ln p}{\ln p}))OPT + 2,$$

*with probability at least $\Omega(\frac{1}{\ln p})$.*

*Proof.* We choose our parameters as in (9), with $d$ replaced by $p$. We claim that in $D_2$ there are no triangles. Assume that $v_1, v_2, v_3$ are vectors corresponding to a triangle in $D_2$. Since $v_i \cdot v_j \leq -1 + 4x$ for every $i \neq j, i, j \in \{1, 2, 3\}$, we get that

$$v_1 \cdot v_2 + v_1 \cdot v_3 + v_1 \cdot v_2 \leq -3 + 4x < -1,$$

which contradicts the constraints of the semidefinite program.

Since $D_2$ does not contain any triangle, and since it is a $p$-claw free graph, we get that the maximum degree in $D_2$ is $p - 1$. One can verify that in the proof of Theorem 1, we only used the property that $G$ is a bounded degree graph when we bounded the number of edges in $D_2$. Thus, we can apply the same proof, replacing $d$ by $p - 1$.

We now prove that when $k$ decreases, the performance ratio of algorithm **part-cover** improves.

**Theorem 3.** *Let $G = (V, E)$ be a graph with a $k$-cover of size $OPT$. For a proper choice of $c, x$ and $\epsilon$, algorithm **part-cover** will produce a $k$-cover $C$ of size at most*

$$|C| \leq (2 - \Omega(\frac{\ln \ln OPT}{\ln OPT}))OPT + 2,$$

*with probability at least $\Omega(\frac{1}{\ln OPT})$.*

*Proof.* We choose our parameters as in (9), with $d$ replaced by $3 \cdot OPT$. Since $\Phi_2 \approx \frac{|S_2|}{2}$, we get that $|S_2| \leq 3 \cdot OPT$. Therefore, the maximum degree in $D_2$ is at most $3 \cdot OPT$. We can now apply the proof of Theorem 1, by replacing $d$ by $3 \cdot OPT$.

Since the optimal $k$-cover of $G$ contains at most $k$ vertices, we get the following corollary.

**Corollary 1.** *Let* $G = (V, E)$ *be a graph. For a proper choice of* $c, x$ *and* $\epsilon$, *algorithm* **part-cover** *will produce a $k$-cover $C$ of size at most*

$$|C| \leq (2 - \Omega(\frac{\ln \ln k}{\ln k}))OPT + 2,$$

*with probability at least* $\Omega(\frac{1}{\ln k})$.

We now mention how the extra "+2" term can be removed from the objective function value from the statements of Theorems 1, 2 and 3, as well as from Corollary 1. First note that our results imply that $OPT \leq 2Z + 2$. Suppose we adopt the following strategy. Let $c_0$ be a sufficiently large constant. If $Z \leq c_0$, we find an optimal solution in polynomial time by exhaustive search, since $OPT$ is bounded by a constant. Next if $Z \geq c_0$ and if $c_0$ has been chosen large enough, then $(2 - \Omega(\frac{\ln \ln OPT}{\ln OPT}))OPT + 2 = (2 - \Omega(\frac{\ln \ln OPT}{\ln OPT}))OPT$. Similarly, consider Theorem 1, where we show an upper bound of the form $(2 - f(d))OPT + 2$, where $f(d) = o(1)$. Again, if $Z \leq c_0$, we find an optimal solution in polynomial time by exhaustive search; if $Z \geq c_0$, we use Theorem 1. Now if $f(d) \cdot OPT \geq 3$, then $(2 - f(d))OPT + 2$ is of the form $(2 - \Omega(f(d)))OPT$. Otherwise if $f(d) \cdot OPT < 3$, showing an upper bound of $2 \cdot OPT - 3$ on the output, will suffice to show that the output is at most $(2 - f(d))OPT$. Such an upper bound of $2 \cdot OPT - 3$ on the output follows from Theorem 3, since $Z \geq c_0$ and $c_0$ is large enough. Similar remarks hold for Theorem 1 and Corollary 1.

## 7   Concluding Remarks

We have presented an improved approximation algorithm for partial vertex cover using semidefinite programming. Our algorithm involves a new rounding technique, which copes with a few different worst-case configurations of the vectors at the same time. Our technique combines rounding methods for semidefinite programming, with additional ingredients such as level-sets distributions. It would be interesting to find other applications of such distributions in the context of semidefinite programming. Finally, we raise the following open problem. Is there a better approximation algorithm for partial vertex cover when restricted to small $k$, and specifically, is it possible to find a $(2 - \Omega(1))$–approximation when $k = o(n)$?

# References

1. R. Bar-Yehuda. Using homogeneous weights for approximating the partial cover problem. In *Proceedings of the 10th Annual ACM-SIAM Symposium on Discrete Algorithms, Baltimore, Maryland*, pages 71–75, 1999.
2. N. Bshouty and L. Burroughs. Massaging a linear programming solution to give a 2-approximation for a generalization of the vertex cover problem. In *Fifteenth Annual Symposium on the Theoretical Aspects of Computer Science*, pages 298–308, 1998.
3. W. Feller. *An Introduction to Probability Theory and its Applications*. John Wiley & Sons, New York, 3 edition, 1968.
4. R. Gandhi, S. Khuller, and A. Srinivasan. Approximation algorithms for partial covering problems. In *Proceedings of the 28th International Colloquium on Automata, Languages and Programming, Crete, Greece*, pages 225–236, 2001.
5. E. Halperin. Improved approximation algorithms for the vertex cover problem in graphs and hypergraphs. In *Proceedings of the 11th Annual ACM-SIAM Symposium on Discrete Algorithms, San Francisco, California*, pages 329–337, 2000.
6. E. Halperin, R. Nathaniel, and U. Zwick. Coloring $k$-colorable graphs using smaller palettes. In *Proceedings of the 12th Annual ACM-SIAM Symposium on Discrete Algorithms, Washington, D.C.*, pages 319–326, 2001.
7. D. S. Hochbaum. The $t$-vertex cover problem: Extending the half integrality framework with budget constraints. In *First International Workshop on Approximation Algorithms for Combinatorial Optimization Problems*, pages 111–122, 1998.
8. D. Karger, R. Motwani, and M. Sudan. Approximate graph coloring by semidefinite programming. *Journal of the ACM*, 45:246–265, 1998.
9. A. Rényi. *Probability Theory*. Elsevier/North-Holland, Amsterdam, London, New York, 1970.
10. A. Srinivasan. Distributions on level-sets with applications to approximation algorithms. In *Proceedings of the 42nd Annual IEEE Symposium on Foundations of Computer Science, Las Vegas, Nevada*, pages 588–597, 2001.

# Minimum Restricted Diameter Spanning Trees

Refael Hassin and Asaf Levin

Department of Statistics and Operations Research,
Tel-Aviv University, Tel-Aviv 69978, Israel,
{hassin,levinas}@post.tau.ac.il

**Abstract.** Let $G = (V, E)$ be a requirements graph. Let $d = (d_{ij})_{i,j=1}^{n}$ be a length metric. For a tree $T$ denote by $d_T(i, j)$ the distance between $i$ and $j$ in $T$ (the length according to $d$ of the unique $i - j$ path in $T$). The restricted diameter of $T$, $D_T$, is the maximum distance in $T$ between pair of vertices with requirement between them. The minimum restricted diameter spanning tree problem is to find a spanning tree $T$ such that the minimum restricted diameter is minimized. We prove that the minimum restricted diameter spanning tree problem is $NP$-hard and that unless $P = NP$ there is no polynomial time algorithm with performance guarantee of less than 2. In the case that $G$ contains isolated vertices and the length matrix is defined by distances over a tree we prove that there exist a tree over the non-isolated vertices such that its restricted diameter is at most 4 times the minimum restricted diameter and that this constant is at least $3\frac{1}{2}$. We use this last result to present an $O(log(n))$-approximation algorithm.

## 1 Introduction

Let $G = (V, E)$ be a *requirements graph* with $|V| = n$, $|E| = m$. Let $d = (d_{ij})_{i,j=1}^{n}$ be a length metric.

For a tree $T$ denote by $d_T(i, j)$ the distance between $i$ and $j$ in $T$ (the length according to $d$ of the unique $i - j$ path in $T$).

For a spanning tree $T$, define the *restricted diameter of* $T$ as

$$D_T = Max_{(i,j)\in E} d_T(i, j).$$

The MINIMUM RESTRICTED DIAMETER SPANNING TREE PROBLEM is to find a spanning tree $T$ that minimizes $D_T$.

When $G$ is a complete graph the problem is that of finding minimum diameter spanning tree. This problem is solvable in $O(mn + n^2 \log n)$ time (see [7]).

The minimum diameter spanning tree problem is motivated by the following (see for example [9]): we want to find a communication network among the vertices, where the communication delay is measured in terms of the length of a shortest path between the vertices. A desirable communication network is naturally one that minimizes the diameter. To keep the routing protocols simple, often the communication network is restricted to be a spanning tree. The minimum restricted diameter spanning tree problem arises when communication

K. Jansen et al. (Eds.): APPROX 2002, LNCS 2462, pp. 175–184, 2002.

takes place only between a specified collection of pairs of vertices. In such a case it is natural to minimize the maximum communication delay between vertices that need to communicate.

The case, in which $G$ is a clique over a subset $S \subseteq V$ is similarly solved ([13]) by finding the shortest paths tree from the weighted absolute 1-center where the weight of a vertex in $S$ is 1 and the weight of a vertex not in $S$ is 0.

As observed in [1], there are cases where $D_T = \Omega(n) Max_{(i,j) \in E} d_{i,j}$ for any spanning tree $T$. Therefore, in the analysis of an approximation algorithm for the problem we need to use a better lower bound than $Max_{(i,j) \in E} d_{i,j}$.

We prove that the MINIMUM RESTRICTED DIAMETER SPANNING TREE PROBLEM is $NP$-hard and that if $P \neq NP$ there is no polynomial time algorithm with performance guarantee of less than 2.

Suppose that $V = V_R \cup V_S$ and $E \subseteq V_R \times V_R$: In [5] it is shown that the *distortion* of tree metric with respect to a steiner tree metric is bounded by a factor of 8. This means that for every spanning tree $ST$ over $V$ there is a tree $T$ over $V_R$ such that for all $i, j \in V_R$ $d_{ST}(i, j) \leq d_T(i, j) \leq 8 d_{ST}(i, j)$. By this result we conclude that in particular $D_T \leq 8 D_{ST}$. In this paper we will provide a better construction with respect to the restricted diameter criteria, which proves that $D_T \leq 4 D_{ST}$. We also show that the best possible bound is at least $3\frac{1}{2}$.

We use this last result to present an $O(\log n)$-approximation algorithm.

A similar problem, MINIMUM COMMUNICATION SPANNING TREE (MCT), was addressed in other papers. In the MCT problem we are given a requirement graph $G = (V, E)$ and a length matrix $d$ and the goal is to find a spanning tree $T$ that minimizes $\Sigma_{(i,j) \in E} d_T(i, j)$. In [4] a derandomization procedure to Bartal's tree metric construction (see [2], [3]) is used in order to obtain a deterministic $O(\log n \ \log \log n)$-approximation algorithm for the general metric case. In [11] an $O(\log n)$-approximation algorithm is presented for the $k$-dimensional Euclidean complete graphs where $k$ is a constant.

Another similar problem, MINIMUM ROUTING TREE CONGESTION (MRTC), was addressed in other papers. In the MRTC problem we are given a requirement graph $G = (V, E)$ and a weight function $w : E \to N$. We want to find a routing tree $T = (V', E')$ (a tree is a routing tree if the leaves of $T$ correspond to $V$ and each internal vertex has degree 3) that minimizes

$$Max_{e \in E'} \Sigma_{(u,v) \in E, u \in S(e), v \notin S(e)} w(u, v)$$

where $S(e)$ is one of the connected components resulting from $T$ by deleting an edge $e \in E'$. In [12] the MRTC is proved to be $NP$-hard and the special case when $G$ is planar, the problem can be solved optimally in polynomial time. In [10] an $O(\log n)$-approximation algorithm is given for the general case (non-planar graphs).

## 2   NP-Hardness

**Theorem 1.** *Unless $P = NP$ there is no polynomial-time approximation algorithm for the restricted diameter spanning tree problem with performance guarantee of less than 2.*

*Proof.* We describe a reduction from MONOTONE SATISFIABILITY (mSAT) (in which in every clause either all the literals are variables or all the literals are negated variables). mSAT is $NP$-complete (see [6]). Consider an instance of the mSAT problem composed of variables $x_1, x_2, \ldots, x_n$ and of clauses $c_1, c_2, \ldots c_m$. We construct an instance for the minimum restricted diameter spanning tree problem as follows: Define a vertex set $V = V_1 \cup V_2 \cup V_3$ where $V_1 = \{root\}$, $V_2 = \{x_i, \bar{x}_i, x'_i | 1 \leq i \leq n\}$ and $V_3 = \{c_j^1, c_j^2 | 1 \leq j \leq m\}$. Define a length matrix as follows: $d_{root,x_i} = d_{root,\bar{x}_i} = 1 \; \forall i$, $d_{root,c_j^1} = d_{root,c_j^2} = 2 \; \forall j$, $d_{x_i,x'_i} = d_{x'_i,\bar{x}_i} = 2 \; \forall i$, $d_{x_i,c_j^1} = d_{x_i,c_j^2} = 1$ if $x_i \in c_j$, $d_{\bar{x}_i,c_j^1} = d_{\bar{x}_i,c_j^2} = 1$ if $\bar{x}_i \in c_j$ and otherwise $d$ is defined as the shortest path length according to the defined distances. Define a requirement graph $G = (V, E)$ by $E = \{(x_i, x'_i), (x'_i, \bar{x}_i) | 1 \leq i \leq n\} \cup \{(root, c_j^1), (c_j^1, c_j^2), (root, c_j^2) | 1 \leq j \leq m\}$.

The instance for the minimum restricted diameter spanning tree problem has solution which is at most 2 if and only if the mSAT instance is satisfiable. If there is a spanning tree with restricted diameter 2 then for every $i$ the spanning tree must include the edges $(x_i, x'_i), (x'_i, \bar{x}_i)$ as there is no other path of length 2 between $x_i$ and $x'_i$ and between $x'_i$ and $\bar{x}_i$. Therefore, the tree may include only one of the edges $(root, x_i)$ or $(root, \bar{x}_i)$. In order to have a path of length 2 between $root$ and $c_j^1$, a path of length 2 between $root$ and $c_j^2$ and a path of length 2 between $c_j^1$ and $c_j^2$ the tree must have some $i$ such that all of $(root, x_i), (x_i, c_j^1), (x_i, c_j^2)$ belong to the tree and such that $x_i \in c_j$, or the tree must include $(root, \bar{x}_i), (\bar{x}_i, c_j^1), (\bar{x}_i, c_j^2)$ such that $\bar{x}_i \in c_j$. Therefore, if the tree includes the edge $(root, x_i)$ we set $x_i$ to TRUE, and otherwise to FALSE. Since the tree has paths of length 2 between $root, c_j^1$ and $c_j^2$ for every $j$ then $c_j$ must have a literal with TRUE assignment, and therefore the formula is satisfied.

Suppose that the formula can be satisfied and consider the tree composed of the edges $(x_i, x'_i), (x'_i, \bar{x}_i)$ for $i = 1, \ldots, n$, the edge $(root, x_i)$ for every TRUE assigned variable, and the edge $(root, \bar{x}_i)$ for every FALSE assigned variable. Every clause $c_j$ must include a literal with TRUE value. Pick one of them and add an edge between this literal and $c_j^1$, and an edge between this vertex and $c_j^2$. Then the restricted diameter of this spanning tree is 2.

We note that for this problem if there is no solution of cost 2 then the minimum cost is at least 4. Therefore, distinguishing between 2 and 4 is $NP$-complete. Therefore, if $P \neq NP$ there is no polynomial time approximation algorithm with performance guarantee better than 2.

## 3   The Role of Steiner Points

In this section we assume that $V = V_R \cup V_S$ where $V_R \cap V_S = \emptyset$ and $E \subseteq V_R \times V_R$, ($V_S$ the Steiner point set and $V_R$ the regular point set). Denote by $ST$ the minimum restricted diameter spanning tree (a spanning tree over $V$). We will prove that there is a spanning tree $T$ over $V_R$ such that $D_T \leq 4D_{ST}$ and that there are cases where $D_T \geq 3\frac{1}{2}D_{ST}$.

**Theorem 2.** *There exists a spanning tree $T$ that can be computed in polynomial time, such that $D_T \leq 4D_{ST}$.*

*Proof.* It is sufficient to prove the claim under the assumption that $G$ is connected. If $G$ is not connected then we can construct a tree over every component with restricted diameter at most $4D_{ST}$ and connect these trees arbitrary. Therefore, w.l.o.g. assume that $G$ is connected.

W.l.o.g. we assume that all the leaves of $ST$ are in $V_R$. This is so since we can remove any leaf $u \in V_S$ of $ST$ from $G$ without affecting $D_{ST}$.

For a vertex $p \in V$ denote by $t(p) = argmin_{u \in V_R} d_{ST}(u, p)$ ($t(p) = p$ if $p \in V_R$). Define a set of vertices $U$ by the following procedure:

1. Arbitrarily choose $r \in V_S$. Add $r$ to $U$, root $ST$ at $r$. Set $i = 0$, and label $r$ 'unvisited' $level(r) = 0$.
2. While there is an 'unvisited' non-leaf vertex $w$ do:
   - Pick an arbitrary non-leaf vertex $w \in U$ with label 'unvisited' and $level(w) = i$ (if there is no such vertex set $i = i + 1$):
   - Mark $w$ 'visited'.
   - Along every path going down the tree from $w$ to a leaf of $ST$ that does not include a vertex $v \in U$ $level(v) = i$ there is a vertex, $u$, which is the last vertex along this path such that $d_{ST}(t(u), t(w)) \le D_{ST}$. Add $u$ to $U$ with label 'unvisited' and $level(u) = i + 1$.

Step 2 is well defined (the vertex $u$ exists): For every $w \in U$ and $v$ a son of $w$, since $G$ is connected, there is a requirement crossing the cut corresponding to $ST \setminus \{(w, v)\}$, and therefore $d_{ST}(t(v), t(w)) \le D_{ST}$.

We now define a spanning tree $T$. For every $v \in V_R \setminus \{t(r)\}$ there is a vertex $u \in U$ on the path from $v$ to $r$ such that $t(u) \ne v$ and we pick the closest vertex $u$ to $v$ satisfying these conditions. We add to the edge set of $T$ the edge $(v, t(u))$. This defines a spanning tree $T$. This is so as there are $|V_R| - 1$ edges and every vertex $v \in V_R$ is connected to a vertex $t(u)$ and $t(u)$ is either $t(r)$ or it is connected to a vertex $t(w)$ such that $w$ is an ancestor of $u$ in $ST$. Therefore, all the vertices are connected by paths to $t(r)$ and $T$ is connected.

Consider an edge $(v, t(u))$ in $T$. By step 2 $u$ is on the path between $v$ and $r$ and between $v$ and $u$ there is no other vertex $p \in U$ unless $v = t(p)$. By the construction of $U$ $d_{ST}(v, t(u)) \le D_{ST}$.

To complete the proof we prove that if $d_{ST}(u, v) \le D_{ST}$ for $u, v \in V_R$ then the path between $u$ and $v$ in $T$ has at most 4 edges. We assume otherwise. Consider $T$ as a rooted tree at $t(r)$. We use the following observations:

- For $u, v \in V_R$ a path in $T$ between $u$ and $v$ goes up the tree until their common ancestor and then goes down the tree.
- Suppose that the path from $u$ to $v$ in $T$ consists of at least 5 edges. W.l.o.g. assume that it goes up the tree in at least 3 edges $(u, t(a))$, $(t(a), t(b))$, $(t(b), t(c))$.
- In $ST$ the path between $u$ and $v$ goes through $b$. Therefore, $d_{ST}(u, v) = d_{ST}(u, b) + d_{ST}(b, v)$.
- $v \in V_R$ and by the definition of $t(b)$ $d_{ST}(v, b) \ge d_{ST}(t(b), b)$.
- In Step 2 a vertex is added to $U$ only if it is the last vertex on the path to a leaf satisfying the conditions. Therefore, $d_{ST}(t(b), u) > D_{ST}$.

These observations together with the triangle inequality yield a contradiction to the definition of $D_{ST}$:

$$d_{ST}(u,v) = d_{ST}(u,b) + d_{ST}(b,v) \geq d_{ST}(u,b) + d_{ST}(t(b),b) \geq d_{ST}(u,t(b)) > D_{ST}.$$

*Remark 1.* Figure 1 shows that the construction in Theorem 2 cannot lead to a better asymptotic ratio. In this example $V_S = \{a,b,c,d\}$, and $G$ contains the edges $\{(u,t(a)), (u,t(d)), (u,v), (v,t(a)), (v,t(d)), (t(a),t(d)), (t(a),t(b)), (t(d),t(b))\}$. $T$ contains the edges $\{(u,t(a)), (t(a),t(b)), (t(b),t(d)), (t(d),v)\}$, and therefore $D_{ST} = 2 + 2\epsilon$ and $D_T = 8 + 2\epsilon$.

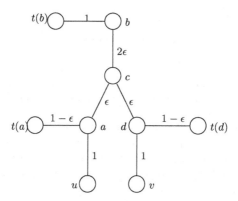

**Fig. 1.** Bad example for the construction in Theorem 2

We now prove a lower bound on the best possible constant in Theorem 2.

**Theorem 3.** *For any $\epsilon > 0$ there exists a requirement graph $G = (V,E)$ such that, $V = V_R \cup V_S$ $E \subseteq V_R \times V_R$, a metric $d$ and a Steiner tree $ST$ over $V$ such that for any tree $T$ over $V_R$, $D_T \geq (3\frac{1}{2} - \epsilon)D_{ST}$.*

*Proof.* Consider the following family of instances: $ST$ contains a rooted (at a vertex $root$) complete binary tree with at least $16K^2$ levels, and the length of these edges is 1. $V_S$ consists of the vertices of this binary tree. Every vertex of $V_S$ is connected in $ST$ to a distinct pair of vertices of $V_R$ by edges of length $K - \frac{\alpha}{2}$. We will start at $root$ with $\alpha = 2$ and every $16K$ levels we will increase $\alpha$ by 2. $D_{ST} = 2K$ and there is a requirement between a pair of vertices from $V_R$ if and only if the distance between them in $ST$ is at most $2K$. The metric will be defined as the distance in $ST$ between the vertices, that is, $d_{ij} = d_{ST}(i,j)$ for every $i,j \in V$. See for example Figure 2.

For $u,v \in V_R$, $(u,v) \in E$ if and only if $d_{ST}(u,v) \leq 2K$. Therefore $D_{ST} = 2K$.

Due to space limitations the proof that for every spanning tree $T$ $D_T > (3\frac{1}{2} - \epsilon)2K$ is omitted.

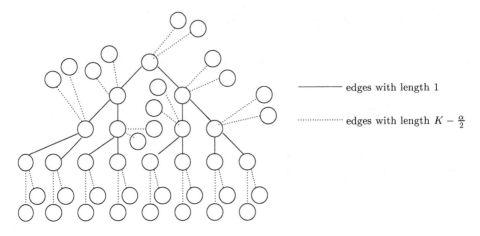

Fig. 2. The structure of $ST$

**Theorem 4.** *Suppose that $G = (V, S \times S)$ for a subset $S \subseteq V$. Then Theorem 2 and Theorem 3 hold with both constants 4 and $3\frac{1}{2}$ replaced by 2.*

*Proof.* The upper bound of 2 is a result of the following argument: Let $r$ be a regular vertex in $ST$. $D_{ST}$ is at least the distance between $r$ and any of the non-steiner vertices in $ST$. Take a tree $T$ which is a star with a center in $r$. Then by the triangle inequality its cost is at most twice the cost of $ST$.

The lower bound is shown by the following example: Let $ST$ be a star with the length of all the edges be 1. Let $S$ be defined as the set of all the star leaves. All the non-$ST$ edges has length 2 and any tree $T$ not containing the star's center has cost at least 4 which is $2D_{ST}$.

## 4    Approximation Algorithm

In this section we provide an $O(\log n)$ approximation algorithm for the MINIMUM RESTRICTED DIAMETER SPANNING TREE PROBLEM.

Denote by $T^*$ a tree that achieves the optimal restricted diameter and denote $D_{T^*} = D^*$.

The following lemma identifies a lower bound and an upper bound on $D^*$.

**Lemma 1.** *Assume that $G$ is connected. Let $MST$ be a minimum spanning tree with respect to the length matrix $d$, and let $(i, j) \in MST$ be the longest edge in $MST$. Then $d_{ij} \leq D^* \leq (n-1)d_{ij}$.*

*Proof.* A path in $MST$ contains at most $n - 1$ edges. Since $(i, j)$ is the longest edge in $MST$ it follows that, $D^* \leq D_{MST} \leq (n-1)d_{ij}$. To see the lower bound let $(I, J)$ be the cut induced by $MST \setminus \{(i, j)\}$. Since $G$ is connected it contains an edge in $(I, J)$, therefore, $D^* \geq min\{d_{k,l} | k \in I, l \in J\} = d_{ij}$.

The legend in the figure reads:

——— edges with length 1

············ edges with length $K - \frac{\alpha}{2}$

**Corollary 1.** *Let* $A = \{d_{ij}, 2d_{ij}, 4d_{ij}, 8d_{ij}, \ldots, 2^{\lceil \log n \rceil} d_{ij}\}$ *then there exists* $\bar{D} \in A$ *such that* $D^* \leq \bar{D} \leq 2D^*$.

We will present an algorithm that for a given test value $D'$ either finds a spanning tree $T'$ such that $D_{T'} = O(D' \log n)$ or concludes that $D^* > D'$. By applying this algorithm for every $D' \in A$ we get an $O(\log n)$-approximation.

We will use the following *decomposition procedure* with a fixed vertex $u$ and test value $D'$. Identify the vertices $V^u(D') = \{v \in V | d_{u,v} \leq D'\}$. To simplify notations we will denote $V^u = V^u(D')$. Let $E(V^u) = E \cap (V^u \times V^u)$. Let $C^1, C^2, \ldots, C^r$ be the connected components of $G \setminus E(V^u)$ which are not singletons (see Figure 3).

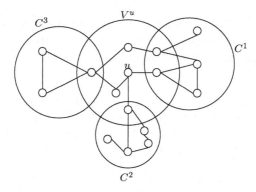

**Fig. 3.** Decomposition procedure at $u$

**Definition 1.** *A $D'$-center is a vertex* $u \in V$ *such that the decomposition procedure with $u$ and $D'$ forms connected components with at most $\frac{n}{2}$ vertices.*

**Lemma 2.** *Assume $D' \geq D^*$, then a $D'$-center $u$ exists.*

*Proof.* Every tree has a centroid, a vertex whose deletion leaves no subtree containing more than $\frac{n}{2}$ vertices (proved by Jordan in 1869, see for example [8]). Let $u$ be a centroid of $T^*$. We will show that $u$ is a $D^*$-center (and therefore it is also a $D'$-center for every $D' \geq D^*$). In $T^*$ every path connecting vertices from $V \setminus V^u$ that belong to different sides of $T^* \setminus \{u\}$ is of length at least $2D^*$. For vertices $w$ and $v$ that belong to distinct sides of $T^* \setminus \{u\}$, $d_{T^*}(w, v) \geq 2D^*$. Therefore, every connected component $C^i$ of $G \setminus E(V^u)$ is fully contained in some side of $T^* \setminus \{u\}$. It follows that $|C^i| \leq \frac{n}{2}$.

We propose to approximate the problem by Algorithm Restricted_Diameter (Figure 4). $l$ is the recursive level of the algorithm ($0 \leq l \leq \lceil \log n \rceil$). The clusters define the connected components in a partial solution (a forest) obtained by previous levels. The application of Algorithm Restricted_Diameter in Figure 4 with parameters $\bar{D}$, $l = 0$, $G$, $d$ and $V_i' = \{v_i\}$ $\forall i$ will result an $O(\log n)$-approximation.

The algorithm first finds a $D'$-center $u$ and adds to the solution the edges connecting $u$ to $V^u \setminus \{u\}$ without closing cycles with previously existing edges. It then solves recursively for every connected component of $G \setminus V^u$. The algorithm uses the information in the partition into clusters to ensure that the solution in each phase doesn't contain cycles. It returns the union of the solutions resulted from the recursive calls and the edges connecting $u$ to $V^u \setminus \{u\}$ without edges that close cycles.

For a formal statement of the algorithm, we need the following notations. For a graph $G' = (V', E')$ and $U \subseteq V'$ the induced subgraph of $G'$ over $U$ will be denoted by $G'(U)$ and the length matrix induced by $U \times U$ will be denoted by $d(U)$.

Denote by $D(V_j')$ the $T$-diameter (regular diameter and not restricted one) of cluster $V_j'$ in the final solution tree $T$ $(D(V_j') = Max_{v,w \in V_j'} d_T(v, w))$. The following holds throughout the algorithm:

**Lemma 3.**

*1) Each component returned by the decomposition procedure at level $l$ of Algorithm Restricted_Diameter has size at most $\frac{n}{2^l}$.*

*2) Let $C$ be a component returned by the decomposition procedure at level $l$. Then,*

$$\Sigma_{j:V_j' \cap C \neq \emptyset} D(V_j') \leq 2lD'.$$

*Proof.* The first property holds by induction because $u$ is chosen in each iteration to be a $D'$-center in a graph induced by a component of the previous level.

The second property holds also by induction over the levels of iterations as follows:

For level $l = 0$ all the clusters are singletons and therefore, have zero diameter and the property holds.

Assume the property holds for the previous levels and we will prove it for $l$:

The only affected clusters in iteration $l$ are the ones that intersect $V^u$. These clusters are all replaced with a new cluster that has diameter of at most $2D'$ plus the sum of all the diameters of the original clusters. Therefore, the property holds for $l$ as well.

**Theorem 5.** *Algorithm Restricted_Diameter applied with $D'$-values which result from binary search over $A$ is an $8 \log n + 4$-approximation algorithm for the minimum restricted diameter spanning tree problem. Its running time is $O(n^3 \log\log n)$.*

*Proof.* Consider a pair of vertices with requirement between them. They both belong to $V^u$ in some level $l$ of the algorithm. Let $C$ be their component in this level.

By Lemma 3 the diameter of their cluster which is an upper bound over their distance in $T$ is at most $2lD$ and $l \leq \log n + 1$.

The presentation of the algorithm assumes the graph is connected but for non-connected graphs vertices from other components may only serve as Steiner points and therefore, by Theorem 2, after multiplying by another factor of 4 one obtains the desired result.

To see the complexity of the algorithm note that using binary search over $A$ we test only $O(\log \log n)$ $D'$ values. It remains to show that each value can be tested in $O(n^3)$ time. To see this note that finding a center can be done in $O(n^3)$ time (by trying all the vertices as candidates to be a center, each candidate $u$ is tested using $BFS$ on $G \setminus E(V^u)$ in $O(n^2)$ time). Denote by $n_i$ the number of vertices in $C^i$ then the running time of a test value satisfies the following recursive

---

*Restricted_Diameter*
    **input**
    *Integers $D'$ and $l$.*
    *Connected graph $G' = (V', E')$.*
    *Length matrix d.*
    *A partition of $V'$ into clusters, $P = \{V'_1, V'_2, \ldots, V'_k\}$.*
    **returns**
    *A spanning tree $T$.*
    **begin**
    $T := (V', \emptyset)$.
    $V'' := \emptyset$.
    **if** $|V'| = 1$
      **then**
          **return** $T$.
      **else**
          $u :=$ *a $D'$-center of $G$ (if it does not exist conclude $D' < D^*$).*
          **for every** $i = 1, 2, \ldots, k$
            **if** $V^u \cap V'_i \setminus \{u\} \neq \emptyset$
            **then**
               *Choose $u_i \in V^u \cap V'_i$, $u_i \neq u$.*
               $T := T \cup \{(u, u_i)\}$.
               $V'' := V'' \cup V'_i$.
               $P := P \setminus V'_i$.
          **end if**
        $P := P \cup V''$.
        $C^1, C^2, \ldots C^r :=$ *components returned by the decomposition procedure when applied to $(G', D', u)$.*
        **for every** $i \in \{1, 2, \ldots, r\}$
            $P_i := \{U \cap C^i | U \in P, \ U \cap C^i \neq \emptyset\}$.
            $T_i :=$ *Restricted_Diameter$(D', l+1, G(C^i), d(C^i), P_i)$.*
        $T := T \cup T_1 \cup \cdots \cup T_r$.
    **end if**
    **return** $T$.
    **end** *Restricted_Diameter*

---

**Fig. 4.** Algorithm Restricted_Diameter

relation: $T(n) \leq cn^3 + \Sigma_{i=1}^r T(n_i)$ where $n_i \leq \frac{n}{2}$ $\forall i$ and $\Sigma_{i=1}^r n_i \leq n$. $\Sigma_{i=1}^r T(n_i)$ is maximized when $r = 2$ and $n_1 = n_2 = \frac{n}{2}$, and therefore $T(n) \leq 2cn^3$.

# References

1. N. Alon, R. Karp, D. Peleg and D. West, "A graph theoretic game and its application to the $k$-server problem", *SIAM J. Comput.*, **24**, 78-100, 1995.
2. Y. Bartal, "Probabilistic approximation of metric spaces and its algorithmic applications", *Proceedings of FOCS 1996*, pages 184-193.
3. Y. Bartal, "On approximating arbitrary metrics by tree metrics", *Proceedings of STOC 1998*, pages 161-168.
4. M. Charikar, C. Chekuri, A. Goel and S. Guha, "Rounding via trees: deterministic approximation algorithms for group steiner trees and $k$-median", *Proceedings of STOC 1998*, pages 114-123.
5. A. Gupta, "Steiner points in tree metrics don't (really) help", *Proceeding of SODA 2001*, pages 220-227.
6. M. R. Garey and D. S. Johnson, "Computers and intractability: a guide to the theory of NP-completeness", W.H. Freeman and Company, 1979.
7. R. Hassin and A. Tamir, "On the minimum diameter spanning tree problem", *IPL*, **53**, 109-111, 1995.
8. F. Harary, "Graph theory", Addison-Wesley, 1969.
9. J.-M. Ho, D. T. Lee, C.-H. Chang, and C. K. Wong, "Minimum diameter spanning trees and related problems", *SIAM J. Comput.*, **20 (5)**, 987-997, 1991.
10. S. Khuller, B. Raghavachari and N. Young, "Designing multi-commodity flow trees", *IPL*, **50**, 49-55, 1994.
11. D. Peleg and E. Reshef, "Deterministic polylog approximation for minimum communication spanning trees", *Proceedings of ICALP 1998*, pages 670-681.
12. P. Seymour and R. Thomas, "Call routing and the rat catcher", *Combinatorica*, **14**, 217-241, 1994.
13. A. Tamir, *Private communication*, 2001.

# Hardness of Approximation for Vertex-Connectivity Network-Design Problems

Guy Kortsarz[1], Robert Krauthgamer[2]*, and James R. Lee[3]

[1] Department of Computer Sciences, Rutgers University, Camden, NJ 08102, USA.
guyk@crab.rutgers.edu
[2] International Computer Science Institute (ICSI) and Computer Science Division,
University of California, Berkeley, CA 94720, USA.
robi@cs.berkeley.edu
[3] Computer Science Division, University of California, Berkeley, CA 94720, USA.
jrl@cs.berkeley.edu

**Abstract.** In the survivable network design problem SNDP, the goal is to find a minimum-cost subgraph satisfying certain connectivity requirements. We study the vertex-connectivity variant of SNDP in which the input specifies, for each pair of vertices, a required number of vertex-disjoint paths connecting them.

We give the first lower bound on the approximability of SNDP, showing that the problem admits no efficient $2^{\log^{1-\epsilon} n}$ ratio approximation for any fixed $\epsilon > 0$ unless $\mathsf{NP} \subseteq \mathsf{DTIME}(n^{\mathrm{polylog}(n)})$. We also show hardness of approximation results for several important special cases of SNDP, including constant factor hardness for the $k$-vertex connected spanning subgraph problem ($k$-VCSS) and for the vertex-connectivity augmentation problem, even when the edge costs are severely restricted.

## 1 Introduction

A basic problem in network design is to find, in an input graph $G = (V, E)$ with nonnegative edge costs, a subgraph of minimum-cost that satisfies certain connectivity requirements, see e.g. the surveys [Fra94,Khu96]. A fundamental problem in this area is the vertex-connectivity variant of the *survivable network design problem* (SNDP). In this problem, the input also specifies a *connectivity requirement* $k_{u,v}$ for every pair of vertices $\{u, v\}$, and the goal is to find a subgraph of minimum-cost such that for every pair of vertices $\{u, v\}$ there are at least $k_{u,v}$ *vertex-disjoint* paths between $u$ and $v$.

Many network design problems (including SNDP) are NP-hard, and a significant amount of research is concerned with *approximation algorithms* for these problems, i.e. polynomial-time algorithms that find a solution whose value is guaranteed to be within some factor (called the *approximation ratio*) of the

---

* Supported in part by NSF grants CCR-9820951 and CCR-0121555 and DARPA cooperative agreement F30602-00-2-0601.

K. Jansen et al. (Eds.): APPROX 2002, LNCS 2462, pp. 185–199, 2002.
© Springer-Verlag Berlin Heidelberg 2002

optimum. A notable success is the 2-approximation of Jain [Jai01] for the edge-connectivity version of SNDP, in which the paths are only required to be *edge-disjoint*. (See also [FJW01] for an extension to a more general version of SNDP.) However, no algorithm is known to achieve a sublinear (in $|V|$) approximation ratio for the vertex-connectivity variant of SNDP.

This disparity between approximating edge-connectivity and vertex-connectivity network design problems might suggest a lack in our understanding of vertex-connectivity or, perhaps, that vertex-connectivity problems are inherently more difficult to approximate. Resolving this question (see e.g. [Vaz01, Section 30.2]) is one of the important open problems in the field of approximation algorithms. We provide an answer by showing that there is a striking difference between the approximability of the edge and vertex-connectivity variants of this problem. Specifically, we show that it is hard to approximate the vertex-connectivity variant of SNDP within a factor of $2^{\log^{1-\epsilon}|V|}$ for any fixed $\epsilon > 0$.

In general, we address the *hardness of approximation* of vertex-connectivity problems by presenting relatively simple variants of SNDP that are nevertheless hard to approximate. Therefore, unless stated otherwise, *connectivity* means vertex-connectivity, *disjoint paths* means vertex-disjoint paths, and all graphs are assumed to be undirected. (For a more in-depth account of approximation algorithms for edge-connectivity problems, see [Fra94,Khu96].) Throughout, let $n$ denote the number of vertices in the input graph $G$.

A classical and well-studied special case of SNDP is the problem of finding a minimum-cost $k$-vertex connected spanning subgraph, i.e., the special case where $k_{u,v} = k$ for all vertex pairs $\{u, v\}$. This is called the *$k$-vertex connected spanning subgraph problem* ($k$-VCSS). A related problem is the *vertex-connectivity augmentation problem* (VCAP$_{\ell,k}$), where the goal is to find a minimum-cost set of edges that augments an $\ell$-connected graph into a $k$-connected graph. We exhibit hardness of approximation results for both of these problems.

Other special cases of SNDP for which we show hardness of approximation are the *subset connectivity problem* and the *outconnectivity to a subset problem* (OSP). In the first problem, the input contains a subset $S$ of the vertices, and the goal is to find a minimum-cost subgraph that contains at least $k$ vertex-disjoint paths between every pair of vertices in $S$. This is SNDP with $k_{u,v} = k$ for all $u, v \in S$ and $k_{u,v} = 0$ otherwise. In the second problem, the input contains a special vertex $r$ (called the *root*) and a subset $S$ of the vertices, and the goal is to find a minimum-cost subgraph that contains at least $k$ vertex-disjoint paths between $r$ and any vertex in $S$. In other words, this is SNDP with $k_{r,v} = k$ for all $v \in S$ and $k_{u,v} = 0$ otherwise.

## 1.1   Previous Work

$k$-VCSS is NP-hard even for $k = 2$ and *uniform costs* (i.e., all edges have the same cost), as this problem already generalizes the Hamiltonian cycle problem (note that a 2-connected subgraph of $G$ has $n$ edges if and only if it is a Hamiltonian cycle). By a similar argument, the outconnectivity to a subset problem is also NP-hard, even for $k = 2$ and $S = V \setminus \{r\}$ [CJN01]. It immediately follows that SNDP (which is a more general problem) is also NP-hard. $VCAP_{0,2}$ is NP-hard by a similar argument [ET76], and $VCAP_{1,2}$ is proved to be NP-hard in [FJ81].

Most previous work on approximating vertex-connectivity problems concentrated on upper bounds, i.e., on designing approximation algorithms. An approximation ratio of $2k$ for $k$-VCSS was obtained in [CJN01] by a straightforward application of [FT89], and the approximation ratio was later improved to $k$ in [KN00]. Recently, Cheriyan, Vempala and Vetta [CVV02] devised improved approximation algorithms for the problem. Their first algorithm achieves approximation ratio $6H(k) = O(\log k)$, where $H(k)$ is the $k$th harmonic number, for the case where $k \leq \sqrt{n/6}$. Their second algorithm achieves an approximation ratio $\sqrt{n/\epsilon}$ for the case where $k \leq (1 - \epsilon)n$. (An approximation ratio of $O(\log k)$ claimed in [RW97] was found to be erroneous, see [RW02].)

Better approximation ratios are known for several special cases of $k$-VCSS. For $k \leq 7$ an approximation ratio of $\lceil (k+1)/2 \rceil$ is known (see [KR96] for $k = 2$, [ADNP99] for $k = 2, 3$, [DN99] for $k = 4, 5$, and [KN00] for $k = 6, 7$). For *metric costs* (i.e., when the costs satisfy the triangle inequality) an approximation ratio $2 + \frac{(k-1)}{n}$ is given in [KN00] (building on a ratio $2 + \frac{2(k-1)}{n}$ previously shown in [KR96]). For uniform costs, an approximation ratio of $1 + 1/k$ is obtained in [CT00]. For a complete Euclidean graph, a polynomial time approximation scheme (i.e., factor $1 + \epsilon$ for any fixed $\epsilon > 0$) is devised in [CL99].

The connectivity augmentation problem has also attracted a lot of attention. A 2-approximation for $VCAP_{1,2}$ is shown in [FJ81,KT93]. In the case where every pair of vertices in the graph forms an augmenting edge of unit cost, $VCAP_{k,k+1}$ is not known to be in P nor to be NP-hard. For the latter problem, a $k - 2$ additive approximation is presented in [Jor95], and optimal algorithms for small values of $k$ are shown in [ET76,WN93,HR91,Hsu92].

The special case of OSP with $S = V \setminus \{r\}$ (called the $k$-outconnectivity problem), can be approximated within ratio 2, see for example [KR96]. Approximation algorithms for related problems are given in [CJN01].

In contrast, there are almost no lower bounds for approximating vertex-connectivity problems. It is shown in [CL99] that 2-VCSS is APX-hard even for bounded-degree graphs with uniform costs and for complete Euclidean graphs in $\mathbb{R}^{\log n}$. No stronger lower bound is known for the more general SNDP.

## 1.2   Our Results

We give hardness of approximation results for several of these vertex-connectivity network design problems. In Section 2 we show that 2-VCSS is APX-*hard*, i.e., there exists some fixed $\epsilon > 0$ such that it is NP-hard to approximate 2-VCSS within ratio $1 + \epsilon$, even in the case of uniform costs. (This result is similar to, but was obtained independently from, [CL99].) We then conclude that $k$-VCSS is APX-hard for any $k \geq 2$, even in the case of 0-1 costs. In particular, the latter is strictly harder to approximate then $k$-VCSS with uniform costs (which can be approximated within ratio $1 + 1/k$), unless P = NP. In addition, we exhibit

APX-hardness for $VCAP_{1,2}$, even in the case where all edges have cost 1 or 2. From this, it follows that $VCAP_{k,k+1}$ with uniform costs is APX-hard for every $k \geq 2$. For fixed $k$, these hardness results match, up to constant factors, the approximation algorithms mentioned in Section 1.1.

In Section 3, we show that the outconnectivity to a subset problem (OSP) cannot be approximated within a ratio of $(\frac{1}{3} - \epsilon) \ln n$ for any fixed $\epsilon > 0$, unless $NP \subseteq DTIME(n^{O(\log \log n)})$. It follows that OSP with a general subset $S$ is much harder to approximate than the special case $S = V \setminus \{r\}$ (which can be approximated within ratio 2). This problem is a special case of SNDP that is already harder to approximate than the edge-connectivity variant of SNDP (which can be approximated within ratio 2), unless $NP \subseteq DTIME(n^{O(\log \log n)})$.

In Section 4, we show that SNDP cannot be approximated within a ratio of $2^{\log^{1-\epsilon} n}$, for any fixed $\epsilon > 0$, unless $NP \subseteq DTIME(n^{\text{polylog}(n)})$. This hardness of approximation result extends also to the subset $k$-connectivity problem which is a special case of SNDP. The lower bound holds for $k = n^\rho$ where $0 < \rho < 1$ is any fixed constant. It then follows that SNDP is provably harder to approximate than $k$-VCSS (which can be approximated within a ratio of $O(\log k)$ for $k \leq \sqrt{n/6}$), unless $NP \subseteq DTIME(n^{\text{polylog}(n)})$.

## 1.3    Preliminaries

For an arbitrary graph $G$, let $V(G)$ denote the vertex set of $G$ and let $E(G)$ denote the edge set of $G$. For a nonnegative cost function $c$ on the edges of $G$ and a subgraph $G' = (V', E')$ of $G$ we use the notation $c(G') = c(E') = \sum_{e \in E'} c(e)$. We denote the set of *neighbors* of a vertex $v$ in $W \subset V$ (namely, the vertices $w \in W$ such that $(v, w) \in E$) by $N(v, W, G)$. When $W = V(G)$ we omit $W$ and write $N(v, G)$, and when $G$ is clear from the context, we use simply $N(v)$.

A set $W$ of $k$ vertices in a graph $G = (V, E)$ is called a *k-vertex-cut* (or just a *vertex-cut*) if the subgraph of $G$ induced on $V \setminus W$ has at least two connected components. A vertex $w \in V$ is called a *cut-vertex* if $W = \{w\}$ is a vertex-cut. We will use the following classical result.

**Theorem 1 (Menger's Theorem, see e.g. [Die00]).**

   a. *A graph $G$ contains at least $k$ vertex-disjoint paths between two of its vertices $u, v$ if and only if every vertex-cut that separates $u$ from $v$ must be of size at least $k$.*
   b. *A graph $G$ is $k$-vertex connected if and only if it has no $(k-1)$-vertex cut.*

## 2    Vertex-Connectivity Spanning Subgraph and Augmentation

### 2.1    The $k$-Vertex Connected Spanning Subgraph Problem

We consider the minimum-cost $k$-vertex connected spanning subgraph problem ($k$-VCSS), both in the uniform cost model, and in models with restricted edge costs. The main result of this section is that for some fixed $\epsilon > 0$ and every $k \geq 2$, it is NP-hard to approximate $k$-VCSS within a factor of $(1 + \epsilon)$, even in the special case where edges have only costs 0 and 1.

Throughout, let $k$-VCSS$(a, b)$ denote the $k$-VCSS problem on the *complete* graph where edges have cost $a$ or $b$. Note that $k$-VCSS with uniform costs is just $k$-VCSS$(1, \infty)$.

**Theorem 2.** *For some fixed $\epsilon > 0$, it is* NP-*hard to approximate the* 2-VCSS$(1, 2)$ *problem within a factor of* $(1 + \epsilon)$.

We prove Theorem 2 by a reduction from the MINTSP$(1, 2)$ problem: Given a complete graph $G = (V, E)$, and a weight function on edges $w : E \to \{1, 2\}$, find a minimum weight tour of $G$. As noted in the introduction, a similar reduction was discovered independently by Czumaj and Lingas [CL99]. We include a proof here for completeness. The following hardness of approximation result for MINTSP$(1, 2)$ follows from [PY93] (and the results of [ALM+98]).

**Theorem 3 ([PY93] and [ALM+98]).** *For some fixed $\epsilon > 0$, it is* NP-*hard to distinguish whether an instance of* MINTSP$(1, 2)$ *on $n$ vertices has an optimal tour of cost $n$ or every tour has cost at least $n(1 + \epsilon)$.*

*Proof (Theorem 2).* We show a gap-preserving reduction from MINTSP$(1, 2)$. Let $G = (V, E)$ be an instance of this problem. Let MINTSP$(G)$ be the weight of an optimal tour of $G$ and let 2-VCSS$(G)$ be the cost of an optimal 2-connected spanning subgraph of $G$, where the cost of an edge is equal to its weight. We will show that for some suitable $\epsilon' > 0$,

$$\text{MINTSP}(G) = n \Longrightarrow \text{2-VCSS}(G) = n$$
$$\text{MINTSP}(G) \geq n(1 + \epsilon) \Longrightarrow \text{2-VCSS}(G) \geq n(1 + \epsilon')$$

Suppose that MINTSP$(G) = n$. Then an optimal tour of $G$ is a cycle with $n$ edges of weight 1. Since a cycle is the smallest possible 2-connected graph on $n$ vertices, it follows that 2-VCSS$(G) = n$.

Now consider a graph $G$ for which MINTSP$(G) \geq n(1 + \epsilon)$ and suppose, for the sake of contradiction, that there exists a 2-connected spanning subgraph $H = (V, E_H)$ such that $\text{cost}(H) < n(1+\epsilon')$. Let $\deg_H(v)$ be the degree of a vertex $v$ in $H$. Since $H$ is 2-connected, $\deg_H(v) \geq 2$ for all $v$. We will repeatedly remove from $H$ edges that are adjacent to vertices whose degree exceeds 2 until no such vertices remain, and then paste the connected components of the resulting graph $H'$ together to obtain a tour of $G$. Let $R$ be the number of edges removed in this process. Note that for any vertex $v$, we remove at most $\deg_H(v) - 2$ edges adjacent to that vertex. It follows that

$$R \leq \sum_{v \in V} (\deg_H(v) - 2) = 2 \cdot |E_H| - 2n \leq 2 \cdot \text{cost}(H) - 2n < 2n\epsilon'.$$

Since we remove $R$ edges from a connected graph $H$, the resulting graph $H'$ has at most $R + 1$ connected components. Each of these components consists of vertices whose degree is at most 2, and hence it is either an isolated vertex, a path, or a cycle. We arbitrarily remove one edge from every cycle without increasing the cost of $H'$. Finally, we add at most $R + 1$ edges to connect the paths (and isolated vertices) from end to end to obtain a tour $T$ of $G$. The cost of each of these additional edges is at most 2, and thus

$$\text{weight}(T) \leq \text{cost}(H) - R + 2(R + 1) < n(1 + \epsilon') + 2n\epsilon' + 2 = n(1 + 3\epsilon') + 2.$$

Choosing $\epsilon' < \epsilon/3$ yields the desired contradiction for sufficiently large $n$. The theorem follows. □

It is straightforward to extend Theorem 2 to 2-VCSS with uniform costs. Indeed, restrict the above instance $G$ of 2-VCSS(1,2) to just the edges of cost 1. Let $\bar{G}$ denote the resulting instance of 2-VCSS with uniform costs. It is easy to see that 2-VCSS($\bar{G}$) $\geq$ 2-VCSS($G$), and that if 2-VCSS($G$) $= n$ then also 2-VCSS($\bar{G}$) $= n$. We thus obtain the following.

**Corollary 1.** *For some fixed $\epsilon > 0$, it is NP-hard to approximate 2-VCSS with uniform costs within a factor of $(1 + \epsilon)$.*

The main result of this section now follows easily from Corollary 1.

**Theorem 4.** *For every $k \geq 2$ and some fixed $\epsilon > 0$ (which is independent of $k$), it is NP-hard to approximate $k$-VCSS with edge costs 0 and 1 within a factor of $(1 + \epsilon)$.*

*Proof.* Given a $(0,1)$-weighted graph $G = (V, E)$ which is $k$-connected, we construct a graph $G'$ by adding a new vertex $v'$ that is connected to every vertex in $V$ by an edge of cost 0. Call this set of edges $E_{new} = \{(v', v) : v \in V\}$. We claim (below) that $k$-VCSS($G$) $= (k + 1)$-VCSS($G'$) (the gap is trivially preserved). The theorem then follows by induction with hardness of the base case, $k = 2$, following from Corollary 1.

To prove the claim, suppose first that $H = (V, E_H)$ is a $k$-connected subgraph of $G$. Then $H' = (V \cup \{v'\}, E_H \cup E_{new})$ is a $(k+1)$-connected subgraph of $G'$ and has the same cost as $H$. Hence, $(k + 1)$-VCSS($G'$) $\leq k$-VCSS($G$). Conversely, let $H'$ be a $(k+1)$-connected subgraph of $G'$. Obtain a graph $H$ by removing $v'$ and all incident edges from $H'$. $H$ is a $k$-connected subgraph of $G$ and has the same cost as $H'$. It follows that $k$-VCSS($G$) $\leq (k + 1)$-VCSS($G'$). □

It is shown in [CT00] that $k$-VCSS with uniform costs is approximable within a factor of $1 + 1/k$, for any fixed $k$. Notice that this ratio approaches 1 as $k \to \infty$. Together with Theorem 4, this yields the following.

**Corollary 2.** *For a sufficiently large fixed $k > 0$, $k$-VCSS with uniform costs is strictly easier to approximate than $k$-VCSS with edge costs 0 and 1, unless P = NP.*

Finally, the analysis in Theorem 2 and the preceding corollary extend also to the corresponding edge-connectivity problem. Such a hardness of approximation result was previously shown in [Fer98] but our proof is considerably simpler.

**Theorem 5.** *For some fixed $\epsilon > 0$, it is NP-hard to approximate the minimum-cost 2-edge-connected spanning subgraph (2-ECSS) problem with uniform costs within a factor of $(1 + \epsilon)$.*

## 2.2 The Vertex-Connectivity Augmentation Problem

The vertex-connectivity augmentation problem ($\text{VCAP}_{k,\ell}$) is defined as follows. Given a $k$-connected graph $G_0 = (V, E_0)$ and a cost function $c : V \times V \to \mathbb{N}$, find a set $E_1 \subseteq V \times V$ of minimum cost so that $G_1 = (V, E_0 \cup E_1)$ is $\ell$-connected. Since all graphs considered here are simple, we will not allow $G_1$ to contain self-loops. $\text{VCAP}_{k,\ell}(a, b)$ will represent a version of the problem where edges have only cost $a$ or $b$ (so $c : V \times V \to \{a, b\}$). The main result of this section is that for some fixed $\epsilon > 0$ and for every $k \geq 1$, it is NP-hard to approximate $\text{VCAP}_{k,k+1}(1, 2)$ within a factor of $(1 + \epsilon)$.

It is possible to convert any instance of $\text{VCAP}_{k_0,k_0+\alpha}$ to an "equivalent" instance of $\text{VCAP}_{k_0+1,k_0+1+\alpha}$ by adding to $G_0$ a new vertex that is connected to every old vertex (similar to the proof of Theorem 4). This yields the following easy lemma.

**Lemma 1.** *If it is NP-hard to approximate $\text{VCAP}_{k_0,k_0+\alpha}$ with edge costs from a set $S$ within a factor of $(1 + \epsilon)$ for some $\alpha \geq 1$, then the same is true of $\text{VCAP}_{k,k+\alpha}$ with edge costs from $S$ for any $k \geq k_0$.*

It is straightforward that $\alpha$-VCSS with edge costs 0 and 1 can be thought of as $\text{VCAP}_{0,\alpha}(0, 1, \infty)$, by setting the cost of each $(u, v) \notin E$ to $\infty$. (Note that we need not use $\infty$; instead we can make each edge not in $E$ have cost $n^2$ so that no optimal solution uses such an edge). Therefore, combining Theorem 4 and Lemma 1 yields the following result.

**Corollary 3.** *For any $k \geq 2$, for any $\alpha \geq 2$, and for some fixed $\epsilon > 0$ (independent of $\alpha$ and $k$), it is NP-hard to approximate $\text{VCAP}_{k,k+\alpha}(0, 1, \infty)$ within a factor of $(1 + \epsilon)$.*

Notice that the above corollary is not applicable to the case $\alpha = 1$. Furthermore, one may wonder about the difficulty of the problem when the edge weights are severely restricted, say to only 1 and 2, or in a model with uniform edge costs. We resolve both of these concerns in the following theorem.

**Theorem 6.** *For any $k \geq 1$ and some fixed $\epsilon > 0$ (independent of $k$), it is NP-hard to approximate $\text{VCAP}_{1,2}(1, 2)$ within a factor of $(1 + \epsilon)$.*

The proof of theorem 6 employs a reduction from 3-dimensional matching (3DM) that was used in [FJ81] to prove that solving $\text{VCAP}_{1,2}(1, 2)$ (optimally) is NP-hard. We obtain a stronger result (i.e., hardness of approximation) by a more involved analysis of the reduction (and by relying on the hardness of approximating a bounded version of the 3-dimensional matching problem, shown in [Pet94]). Due to space considerations, the proof is deferred to the full version of this paper. The main result of this section follows by applying Lemma 1 to the above theorem.

**Corollary 4.** *For any $k \geq 1$ and some fixed $\epsilon > 0$ (independent of $k$), it is* NP-*hard to approximate* $\mathrm{VCAP}_{k,k+1}(1,2)$ *within a factor of $(1 + \epsilon)$. The same result holds for* $\mathrm{VCAP}_{k,k+1}(1,\infty)$, *i.e.* VCAP *with uniform costs.*

**Remark.** The techniques used to prove Theorem 6 can be used to show that a number of other connectivity augmentation problems are APX-hard (including the edge-connectivity variant of VCAP). Also, both Theorem 4 and Corollary 4 remain true even when $k = k(n)$ grows with $n$ and $k(n)$ is computable in polynomial time.

# 3   The Outconnectivity to a Subset Problem

We prove the following theorem in the full version of this paper. The proof is based on a hardness result shown by Feige et al. [FHKS02] for the problem of packing set-covers (in a set-system).

**Theorem 7.** *The outconnectivity to a subset problem cannot be approximated within a ratio of $(\frac{1}{3} - \epsilon) \ln n$ for any fixed $\epsilon > 0$, unless* NP $\subseteq$ DTIME$(n^{O(\log \log n)})$.

# 4   Survivable Network Design and the Subset Connectivity Problem

In this section we show hardness results for approximating the subset connectivity problem, and thus also SNDP (which is a more general problem), within a ratio of $2^{\log^{1-\epsilon} n}$ for any fixed $\epsilon > 0$. These lower bounds are proven by a reduction from a graph-theoretic problem, called MINREP that is defined in [Kor01]. This problem is closely related to the LABELCOVER$_{\max}$ problem of [AL96] and to the parallel repetition theorem of [Raz98]. We first describe the MINREP problem and the hardness results known for it in Section 4.1. Then, in Section 4.2, we give a reduction from MINREP to SNDP. Finally, we adapt this reduction to the subset connectivity problem in Section 4.3.

## 4.1   The MINREP Problem

Arora and Lund [AL96] introduced the LABELCOVER$_{\max}$ problem as a graph-theoretic description of one-round two-prover proof systems, for which the parallel repetition theorem of Raz [Raz98] applies. The MINREP problem described below is closely related to LABELCOVER$_{\max}$ and was defined in [Kor01] for the same purpose.

The input to the MINREP problem consists of a bipartite graph $G(A, B, E)$, with an explicit partitioning of each of $A$ and $B$ into equal-sized subsets, namely, $A = \bigcup_{i=1}^{q_A} A_i$, $B = \bigcup_{j=1}^{q_B} B_j$ where all the sets $A_i$ have the same size $m_A$ and all the sets $B_j$ have the same size $m_B$. The bipartite graph $G$ induces a *super-graph* $H$, as follows. The *super-vertices* (i.e., the vertices of $H$) are the $q_A + q_B$ sets $A_i$

and $B_j$. A *super-edge* (i.e., an edge in $H$) connects two super-vertices $A_i$ and $B_j$ if there exist $a \in A_i$ and $b \in B_j$ that are adjacent in $G$ (i.e., $(a, b) \in E$).

A pair $(a, b)$ *covers* a super-edge $(A_i, B_j)$ if $a \in A_i$ and $b \in B_j$ are adjacent in $G$ (i.e., $(a, b) \in E$). Let $A'_i \subseteq A_i$ and $B'_j \subseteq B_j$. (The vertices of $A'_i$ and $B'_j$ can be thought of as *representatives* of $A_i$ and $B_j$, respectively.) We say that $A'_i \cup B'_j$ *covers* the super-edge $(A_i, B_j)$ if there exists a pair $(a, b)$ that covers the super-edge $(A_i, B_j)$, namely, there exists a vertex $a_i \in A'_i$ and a vertex $b_j \in B'_j$ such that $(a_i, b_j) \in E$.

The goal in the MINREP problem is to select representatives from the sets $A_i$ and $B_j$ such that all the super-edges are covered and the total number of representatives selected is minimal. That is, we wish to find subsets $A' \subseteq A$ and $B' \subseteq B$ with minimal total size $|A| + |B|$, such that for every super-edge $(A_i, B_j)$ there exist representatives $a \in A' \cap A_i$ and $b \in B' \cap B_j$ that are adjacent in $G$ (i.e., $(a, b) \in E$).

For our purposes, it is convenient (and possible) to restrict the MINREP problem so that for every super-edge $(A_i, B_j)$, each vertex in $B_j$ is adjacent to at most one vertex in $A_i$. We call this additional property of $G$ the *star property* because it is equivalent to saying that for every super-edge $(A_i, B_j)$ the subgraph of $G$ induced on $A_i \cup B_j$ is a collection of vertex-disjoint stars whose centers are in $A_i$. [1] (See Figure 1 for an illustration.)

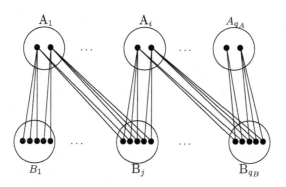

**Fig. 1.** An instance of MINREP with the star property

The next theorem follows by a straightforward application of the parallel repetition theorem of Raz [Raz98], since the MINREP problem is graph-theoretic description of two-prover one-round proof systems. The additional property of the MINREP instance (namely, the star property) is achieved by using a specific proof system for MAX3SAT(5). A description can be found in [Fei98, Section 2.2].

---

[1] A *star* is a graph all of whose vertices have degree 1, except for one vertex that has degree larger than 1. The vertex with degree larger than 1 is called the *center* of the star. The vertices of degree 1 are called *leaves* of the star.

**Theorem 8.** *Let $L$ be any* NP-*complete language and fix $\epsilon > 0$. Then there exists an algorithm (i.e., a reduction), whose running time is quasi-polynomial, namely $n^{\mathrm{polylog}(n)}$, and that given an instance $x$ of $L$ produces an instance $G(A, B, E)$ of the* MinRep *problem with the star property, such that the following holds.*

- *If $x \in L$ then the* MinRep *instance $G$ has a solution of value $q_A + q_B$ (namely, with one representative from each $A_i$ and one from each $B_j$).*
- *If $x \notin L$ then the value of any solution of the* MinRep *instance $G$ is at least $(q_A + q_B) \cdot 2^{\log^{1-\epsilon} n}$, where $n$ is the number of vertices in $G$.*

*Hence, the* MinRep *problem cannot be approximated within ratio $2^{\log^{1-\epsilon} n}$, for any fixed $\epsilon > 0$, unless* NP $\subseteq$ DTIME($n^{\mathrm{polylog}(n)}$).

## 4.2   The Survivable Network Design Problem

We prove the following theorem by a reduction whose starting point is Theorem 8.

**Theorem 9.** *The* SNDP *problem cannot be approximated within ratio $2^{\log^{1-\epsilon} n}$, for any fixed $\epsilon > 0$, unless* NP $\subseteq$ DTIME($n^{\mathrm{polylog}(n)}$).

*The reduction.* Given the instance $G(A, B, E)$ of the MinRep problem as described in Section 4.1, create an instance $\bar{G}(\bar{V}, \bar{E})$ of SNDP as follows. (See Figure 2 for illustration.)

1. Take $G$ and let all its edges have cost 0.
2. For each $i = 1, \ldots, q_A$ create a new vertex $u_i$ that is connected to every vertex in $A_i$ by an edge of cost 1. Similarly, for each $j = 1, \ldots, q_B$ create a new vertex $w_j$ that is connected to every vertex in $B_j$ by an edge of cost 1. Let $U = \{u_1, \ldots, u_{q_A}\}$ and $W = \{w_1, \ldots, w_{q_B}\}$.
3. For each super-edge $(A_i, B_j)$ create two new vertices $x_i^j$ and $y_j^i$. For every $i$, let $X_i = \{x_i^j\}$ consist of the $d$ vertices $x_i^j$ (note that the values of $j$ need not be consecutive), and connect every vertex of $X_i$ to $u_i$ by edges of cost 0. Similarly, for every $j$ let $Y_j = \{y_j^i\}$ and connect every vertex in $Y_j$ to $w_j$ by an edge of cost 0.
4. For every super-edge $(A_i, B_j)$ connect every vertex in $\{x_i^j, y_j^i\}$ to every vertex in $(A \setminus A_i) \cup (B \setminus B_j)$ by an edge of cost 0.
5. Let $X = \cup_{i=1}^{q_A} X_i$ and $Y = \cup_{j=1}^{q_B} Y_j$. Connect every two vertices in $X \cup Y$ by an edge of cost 0.
6. Finally, require $k = |X| + |Y| + (q_A - 1)m_A + (q_B - 1)m_B$ vertex-disjoint paths from $x_i^j$ to $y_j^i$ for every super-edge $(A_i, B_j)$. (Recall that $|X| = |Y| = q_A \cdot d_A = q_B \cdot d_B$.)

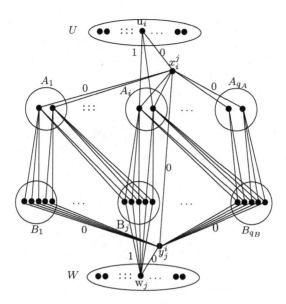

**Fig. 2.** The vertices $x_i^j, y_j^i$ in the SNDP instance $\bar{G}$

*The Analysis.* Suppose $x \in L$ and then by Theorem 8 there exists a choice of $q_A + q_B$ representatives (one representative from each $A_i$ and one from each $B_j$) that cover all the super-edges. Let $G'$ be the subgraph of $\bar{G}$ that contains an edge between each $u_i$ and the representative chosen in $A_i$, an edge between each $w_j$ and the representative chosen in $B_j$, and all the edges of cost 0 in $\bar{G}$. Let us now show that $G'$, whose cost is clearly $2q$, is a solution to the instance $\bar{G}$ of the SNDP problem. Consider a pair of vertices $x_i^j, y_j^i$ such that $(A_i, B_j)$ is a super-edge in $G$. Each vertex in $\mathcal{F}_{i,j} = (X \setminus \{x_i^j\}) \cup (Y \setminus \{y_j^i\}) \cup (A \setminus A_i) \cup (B \setminus B_j)$ defines a path of length 2 in $G'$ between $x_i^j$ and $y_j^i$, and the edge $(x_i^j, y_j^i)$ defines a path of length 1, so we get a total of $|X| - 1 + |Y| - 1 + (q_A - 1)m_A + (q_B - 1)m_B + 1 = k - 1$ vertex-disjoint paths between $x_i^j$ and $y_j^i$. There is an additional path that goes through $V \setminus \mathcal{F}_{i,j} = U \cup W \cup A_i \cup B_j \cup \{x_i^j, y_j^i\}$, namely, $x_i^j - u_i - a_i - b_j - w_j - y_j^i$ where $a_i$ and $b_j$ are the representatives chosen from $A_i$ and $B_j$, respectively. This additional path is clearly vertex-disjoint from the others, yielding $k$ vertex-disjoint paths between $x_i^j$ and $y_j^i$.

The next lemma will be used to complete the proof of Theorem 9. Let $G'$ be a feasible solution to this instance of the SNDP problem, i.e. a subgraph of $\bar{G}$ in which $x_i^j$ is $k$-vertex connected to $y_j^i$ for every super-edge $(A_i, B_j)$.

**Lemma 2.** *For every super-edge* $(A_i, B_j)$, *the subgraph* $G'$ *contains an edge connecting* $u_i$ *to some* $a_i \in A_i$, *and an edge connecting* $w_j$ *to some* $b_j \in B_j$, *such that* $(a_i, b_j) \in E$ *(i.e., the pair* $(a_i, b_j)$ *covers the super-edge).*

*Proof.* Since $G'$ is a feasible solution, it contains $k$ vertex-disjoint paths between $x_i^j$ and $y_j^i$. Let $\mathcal{F}_{i,j} = (X \setminus \{x_i^j\}) \cup (Y \setminus \{y_j^i\}) \cup (A \setminus A_i) \cup (B \setminus B_j)$. At most

$|\mathcal{F}_{i,j}| = |X| - 1 + |Y| - 1 + (q_A - 1)m_A + (q_B - 1)m_B = k - 2$ of these $k$ paths can visit vertices of $\mathcal{F}_{i,j}$, and at most one of these paths can use the edge $(x_i^j, y_j^i)$. (This is essentially a "mixed" cut that contains both vertices and edges; see Figure 3.) Hence, $G'$ contains a path between $x_i^j$ and $y_j^i$, that visits only vertices of $\bar{V} \setminus \mathcal{F}_{i,j} = U \cup W \cup A_i \cup B_j \cup \{x_i^j, y_j^i\}$ and whose length is at least two.

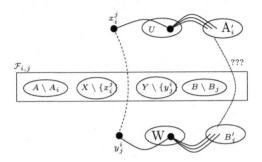

**Fig. 3.** A mixed cut with $k - 2$ vertices and one edge.

Observe that in the subgraph of $G'$ induced on $\bar{V} \setminus \mathcal{F}_{i,j}$ the following holds. (Assume without loss of generality that $G'$ contains all the edges of cost 0 in $\bar{G}$.) The only neighbor of $x_i^j$ is $u_i$, so the vertices at distance 2 from $x_i^j$ (i.e., the neighbors of $u_i$ except for $x_i^j$) form a subset $A_i'$ of $A_i$. Thus, the vertices at distance 3 from $x_i^j$ (i.e. all the neighbors of $A_i'$ except for $u_i$) are all from $B_j$. Similarly, the only neighbor of $y_j^i$ is $w_j$, so vertices at distance 2 from $y_j^i$ form a subset $B_j'$ of $B_j$, and all vertices at distance 3 from $y_j^i$ are from $A_i$. Note that the subgraph of $\bar{G}$ induced on $A_i \cup B_j$ is a collection of vertex-disjoint stars, whose centers are in $A_i$ and whose leaves are in $B_j$. The aforementioned path in $G'$ between $x_i^j$ and $y_j^i$ (that visits only vertices of $\bar{V} \setminus \mathcal{F}_{i,j}$) then must be have the form $x_i^j - u_i - a_i - b_j - w_j - y_j^i$ with $a_i \in A_i'$ and $b_j \in B_j'$ (note that the other vertices of $U \cup W$ are unreachable), and the lemma follows.                □

We now complete the proof of Theorem 9. Suppose that $x \notin L$ and let $G'$ be a feasible solution to the instance $\bar{G}$ of the SNDP problem. Let $A_i'$ be the set of neighbors of $u_i$ among $A_i$ (in $G'$), and let $B_j'$ be the set of neighbors of $w_j$ among $B_j$ (in $G'$). By Lemma 2 the representatives $A' = \cup_i A_i'$ and $B' = \cup_j B_j'$ cover all the super-edges $(A_i, B_j)$, thus forming a feasible solution to the MINREP instance $G$. By Theorem 8 the value of this MINREP solution is $|A'| + |B'| \geq (q_A + q_B) \cdot 2^{\log^{1-\epsilon} n}$, where $n$ denotes the number of vertices in $G$. Observe that the cost of $G'$ is also $|A'| + |B'|$. Since the number of vertices in $\bar{G}$ is $n^{O(1)}$, the cost of $G'$ is at least $(q_A + q_B) \cdot 2^{\log^{1-\epsilon} |V(\bar{G})|}$, and the theorem follows.

### 4.3   The Subset $k$-Connectivity Problem

We adapt the reduction of Theorem 9 to the subset $k$-connectivity problem, as follows. We require that the subset $S = X \cup Y$ is $k$-vertex connected. For this $S$ to be $k$-vertex connected in the case $x \in L$, we add, for every $z, z' \in S$ that are not a pair $x_i^j, y_j^i$, a set $Q_{z,z'}$ of $k$ new vertices that are all connected to $z$ and to $z'$ by edges of cost 0. The analysis of the case $x \notin L$ (including the proof of Lemma 2) remains valid. Specifically, in the mixed cut defined by the vertices $\mathcal{F}_{i,j}$ and the edge $(x_i^j, y_j^i)$ (see Figure 3), the sets $Q_{x_i^j, z}$ are on the side of $x_i^j$ and similarly for $y_j^i$, and all other sets $Q_{z,z'}$ are isolated cliques. We thus obtain the following hardness of approximation result for the subset $k$-connectivity problem.

**Theorem 10.** *The subset $k$-connectivity problem cannot be approximated within ratio $2^{\log^{1-\epsilon} n}$, for any fixed $\epsilon > 0$, unless* $\mathsf{NP} \subseteq \mathsf{DTIME}(n^{\mathrm{polylog}(n)})$.

Note that in our reduction above $k$ is $\Theta(|E(\bar{G})|)$ while the number of vertices is $\Theta(k|E(\bar{G})|^2)$. Therefore, our hardness result for the subset $k$-connectivity applies for $k = \Theta(n^{1/3})$, where $n$ denotes the number of vertices in the input graph. In this problem, it is straightforward to achieve $k = n^\alpha$ for any fixed $0 < \alpha < 1$ by adding sufficiently many vertices that are either isolated or connected to all other vertices by edges of cost 0.

It follows that the min-cost subset $k$-connectivity problem is provably harder to approximate than the min-cost $k$-connectivity problem (for values of $k$ as above). Indeed, it is shown in [CVV02] that the latter problem can be approximated within ratio $O(\log k)$ for $k \leq \sqrt{n/6}$.

# References

ADNP99.    V. Auletta, Y. Dinitz, Z. Nutov, and D. Parente. A 2-approximation algorithm for finding an optimum 3-vertex-connected spanning subgraph. *J. Algorithms*, 32(1):21–30, 1999.

AL96.    S. Arora and C. Lund. Hardness of approximations. In D. Hochbaum, editor, *Approximation Algorithms for NP-Hard Problems*. PWS Publishing Company, 1996.

ALM+98.    S. Arora, C. Lund, R. Motwani, M. Sudan, and M. Szegedy. Proof verification and the hardness of approximation problems. *J. ACM*, 45(3):501–555, 1998.

CJN01.    J. Cheriyan, T. Jordán, and Z. Nutov. On rooted node-connectivity problems. *Algorithmica*, 30(3):353–375, 2001.

CL99.    A. Czumaj and A. Lingas. On approximability of the minimum-cost $k$-connected spanning subgraph problem. In *Proceedings of the 10th Annual ACM-SIAM Symposium on Discrete Algorithms*, pages 281–290. ACM, 1999.

CT00.    J. Cheriyan and R. Thurimella. Approximating minimum-size $k$-connected spanning subgraphs via matching. *SIAM J. Comput.*, 30(2):528–560, 2000.

CVV02.    J. Cheriyan, S. Vempala, and A. Vetta. Approximation algorithms for minimum-cost $k$-vertex connected subgraphs. In *34th Annual ACM Symposium on the Theory of Computing*, 2002. To appear.

Die00.    R. Diestel. *Graph theory*. Springer-Verlag, New York, second edition, 2000.

DN99.    Y. Dinitz and Z. Nutov. A 3-approximation algorithm for finding optimum 4, 5-vertex-connected spanning subgraphs. *J. Algorithms*, 32(1):31–40, 1999.

ET76.       K. P. Eswaran and R. E. Tarjan. Augmentation problems. *SIAM J. Comput.*, 5(4):653–665, 1976.

Fei98.      U. Feige. A threshold of ln $n$ for approximating set cover. *J. ACM*, 45(4):634–652, 1998.

Fer98.      C. G. Fernandes. A better approximation ratio for the minimum size $k$-edge-connected spanning subgraph problem. *J. Algorithms*, 28(1):105–124, 1998.

FHKS02.     U. Feige, M. M. Halldórsson, G. Kortsarz, and A. Srinivasan. Approximating the domatic number. *SIAM J. Comput.*, 2002. To appear.

FJ81.       G. N. Frederickson and J. JáJá. Approximation algorithms for several graph augmentation problems. *SIAM J. Comput.*, 10(2):270–283, 1981.

FJW01.      L. Fleischer, K. Jain, and D. P. Williamson. An iterative rounding 2-approximation algorithm for the element connectivity problem. In *42nd Annual IEEE Symposium on Foundations of Computer Science*, pages 339–347, 2001.

Fra94.      A. Frank. Connectivity augmentation problems in network design. In J. R. Birge and K. G. Murty, editors, *Mathematical Programming: State of the Art 1994*, pages 34–63. The University of Michigan, 1994.

FT89.       A. Frank and É. Tardos. An application of submodular flows. *Linear Algebra Appl.*, 114/115:329–348, 1989.

HR91.       T. Hsu and V. Ramachandran. A linear time algorithm for triconnectivity augmentation. In *32nd Annual IEEE Symposium on Foundations of Computer Science*, pages 548–559, 1991.

Hsu92.      T. Hsu. On four-connecting a triconnected graph. In *33nd Annual IEEE Symposium on Foundations of Computer Science*, pages 70–79, 1992.

Jai01.      K. Jain. A factor 2 approximation algorithm for the generalized Steiner network problem. *Combinatorica*, 21(1):39–60, 2001.

Jor95.      T. Jordán. On the optimal vertex-connectivity augmentation. *J. Combin. Theory Ser. B*, 63(1):8–20, 1995.

Khu96.      S. Khuller. Approximation algorithms for finding highly connected subgraphs. In D. Hochbaum, editor, *Approximation Algorithms for NP-Hard Problems*. PWS Publishing Company, 1996.

KN00.       G. Kortsarz and Z. Nutov. Approximating node connectivity problems via set covers. In *3rd International workshop on Approximation algorithms for combinatorial optimization (APPROX)*, pages 194–205. Springer, 2000.

Kor01.      G. Kortsarz. On the hardness of approximating spanners. *Algorithmica*, 30(3):432–450, 2001.

KR96.       S. Khuller and B. Raghavachari. Improved approximation algorithms for uniform connectivity problems. *J. Algorithms*, 21(2):434–450, 1996.

KT93.       S. Khuller and R. Thurimella. Approximation algorithms for graph augmentation. *J. Algorithms*, 14(2):214–225, 1993.

Pet94.      E. Petrank. The hardness of approximation: gap location. *Comput. Complexity*, 4(2):133–157, 1994.

PY93.       C. H. Papadimitriou and M. Yannakakis. The traveling salesman problem with distances one and two. *Math. Oper. Res.*, 18(1):1–11, 1993.

Raz98.      R. Raz. A parallel repetition theorem. *SIAM J. Comput.*, 27(3):763–803, 1998.

RW97.       R. Ravi and D. P. Williamson. An approximation algorithm for minimum-cost vertex-connectivity problems. *Algorithmica*, 18(1):21–43, 1997.

RW02.     R. Ravi and D. P. Williamson. Erratum: An approximation algorithm for minimum-cost vertex-connectivity problems. In *Proceedings of the 13th Annual ACM-SIAM Symposium on Discrete Algorithms*, pages 1000–1001, 2002.

Vaz01.     V. V. Vazirani. *Approximation algorithms*. Springer-Verlag, Berlin, 2001.

WN93.     T. Watanabe and A. Nakamura. A minimum 3-connectivity augmentation of a graph. *J. Comput. System Sci.*, 46(1):91–128, 1993.

# Non-abusiveness Helps:
# An $\mathcal{O}(1)$-Competitive Algorithm
# for Minimizing the Maximum Flow Time
# in the Online Traveling Salesman Problem

Sven O. Krumke[1,*], Luigi Laura[2], Maarten Lipmann[3,**],
Alberto Marchetti-Spaccamela[2], Willem E. de Paepe[3,***],
Diana Poensgen[1,†], and Leen Stougie[4,‡]

[1] Konrad-Zuse-Zentrum für Informationstechnik Berlin, Germany.
{krumke,poensgen}@zib.de
[2] Dipartimento di Informatica e Sistemistica,
Universita di Roma "La Sapienza", Italy.
{alberto,laura}@dis.uniroma1.it
[3] Technical University of Eindhoven, The Netherlands.
m.lipmann@tue.nl,
w.e.d.paepe@tm.tue.nl
[4] Technical University of Eindhoven,
and Centre for Mathematics and Computer Science (CWI),
Amsterdam, The Netherlands.
leen@win.tue.nl

**Abstract.** In the online traveling salesman problem OLTSP requests for visits to cities arrive online while the salesman is traveling. We study the $F_{\max}$-OLTSP where the objective is to minimize the maximum flow time. This objective is particularly interesting for applications. Unfortunately, there can be no competitive algorithm, neither deterministic nor randomized. Hence, competitive analysis fails to distinguish online algorithms. Not even resource augmentation which is helpful in scheduling works as a remedy. This unsatisfactory situation motivates the search for alternative analysis methods.

We introduce a natural restriction on the adversary for the $F_{\max}$-OLTSP on the real line. A *non-abusive adversary* may only move in a direction if there are yet unserved requests on this side. Our main result is an algorithm which achieves a constant competitive ratio against the non-abusive adversary.

* Research supported by the German Science Foundation (DFG, grant GR 883/10)
** Supported by the TMR Network DONET of the European Community ERB TMRX-CT98-0202
*** Partially supported by Algorithmic Methods for Optimizing the Railways in Europe (AMORE) grant HPRN-CT-1999-00104
† Partially supported by Algorithmic Methods for Optimizing the Railways in Europe (AMORE) grant HPRN-CT-1999-00104
‡ Supported by the TMR Network DONET of the European Community ERB TMRX-CT98-0202

K. Jansen et al. (Eds.): APPROX 2002, LNCS 2462, pp. 200–214, 2002.
© Springer-Verlag Berlin Heidelberg 2002

# 1    Introduction

In the online traveling salesman problem (OLTSP) requests for visits to cities arrive online while the salesman is traveling. An online algorithm learns from the existence of a request only at its release time. The OLTSP has been studied for the objectives of minimizing the makespan [2,1,?], the weighted sum of completion times [5,9], and the maximum/average flow time [6]. In view of applications, the maximum flow time is of particular interest. For instance, it can be identified with the maximal dissatisfaction of customers. Alas, there can be no competitive algorithm, neither deterministic nor randomized [6]. Moreover, in contrast to scheduling [11] resource augmentation, e.g. providing the online algorithm with a faster server, does not help, the crucial difference being that servers move in space.

The only hope to overcome the weaknesses of standard competitive analysis in the context of the $F_{\max}$-OLTSP is to restrict the powers of the adversary. In this paper we consider the $F_{\max}$-OLTSP on the real line and introduce a natural restriction on the adversary: a *non-abusive* adversary may move its server only in a direction, if yet unserved requests are pending on that side. We construct an algorithm, called DETOUR which achieves a competitive ratio of eight against the non-abusive adversary.

**Related Work** Koutsoupias and Papadimitriou introduced the concept of *comparative analysis* for restricting the adversary [8]. The *fair adversary* of Blom et al. [3] implements this concept in the context of the OLTSP as follows: a fair adversary may only move within the convex hull of all requests released so far. While one can obtain improved competitiveness results for the minimization of the makespan against a fair adversary [3], still a constant competitive ratio for the maximum flow time is out of reach (see Theorem 1). The non-abusive adversary presented in this paper can be viewed as a refinement of the fair adversary. In [7] it is shown that minimizing the average flow time in scheduling is obnoxiously hard, both online and offline.

**Paper Outline.** In Section 2 we formally define the $F_{\max}$-OLTSP and the non-abusive adversary. We also show lower bound results for the competitive ratio against a fair and non-abusive adversary, respectively. Section 3 presents our algorithm DETOUR, the proof of its performance is sketched in Section 4.

# 2    Preliminaries

An instance of the *Online Traveling Salesman Problem* ($F_{\max}$-OLTSP) consists of a metric space $M = (X, d)$ with a distinguished origin $o \in X$ and a sequence $\sigma = r_1, \ldots, r_m$ of requests. A server is located at the origin $o$ at time 0 and can move at most at unit speed. In this paper we are concerned with the special case that $M$ is $\mathbb{R}$, the real line endowed with the Euclidean metric $d(x, y) = |x - y|$;

the origin $o$ equals the point 0. Each *request* is a pair $r_i = (t_i, x_i)$, where $t_i \in \mathbb{R}$ is the time at which request $r_i$ is released, and $x_i \in X$ is the point in the metric space to be visited. We assume that the sequence $\sigma = r_1, \ldots, r_m$ of requests is given in order of non-decreasing release times. For a real number $t$, we denote by $\sigma_{\leq t}$ ($\sigma_{<t}$) the subsequence of requests in $\sigma$ released up to time $t$ (strictly before time $t$).

An online algorithm ALG gets to know request $r_j$ only at its release time $t_j$. In particular, ALG has neither information about the release time of the last request nor about the total number of requests. Hence, at any moment in time $t$, ALG must make its decisions only knowing the requests in $\sigma_{\leq t}$. An offline algorithm has complete knowledge about the sequence $\sigma$ already at time 0.

Given a sequence $\sigma$ of requests, an algorithm ALG for the $F_{\max}$-OLTSP must find a route for the server which starts in the origin and visits each point in $\sigma$, but not earlier than its release time. By $C_j^{\text{ALG}}$ and $F_j^{\text{ALG}} = C_j^{\text{ALG}} - t_j$ we denote the completion time and flow time of request $r_j$, respectively, in the solution produced by ALG. The goal in the $F_{\max}$-OLTSP is to minimize the maximum flow time $\text{ALG}(\sigma) := \max_j F_j^{\text{ALG}}$.

Let OPT denote an optimal offline algorithm. A deterministic online algorithm ALG for the $F_{\max}$-OLTSP is *c-competitive*, if there exists a constant $c$ such that for any request sequence $\sigma$, $\text{ALG}(\sigma) \leq c \cdot \text{OPT}(\sigma)$. If ALG is randomized, then $\text{ALG}(\sigma)$ is replaced by the expected solution value (this corresponds to the oblivious adversary, see [4]). The *competitive ratio* of ALG is the infimum over all $c$ such that ALG is $c$-competitive.

The following lower bound result shows that the fairness restriction on the adversary introduced in [3] is still not strong enough to allow for competitive algorithms in the $F_{\max}$-OLTSP.

**Theorem 1.** *No randomized algorithm for the $F_{\max}$-OLTSP on $\mathbb{R}$ can achieve a constant competitive ratio against an oblivious adversary. This result still holds, even if the adversary is fair, i.e., if at any moment in time $t$ the server operated by the adversary is within the convex hull of the origin and the requested points from $\sigma_{\leq t}$.*

*Proof.* Let $\varepsilon > 0$ and $k \in \mathbb{N}$ be arbitrary. We present two request sequences $\sigma_1 = (\varepsilon, \varepsilon), (2\varepsilon, 2\varepsilon), \ldots, (k\varepsilon, k\varepsilon), (T, 0)$ and $\sigma_2 = (\varepsilon, \varepsilon), (2\varepsilon, 2\varepsilon), \ldots, (k\varepsilon, k\varepsilon), (T, k\varepsilon)$, each with probability $1/2$, where $T = 4k\varepsilon$. The expected cost of an optimal fair offline solution is at most $\varepsilon$, while any deterministic online algorithm has cost at least $k\varepsilon/2$. The claim now follows by applying Yao's principle [4,10].     □

The fair adversary is still too powerful in the sense that it can move to points where it knows that a request will pop up without revealing any information to the online server before reaching the point. A *non-abusive* adversary is stripped of this power.

**Definition 1 (Non-abusive Adversary).** *An adversary ADV for the OLTSP on $\mathbb{R}$ is non-abusive, if the following holds: At any moment in time $t$, where the adversary moves its server from its current position $p^{\text{ADV}}(t)$ to the right (left),* *there is a request from $\sigma_{\leq t}$ to the right (left) of $p^{\mathrm{ADV}}(t)$ which ADV has not served yet.*

In the sequel we slightly abuse notation and denote by OPT($\sigma$) the maximal flow time in an optimal *non-abusive* offline solution for the sequence $\sigma$. The following result shows that the $F_{\max}$-OLTSP is still non-trivial against a non-abusive adversary.

**Theorem 2.** *No deterministic algorithm for the $F_{\max}$-OLTSP on $\mathbb{R}$ can achieve a competitive ratio less than 2 against a non-abusive adversary.*

*Proof.* Let ALG be any deterministic online algorithm. The adversary first presents the following $2m$ requests: $(0, \pm 1), (3, \pm 2), \ldots, (\sum_{k=1}^{m-1}(1 + 2k), \pm m)$. W.l.o.g., let ALG serve the request in $-m$ later than the one in $+m$, and let $T$ be the time it reaches $-m$. Clearly, $T \geq \sum_{k=1}^{m}(1 + 2k)$. At time $T$, the adversary presents one more request in $+(3m + 1)$ which results in a flow time of at least $4m + 1$ for ALG. On the other hand, a non-abusive offline algorithm can serve all of the first $2m$ requests with maximum flow time $2m+1$ by time $\sum_{k=1}^{m}(1+2k)$, ending with the request at $+m$. From there it can easily reach the last request with flow time $2m + 1$. The theorem follows by letting $m \to \infty$. □

# 3   The Algorithm detour

We denote by $p^{\mathrm{DTO}}(t)$ the position of the server operated by DETOUR (short DTO) at time $t$. The terms *ahead of* and *in the back of* the server refer to positions on the axis w.r.t. the direction the server currently moves: if it is moving from left to right on the $\mathbb{R}$-axis, "ahead" means to the right of the server's current position, while a request "in the back" of the server is to its left. The other case is defined analogously.

Given a point $p \in \mathbb{R}$, we call the pair $(t_i, x_i)$ *more critical* than the request $r_j = (t_j, x_j)$ w.r.t. $p$ if both $x_i$ and $x_j$ are on the same side of $p$, and $d(p, x_i) - t_i > d(p, x_j) - t_j$. If at time $t$, request $r_i$ is more critical than $r_j$ w.r.t. DTO's position $p^{\mathrm{DTO}}(t)$, this yields that DTO cannot serve $r_i$ with the same (or a smaller) flow time than $r_j$. Moreover, $r_i$ remains more critical than $r_j$ after time $t$ as long as both requests are unserved. Conversely, we have the following observation.

**Observation 3.** *If, at time $t$, request $r_i$ is more critical than $r_j$ w.r.t. $p^{\mathrm{DTO}}(t)$, and DTO moves straight ahead to the more distant request after having served the one closer to $p^{\mathrm{DTO}}(t)$, then DTO's flow time for $r_j$ is at most the flow time it achieves for $r_i$.* □

The *critical region* $\vee(r_j, p, G)$ w.r.t. a request $r_j$, a point $p \in \mathbb{R}$ and a bound $G$ for the allowed maximal flow time contains all those pairs $(t, x) \in \mathbb{R}_+ \times \mathbb{R}$ with the property that (i) $(t, x)$ is more critical than $r_j$ w.r.t. $p$, and (ii) $t + d(x, x_j) - t_j \leq G$.

In the setting of DTO, $p$ will be the position of the online server at a certain time $t'$. Condition (ii) has the meaning that a request in $(t, x)$ could be served before $r_j$ in an offline tour serving both $r_j$ and $(t, x)$ with maximal flow time at most $G$.

DTO's decisions at time $t$ are based on an approximation, called the *guess* $G(t)$, of $\text{OPT}(\sigma_{\leq t})$. DTO's rough strategy is to serve all requests in a first-come-first-serve (FCFS) manner. However, blindly following an FCFS-scheme makes it easy for the adversary to fool the algorithm. DTO enforces the offline cost in a malicious sequence to increase by making a detour on its way to the next target: it first moves its server in the "wrong direction" as long as it can serve the target with flow time thrice the guess. If the guess changes, the detour, and possibly the target, are adjusted accordingly (this requires some technicalities in the description of the algorithm).

Our algorithm DTO can assume three modes:

**idle** In this mode, DTO's server has served all unserved requests, and is waiting for new requests at the point at which it served the last request.

**focus** Here, the server is moving in one direction serving requests until a request in its back becomes the oldest unserved one or all requests have been served.

**detour** In this case, the server is moving away from its current target (possibly serving requests on the way), thus making a "detour".

At any time, at most one unserved request is marked as a *target* by DTO. Moreover, there it keeps at most one critical region, denoted by $\vee$. Before we formalize the behavior of DTO we specify important building blocks for the algorithm.

- *Guess Update:* Replace the current guess value $G$ by $G'$, defined as follows: If $G = 0$, then $G' := \text{OPT}(\sigma_{\leq t})$. If $G > 0$, then $G' := 2^a G$, where $a$ is the smallest integer $k$ such that $\text{OPT}(\sigma_{\leq t}) \leq 2^k G$.
- *Target Selection:* Given a candidate set $C$ and the current time $t$, let $s_0 = (t_0, x_0)$ be the most critical request from $C$ w.r.t. $p^{\text{DTO}}(t)$ with the property that $s_0$ is *feasible* in the following sense:
  Let $X_0$ be the point ahead of the server such that $t + d(p^{\text{DTO}}(t), X_0) + d(X_0, x_0) = t_0 + 3G$, provided such a point exists, otherwise let $X_0 := p^{\text{DTO}}(t)$. Define the *turning point* $\text{TP}_0 = (T_0, X_0)$, where $T_0 := t + d(p^{\text{DTO}}(t), X_0)$. There is no unserved request ahead of the server further away from $p^{\text{DTO}}(t)$ than $\text{TP}_0$ and older than $s_0$.
  If necessary, unmark the current target and turning point. Mark $s_0$ as a target and set $\text{TP}_0$ to be the current turning point.
- *Mode Selection:* If $\text{TP}_0 \neq p^{\text{DTO}}(t)$, then set $\vee := \vee(s_0, p^{\text{DTO}}(t), G)$ and enter the detour mode. Otherwise, set $\vee := \emptyset$ and unmark $s_0$ as a target. Change the direction and enter the focus mode.

We now specify for each of the three states how DTO reacts to possible events. All events not mentioned are ignored. In the beginning, the guess value is set to $G := 0$, $\vee := \emptyset$ and the algorithm is in the idle mode.

*Idle Mode* In the idle mode DTO waits for the next request to occur.

- A new request is released at time $t$.

The pair $(t, p^{\mathrm{DTO}}(t))$ is called a *selection point*. DTO performs a *guess update*. The direction of the server is defined to be such that the oldest unserved request is in its back. In case that there is one oldest request on both sides of the server chooses to have the most critical one in the server's back.

DTO defines $C$ to be the set of unserved requests in the back of the server and performs a *target selection*, followed by a *mode selection*.

*Detour Mode* In this mode, DTO has a current target $s_m = (t_m, x_m)$, a critical region $\vee \neq \emptyset$ and a turning point TP. Let $T$ be the time DTO entered the detour mode and $s_0$ the then chosen target. The server moves towards TP until one of following two events happens, where the first one has a higher priority than the second one (i.e., the first event is always checked for first).

- A new request is released at time $t$ or an already existing request has just been served.

  DTO performs a *guess update*. Then it enlarges the critical region to $\vee :=$ $\vee(s_0, p^{\mathrm{DTO}}(T), G)$ where $G$ is the updated guess value. Replace the old turning point by a new point TP which satisfies $t + d(p^{\mathrm{DTO}}(t), \mathrm{TP}) + d(\mathrm{TP}, x_m) = t_m + 3G$ for the updated guess value $G$.

  DTO defines $C$ to be the set of unserved requests which are in $\vee$ and more critical than the current target $s_m$. If $C \neq \emptyset$, it executes the *target selection* and the *mode selection*. If $C = \emptyset$, DTO remains in the detour mode.
- The turning point TP is reached at time $t$.

  DTO unmarks the current target, sets $\vee := \emptyset$ and clears the turning point. The server reverses direction and enters the focus mode.

*Focus Mode* When entering the focus mode, DTO's server has a direction. It moves in this direction, reacting to the following events:

- A new request is released at time $t$.

  A *guess update* is performed, and the server remains in the focus mode.
- The last unserved request has been served.

  The server stops, forgets its direction and enters the idle mode.
- A request in the back of the server becomes the oldest unserved request.

  If this happens at time $t$, the pair $(t, p^{\mathrm{DTO}}(t))$ is also called a *selection point*. DTO defines $C$ to be the set of unserved requests in the back of the server and performs a *target selection*, followed by a *mode selection* (exactly as in the idle mode).

# 4   Analysis of **detour**

For the analysis of DTO it is convenient to compare intermediate solutions $\mathrm{DTO}(\sigma_{\leq t})$ not only to the optimal non-abusive solution on $\sigma_{\leq t}$, but to a whole class of offline solutions $\mathcal{ADV}(t)$ which depend on the guess value $G(t)$ maintained by DTO. Notice that for any time $t$ we have $\mathrm{OPT}(\sigma_{\leq t}) \leq G(t) \leq 2\mathrm{OPT}(\sigma_{\leq t})$.

**Definition 2.** *By $\mathcal{ADV}(t)$, we denote the set of all non-abusive offline solutions for the sequence $\sigma_{\leq t}$ with the property that the maximum flow time of any request in $\sigma_{\leq t}$ is bounded from above by $G(t)$.*

*Let $r_i \in \sigma_{\leq t}$ . The smallest achievable flow time $\alpha_i(t)$ is defined to be minimum flow time of $r_i$ taken over all solutions in $\mathcal{ADV}(t)$.*

Note that, by definition, $\alpha_i(t) \leq G(t)$. In general, $\alpha_i(t) \geq \alpha_i(t')$ for $t' > t$, as an increase in the allowed flow time can help an adversary to serve a request $r_i$ earlier. On the other hand, we have the following property.

**Observation 4.** *If $t \leq t'$ and $G(t) = G(t')$, then $\alpha_i(t) \leq \alpha_i(t')$ for any request $r_i = (t_i, x_i)$ with $t_i \leq t$.*

To derive bounds on the flow times for DTO we would like to conclude as follows: if request $r_i$ is served by DTO "in time" and $r_j$ is served directly after $r_i$ (i.e., without any detour in between), then $r_j$ is also served "in time".

**Definition 3 (Served in time).** *Given $r_i = (t_i, x_i)$, we define $\tau_i := \max\{t_i, T\}$, where $T$ is the last time DTO reverses direction before serving $r_i$. Furthermore, we say that $r_i$ is served in time by DTO, if $C_i^{\mathrm{DTO}} \leq t_i + 3G(\tau_i) + \alpha_i(\tau_i)$.*

Notice that any request $r_i$ served in time is served with a flow time of at most $4G(\tau_i) \leq 4G(C_i^{\mathrm{DTO}})$ since $\alpha_i(\tau_i) \leq G(\tau_i)$ by definition.

**Lemma 1.** *Let $r_i = (t_i, x_i)$ and $r_j = (t_j, x_j)$ be two requests such that: (i) $r_i$ is served in time by DTO, (ii) $C_j^{\mathrm{DTO}} \leq C_i^{\mathrm{DTO}} + d(x_i, x_j)$, (iii) $\tau_i \leq \tau_j$, and (iv) $r_j$ is served after $r_i$ by all ADV $\in \mathcal{ADV}(\tau_j)$. Then, DTO serves $r_j$ in time.*

*Proof.* From (iv) we have

$$t_j + \alpha_j(\tau_j) \geq t_i + \alpha_i(\tau_j) + d(x_i, x_j). \tag{1}$$

In particular, this means that

$$t_i + d(x_i, x_j) \leq t_j + G(\tau_j). \tag{2}$$

By (i), (ii), and the definition of in time, we get:

$$C_j^{\mathrm{DTO}} \leq t_i + \alpha_i(\tau_i) + 3G(\tau_i) + d(x_i, x_j). \tag{3}$$

If $G(\tau_i) = G(\tau_j)$, we have that $\alpha_i(\tau_i) \leq \alpha_i(\tau_j)$ by Observation 4. In this case, inequality (3) yields that

$$C_j^{\mathrm{DTO}} \leq t_i + \alpha_i(\tau_j) + 3G(\tau_j) + d(x_i, x_j) \overset{(1)}{\leq} t_j + \alpha_j(\tau_j) + 3G(\tau_j).$$

If $G(\tau_i) < G(\tau_j)$, it must hold that $2G(\tau_i) \leq G(\tau_j)$, and inequality (3) yields

$$C_j^{\mathrm{DTO}} \leq t_i + 4G(\tau_i) + d(x_i, x_j) \overset{(2)}{\leq} t_j + G(\tau_j) + 4G(\tau_i) \leq t_j + 3G(\tau_j).$$

$\square$

An easy but helpful condition which ensures that assumption (iv) of Lemma 1 holds, is that $t_j + d(x_j, x_i) > t_i + G(t)$. This yields the following observation which will be used frequently in order to apply Lemma 1:

**Observation 5.** *(i)* *If $d(x_i, x_j) > G(t)$ and $t_i \leq t_j \leq t$, then $r_i$, the older request, must be served before $r_j$ in any offline solution in $\mathcal{ADV}(t)$.*
*(ii)* *If a request $r_j$ is outside the critical region $\vee(r_i, p, G(t))$ valid at time $t$, request $r_j$ is served after $r_i$ in any offline solution in $\mathcal{ADV}(t)$.*

We define a *busy period* to be the time between the moment in time when DTO leaves the idle mode and the next time the idle mode is entered again.

**Lemma 2.** *Suppose that at the beginning of a busy period at time $t$ each request $r_j$ served in one of the preceeding busy periods was served by DTO with a flow time at most $4\,G(C_j^{\mathrm{DTO}})$. Then, $d(p^{\mathrm{DTO}}(t), p^{\mathrm{ADV}}(t)) \leq 5/2 G(t)$ for any ADV $\in \mathcal{ADV}(t)$.*

*Proof.* The claim of the lemma is trivially true for the first busy period, since a non-abusive adversary must keep its server in the origin until the first request is released. Hence, it suffices to consider the later busy periods.

Let $r_l$ be last request served by DTO in the preceeding busy period, so DTO enters the idle mode at time $C_l^{\mathrm{DTO}}$ again. Consider that ADV $\in \mathcal{ADV}(t)$ in which the adversary's server is furthest away from $p^{\mathrm{DTO}}(t) = x_l$.

**Case 1:** At time $t$, ADV has served all requests in $\sigma_{<t}$.
In this case, since ADV is non-abusive, its server satisfies $p^{\mathrm{ADV}}(t) = x_k$ for the request $r_k$ it served last. DTO must have served $r_k$ in the preceeding busy period, hence no later than $r_l$. This gives $C_k^{\mathrm{DTO}} \leq C_l^{\mathrm{DTO}} - d(x_k, x_l)$ and we obtain that

$$t_k \leq C_l^{\mathrm{DTO}} - d(x_k, x_l) \leq t_l + 4G(C_l^{\mathrm{DTO}}) - d(x_k, x_l),$$

because $r_l$ was served with a flow time of at most $4G(C_l^{\mathrm{DTO}})$ according to the assumption of the lemma. On the other hand, ADV serves $r_l$ no later than $r_k$, which implies that $t_l + d(x_k, x_l) \leq t_k + G(C_l^{\mathrm{DTO}})$. The two inequalities together yield

$$d(x_k, x_l) \leq t_k - t_l + G(C_l^{\mathrm{DTO}}) \leq 5G(C_l^{\mathrm{DTO}}) - d(x_k, x_l).$$

hence the claim, since $d(p^{\mathrm{DTO}}(t), p^{\mathrm{ADV}}(t)) = d(x_k, x_l)$.

**Case 2:** At time $t$, there is a request from $\sigma_{<t}$ which has not been served yet by ADV.
If $r_l$ has not been served by ADV at time $t$, then the distance $d(p^{\mathrm{ADV}}(t), x_l) = d(p^{\mathrm{ADV}}(t), p^{\mathrm{DTO}}(t))$ is at most $G(t)$, because otherwise the adversary's flow time for $r_l$ would be greater than $G(t)$. Otherwise, $r_l$ has been served, but another request in $\sigma_{<t}$ is yet unserved by ADV. Let $r_k$ be the request in $\sigma_{<t}$ which is furthest away from $r_l$ and yet unserved by ADV. The same reasoning as in Case 1 shows that $d(x_k, x_l) \leq \frac{5}{2}G(C_l^{\mathrm{DTO}})$. So, if the adversary's server is between $r_k$ and $r_l$, the claim holds true. Assume that the adversary's server is further away

from $r_l$ than $r_k$. Since the adversary is non-abusive, there must be a request $r_j$ even further away from $r_l$ than $p^{\text{ADV}}(t)$, which ADV served last before (or at) time $t$. In particular, ADV served $r_l$ before $r_j$. Thus, the same arguments as in Case 1 apply to $r_j$ in place of $r_k$, showing that $d(x_j, x_l) \leq \frac{5}{2}G(C_l^{\text{DTO}})$.     □

We further subdivide each busy period into *phases*, where a phase is defined to be the time between two subsequent selection points of DTO. Remember that DTO reaches a selection point whenever it leaves the idle mode, and each time at which a request in the server's back becomes the oldest unserved one. The following statement is the key theorem of our analysis.

**Theorem 6.** *The following is true for any phase $\rho \geq 1$:*

(a) *At any time $t$ in phase $\rho$ at which DTO is in the detour mode, it holds that $d(X_i, x_i) \geq G(t)$ for the turning point $TP_i = (X_i, T_i)$ valid at that time and its corresponding target $r_i = (t_i, x_i)$. Moreover, if at some time $t$ during phase $\rho$, a request $r_i$ failed to become a new target only because it was infeasible, the above inequality holds as well for $r_i$ and its hypothetical turning point $TP_i$.*

(b) *Any request $r_j$ served in phase $\rho$ is served with a flow time of at most $4G(C_j^{\text{DTO}})$.*

(c) *The last request served in phase $\rho$ is served in time.*

*Proof.* We prove the statement by induction on the total number of phases. In the inductive step we distinguish whether phase $\rho$ is the first phase of a busy period or not. The former case includes the induction base ($\rho = 1$), i.e., the first phase of the first busy period, as a special case.

Let $\rho \geq 1$ be the number of the phase under consideration and assume that the three statements of the theorem all hold true for each phase $\rho' < \rho$, provided such a phase exists. Note that in the case that phase $\rho$ is the first phase of a busy period, Lemma 2 can be applied.

At the beginning of phase $\rho$, DTO determines a turning point which might be replaced later on in the phase. We call the turning point $TP^\rho = (T^\rho, X^\rho)$ at which the server actually reverses direction the *realized turning point* of the phase $\rho$. Each turning point ever considered has a corresponding target. Note that both the realized turning point $TP^\rho$ and its corresponding target $s^\rho = (t^\rho, x^\rho)$ are reached by DTO in the same phase. Moreover, at any time when DTO is in the detour mode, the algorithm has a valid turning point and a corresponding target.

Throughout the whole proof we assume without loss of generality that the realized turning point is to the right of the final target, that is, in phase $\rho$, DTO moves to the right while in the detour mode and to the left after entering the focus mode. We may also assume without loss of generality that at time 0 a request appears in the origin since this request does not increase the offline cost.

**Proof of Statement (a):** Let $TP_0 = (T_0, X_0)$ be the first turning point chosen in phase $\rho$, $TP_1$ the next one, etc. until $TP^\rho$, the realized turning point of the phase. Let $s_i = (t_i, x_i)$ be the target corresponding to $TP_i$ ($s_i$ is released at time $t_i$). Part (a) is proven by induction on the number of turning points in the considered phase $\rho$.

*1. Phase $\rho$ is the first phase of a busy period.*

The first target $s_0 = (t_0, x_0)$ must be among the requests whose release initiates the start of the busy period at time $t_0$, and $\text{TP}_0 = (T_0, X_0)$ is chosen such that $t_0 + d(p^{\text{DTO}}(t_0), X_0) + d(x_0, X_0) \geq t_0 + 3G(t_0)$. Since $d(p^{\text{DTO}}(t_0), X_0) < d(x_0, X_0)$, it readily follows that $d(X_0, x_0) > \frac{3}{2} G(t_0)$.

Assume that (a) holds for the turning points $\text{TP}_0, \ldots, \text{TP}_{i-1}$ of phase $\rho$. We prove (a) for the next turning point $\text{TP}_i = (T_i, X_i)$. Assume that $\text{TP}_i$ replaces $\text{TP}_{i-1}$ at time $t$ of phase 1, and let $s_i = (t_i, x_i)$ be $\text{TP}_i$'s corresponding target.

If $s_i$ was released at time $t_0$, then the turning point $\text{TP}_i$ planned by DTO at time $t$ is exactly the same as if the guess value at time $t_0$ had already been $G(t)$ and $s_i$ had been selected as a target at time $t_0$. Exactly as before we can conclude that $d(X_i, x_i) \geq \frac{3}{2} G(t)$.

If $s_i$ was released later than $t_0$, the detour taken by DTO is only longer as the one chosen if $s_i$ had been released already at time $t_0$. Hence, the arguments of above apply again and $d(X_i, x_i) \geq \frac{3}{2} G(t)$.

*2. Phase $\rho$ is not the first phase of a busy period.*

Again, we first consider $\text{TP}_0 = (T_0, X_0)$, the turning point planned first in phase $\rho$. Both $s_0$ and $T_0$ are determined in the selection point SP which marks the end of phase $\rho - 1$ and the start of phase $\rho$. It must hold that $\text{SP} = (C_l^{\text{DTO}}, x_l)$ for some request $r_l$. When DTO serves request $r_l$, the oldest unserved request, call it $r_z$, is in its back. Observe that SP cannot be reached before the final target $s^{\rho-1}$ of phase $\rho - 1$ is served: If that was the case, there would be an unserved request in the back of DTO's server before $s^{\rho-1}$ is reached which is older than $s^{\rho-1}$. But in that case, $s^{\rho-1}$ would have been infeasible at the time it became a target, which is a contradiction.

Assume first that $s_0$ is located between $X^{\rho-1}$ and $x_l$. Then, the release time of $s_0$ satisfies

$$t_0 \geq C_l^{\text{DTO}} - d(x_0, x_l), \tag{4}$$

because otherwise $s_0$ would have been served on the way to $r_l$. Since $\text{TP}_0$ is chosen at time $C_l^{\text{DTO}}$ in such a way that $s_0$ is served not earlier than time $t_0 + 3G(C_l^{\text{DTO}})$, we have $C_l^{\text{DTO}} + d(x_l, X_0) + d(X_0, x_0) \geq t_0 + 3G(C_l^{\text{DTO}})$. It follows that

$$d(X_0, x_0) \geq t_0 + 3G(C_l^{\text{DTO}}) - C_l^{\text{DTO}} - d(x_l, X_0)$$
$$\overset{(4)}{\geq} 3G(C_l^{\text{DTO}}) - d(x_0, x_l) - d(x_l, X_0) = 3G(C_l^{\text{DTO}}) - d(x_0, X_0).$$

Hence $d(x_0, X_0) \geq \frac{3}{2} G(C_l^{\text{DTO}})$.

We now have to cover the case that $s_0$ is further away from $r_l$ than $\text{TP}^{\rho-1}$, that is, $d(x_0, x_l) > d(X^{\rho-1}, x_l)$. Notice that $r_l$ must be older than $s_0$: If $s_0$ was older, the oldest unserved request would have been in DTO's back before reaching $r_l$, and $(C_l^{\text{DTO}}, x_l)$ would not have been the selection point. Observe also that $d(X^{\rho-1}, x_l)$ is at least the distance between the realized turning point $\text{TP}^{\rho-1}$ and the corresponding target, as the final target of a phase is always served within that phase, as shown above. From the inductive hypothesis for phase $\rho - 1$, Statement (a), we obtain that

$$d(x_l, x_0) > G(T^{\rho-1}). \tag{5}$$

As $d(X_0, x_0) \geq d(x_l, x_0)$, the only interesting case to consider is that $G(T^{\rho-1}) < G(C_l^{\mathrm{DTO}})$. So let $2^a G(T^{\rho-1}) = G(C_l^{\mathrm{DTO}})$ for some integer $a \geq 1$. We need to distinguish two cases.

**Case 1:** $t_0 \geq T^{\rho-1}$. In this case we have $T^{\rho-1} + d(X^{\rho-1}, X_0) + d(X_0, x_0) \geq t_0 + 3G(C_l^{\mathrm{DTO}})$, as DTO's server started from $X^{\rho-1}$ at time $T^{\rho-1}$ and chooses the turning point $\mathrm{TP}_0$ at time $C_l^{\mathrm{DTO}}$ in such that the corresponding target $s_0$ is not served with a smaller flow time than $3G(C_l^{\mathrm{DTO}})$. But since we consider the case in which $d(x_0, X_0) \geq d(X^{\rho-1}, X_0)$, this yields

$$2d(x_0, X_0) \geq t_0 - T^{\rho-1} + 3G(C_l^{\mathrm{DTO}}) \geq 3G(C_l^{\mathrm{DTO}}),$$

as $t_0 \geq T^{\rho-1}$. This implies the claim.

**Case 2:** $t_0 < T^{\rho-1}$. Here we get $t_l \leq t_0 < T^{\rho-1}$ since $r_l$ is older than $s_0$, as reasoned above. By (5) and Observation 5 (i), request $s_0$ is served after $r_l$ by every $\mathrm{ADV} \in \mathcal{ADV}(T^{\rho-1})$. Hence,

$$t_0 + \alpha_0(T^{\rho-1}) \geq t_l + \alpha_l(T^{\rho-1}) + d(x_0, x_l). \tag{6}$$

From the assumption that Statement (c) holds true for phase $\rho - 1$, we know that $r_l$ is served in time, i.e.,

$$C_l^{\mathrm{DTO}} \leq t_l + \alpha_l(T^{\rho-1}) + 3G(T^{\rho-1}), \tag{7}$$

because $T^{\rho-1}$ was the last time DTO turned around before it served $r_l$, and because of $t_l \leq T^{\rho-1}$. Hence, if DTO's server turned around immediately after serving $r_l$, it would hold that

$$C_0^{\mathrm{DTO}} = C_l^{\mathrm{DTO}} + d(x_0, x_l) \overset{(7)}{\leq} t_l + \alpha_l(T^{\rho-1}) + 3G(T^{\rho-1}) + d(x_0, x_l)$$

$$\overset{(6)}{\leq} t_0 + \alpha_0(T^{\rho-1}) + 3G(T^{\rho-1}) \leq t_0 + 4G(T^{\rho-1}) \leq t_0 + 2^{-a+2}G(C_l^{\mathrm{DTO}}).$$

Thus, $s_0$ would be served with a flow time of at most $2\,G(C_l^{\mathrm{DTO}})$, because $a \geq 1$. Since DTO never plans its turning point in such a way that the target is reached with a flow time of less than three times the current guess value, we can deduce that the server does in fact not turn around, but has time of at least $(3 - 2^{-a+2})G(C_l^{\mathrm{DTO}}) > 0$ to spend on a detour, starting at time $C_l^{\mathrm{DTO}}$. Thus, the distance between $x_l$ and the turning point $\mathrm{TP}_0$ planned at time $C_l^{\mathrm{DTO}}$ is at least $\frac{1}{2}(3 - 2^{-a+2})G(C_l^{\mathrm{DTO}})$, and we conclude that

$$d(X_0, x_0) = d(X_0, x_l) + d(x_l, x_0) \geq \frac{1}{2}\left(3 - 2^{-a+2}\right) G(C_l^{\mathrm{DTO}}) + d(x_l, x_0)$$

$$\overset{(5)}{\geq} \left(1 - 2^{-a}\right) G(C_l^{\mathrm{DTO}}) + G(T^{\rho-1}) = \left(1 - 2^{-a}\right) G(C_l^{\mathrm{DTO}}) + 2^{-a}G(C_l^{\mathrm{DTO}})$$

$$\geq G(C_l^{\mathrm{DTO}}),$$

which was our claim for $\mathrm{TP}_0$.

Assume now that (a) holds for the turning points $\text{TP}_0, \ldots, \text{TP}_{i-1}$ of the considered phase $\rho$. We have to prove (a) for the next turning point $\text{TP}_i = (T_i, X_i)$. Assume that $\text{TP}_i$ replaces $\text{TP}_{i-1}$ at time $t$ of phase $\rho$, and let $t' < t$ be the time when $\text{TP}_{i-1}$ was valid for the first time. Denote by $s_i = (t_i, x_i)$ the target corresponding to $\text{TP}_i$. Recall that we assumed w.l.o.g. that $\text{TP}_i$ is to the right of $s_i$.

In the case that $s_i = s_0$, we can deduce that $G(t) \geq 2G(t')$ because the turning point changes but the target does not. That is, in order to serve the target with a flow time of $3G(t)$ instead of $3G(t')$, DTO has now at least $3(G(t) - G(t'))$ additional time units to spend on the way to the target. Hence, it can enlarge its detour by $d(X_i, X_{i-1}) \geq \frac{3}{2}(G(t) - G(t')) \geq G(t) - G(t')$. By the inductive assumption, $d(X_{i-1}, x_0) \geq G(t')$. Consequently,

$$d(X_i, x_0) = d(X_i, X_{i-1}) + d(X_{i-1}, x_0) \geq G(t) - G(t') + G(t') = G(t).$$

If $s_i \neq s_0$ and the new target $s_i$ was released later than time $T^{\rho-1}$, then independent of whether $s_i$ is between $T^{\rho-1}$ and $r_l$ or further away from $r_l$, we can conclude as in the proof for $\text{TP}_0$, that $d(X_i, x_i) \geq \frac{3}{2}G(t)$.

If $s_i \neq s_0$ and $s_i$ had already been released at time $T^{\rho-1}$, it follows that $s_i$ must have been infeasible at time $C_l^{\text{DTO}}$ when the initial target was chosen, as $s_i$ was more critical than $s_0$ at that time: otherwise it could not have become a target later on. Hence, there exists a request $r_j$ older than $s_i$ and to the right of $s_i$'s hypothetical turning point $\text{TP}_i' = (T_i', X_i')$ considered at time $C_l^{\text{DTO}}$.

By exactly the same arguments used for $\text{TP}_0$ above, we can deduce for the hypothetical turning point $\text{TP}_i'$ considered at time $C_l^{\text{DTO}}$ that

$$d(X_i', x_i) \geq G(C_l^{\text{DTO}}). \tag{8}$$

If $G(C_l^{\text{DTO}}) = G(t)$, we obtain that $d(X_i, x_i) > d(X_i', x_i) \geq G(C_l^{\text{DTO}}) = G(t)$. On the other hand, if $G(t) \geq 2G(C_l^{\text{DTO}})$, DTO has at least $3(G(t) - G(C_l^{\text{DTO}}))$ time units more to spend on its detour to $x_i$ than it would have had if $s_i$ had become the target at time $C_l^{\text{DTO}}$. Hence, the distance of the hypothetical turning point $\text{TP}_i'$ considered at time $C_l^{\text{DTO}}$ and the one chosen at time $t$, namely $\text{TP}_i$, is at least half of that additional time. More precisely, we have that $d(X_i', X_i) \geq \frac{3}{2}(G(t) - G(C_l^{\text{DTO}})) \geq G(t) - G(C_l^{\text{DTO}})$. Together with (8), we obtain that

$$d(X_i, x_i) = d(X_i, X_i') + d(X_i', x_i) \quad \geq G(t) - G(C_l^{\text{DTO}}) + G(C_l^{\text{DTO}}) = G(t),$$

which proves the inductive step.

Notice that exactly the same arguments apply whenever a request is not made a target because it is not feasible: its hypothetical turning point considered at that time $t$ must be at least at distance $G(t)$ to the corresponding target.

**Proof of Statements (b) and (c):** Let SP denote the selection point which defines the end of the previous phase. If no previous phase exists, we define $\text{SP} = (0, 0)$. By definition, $\text{SP} = (C_l^{\text{DTO}}, x_l)$ for some request $r_l$. We distinguish two cases: in Case I we consider the situation that DTO's server immediately turns around in the selection point, thus not entering the detour mode, in Case II, we assume that it enters the detour mode at the selection point. Furthermore, we partition the set of requests served in phase $\rho$ into three classes, depending on which part of DTO's route they are served in: Class 1 contains all requests served between the realized turning point and its corresponding target. Class 2 consists of those requests served in phase $\rho$ after the target; whereas all requests served between the selection point and the realized turning point belong to Class 3.

Let $r_z$ be the oldest unserved request at time $C_l^{\text{DTO}}$. Denote by $\text{TP}_0 = (T_0, X_0)$ be the turning point chosen at time $C_l^{\text{DTO}}$, and by $s_0 = (t_0, x_0)$ its corresponding target.

**Case I:** DTO turns around in the selection point

*1. Phase $\rho$ is the first phase of a busy period*

Hence, all requests served in that phase are released at time $t_0$ or later, and since DTO immediately enters the focus mode at the beginning of the phase, they are all served without any detour. By 2, we know that $d(p^{\text{DTO}}(t_0), p^{\text{ADV}}(t_0)) \leq \frac{5}{2}G(t_0)$. Therefore, DTO reaches all requests served in phase 1 at most $\frac{5}{2}G(t_0)$ time units later than the adversary, so all requests are served in time.

*2. Phase $\rho$ is not the first phase of a busy period*

We have $\text{TP}_0 = \text{SP} = (C_l^{\text{DTO}}, x_l)$, because the server turns around immediately as it cannot serve $s_0$ with a flow time of $3G(C_l^{\text{DTO}})$ or less. Thus, $\text{TP}_0$ equals the realized turning point $\text{TP}^\rho = (T^\rho, X^\rho)$, request $s_0$ is the final target and $C_0^{\text{DTO}} = C_l^{\text{DTO}} + d(x_0, x_l)$. Note that Class 3 is empty in this case. First of all, we know that $s_0$ cannot be older than $r_l$: If $s_0$ was older, DTO would not have served $r_l$ anymore, as it only remains in the focus mode until the oldest unserved request is in its back. Furthermore, by part (a) it holds that $d(x_l, x_0) = d(X^\rho, x_0) \geq G(T^\rho) = G(C_l^{\text{DTO}})$.

Therefore, by Observation 5 (i), in all offline solutions in $\mathcal{ADV}(T^\rho)$, request $r_l$ must be served before $s_0$. It is easy to see that $\tau_l \leq \tau_0 = T^\rho$, and since Statement (c) for phase $\rho - 1$ tells us that $r_l$ is served in time, we can apply Lemma 1 and conclude that $s_0$ is also served in time. Notice that exactly the same arguments apply to all requests in Class 2. This ensures Statement (c).

It remains to consider all other requests of Class 1. To this end, let $r_j$ be a request served between $r_l$ and $s_0$ by DTO. If $r_j$ was more critical than $s_0$, it must be at least as old as $s_0$, because it is to the right of $s_0$. Since $r_j$ was not chosen as target at time $C_l^{\text{DTO}}$, it must have been infeasible, that is, there is a request $r_b$ older than $r_j$ and to the right of $r_j$'s hypothetical turning point $\text{TP}'_j$. But as $r_j$ is more critical than $s_0$, and DTO turns around immediately for $s_0$, we have that $\text{TP}'_j = \text{TP}^\rho$, which implies that also $s_0$ cannot have been feasible at time $C_l^{\text{DTO}}$, a contradiction. Thus, $r_j$ must be less critical than $s_0$. This means that it is served with the same or a smaller flow time than $s_0$, hence with flow time of at most $4G(T^\rho) \leq 4G(C_j^{\text{DTO}})$. This proves Case I.

**Case II:** DTO enters the detour mode at the selection point

The proof of this case is omitted due to lack of space and can be found in the full version of the paper. The key arguments used are similar to the ones in Case I, but more involved: We first show that all requests ever marked as a target, excluding the final target (set $S$), are served with a flow time at most $3G(T^\rho)$. This lets us conclude that requests which are less critical than those in $S$ are served with smaller flow times. In order to show that other requests $r_j$ are served with the desired flow time, we apply Lemma 1 with a careful choice of the request $r_i$ which is served before $r_j$ by ADV. Observation 5 will be used to determine a suitable $r_i$. Another helpful ingredient is the following: if $d(p^{\mathrm{DTO}}(T^\rho), p^{\mathrm{ADV}}(T^\rho)) \le 3G(T^\rho)$, then all requests served by both servers after time $T^\rho$ are served in time (ADV $\in \mathcal{ADV}(T^\rho)$). □

**Theorem 7.** DTO *is 8-competitive against a non-abusive adversary for the* $F_{\max}$-OLTSP.

*Proof.* By Theorem 6 we have that any request $r_i$ is served with flow time at most $4G(C_i^{\mathrm{DTO}})$. If $C_{\mathrm{last}}^{\mathrm{DTO}}$ is the time at which the last request is served by DTO, all requests are served with flow time at most $4G(C_{\mathrm{last}}^{\mathrm{DTO}})$, which, by construction, is bounded by $2\mathrm{OPT}(\sigma)$, thence the claim. □

# References

1. N. Ascheuer, S. O. Krumke, and J. Rambau. Online dial-a-ride problems: Minimizing the completion time. In *Proceedings of the 17th International Symposium on Theoretical Aspects of Computer Science*, volume 1770 of *Lecture Notes in Computer Science*, pages 639–650. Springer, 2000.
2. G. Ausiello, E. Feuerstein, S. Leonardi, L. Stougie, and M. Talamo. Algorithms for the on-line traveling salesman. *Algorithmica*, 29(4):560–581, 2001.
3. M. Blom, S. O. Krumke, W. E. de Paepe, and L. Stougie. The online-TSP against fair adversaries. *Informs Journal on Computing*, 13(2):138–148, 2001.
4. A. Borodin and R. El-Yaniv. *Online Computation and Competitive Analysis*. Cambridge University Press, 1998.
5. E. Feuerstein and L. Stougie. On-line single server dial-a-ride problems. *Theoretical Computer Science*, 2001. To appear.
6. D. Hauptmeier, S. O. Krumke, and J. Rambau. The online dial-a-ride problem under reasonable load. *Theoretical Computer Science*, 2001. A preliminary version appeared in the Proceedings of the 4th Italian Conference on Algorithms and Complexity, 2000, vol. 1767 of Lecture Notes in Computer Science.
7. H. Kellerer, Th. Tautenhahn, and G. J. Woeginger. Approximability and non-approximability results for minimizing total flow time on a single machine. In *Proceedings of the 28th Annual ACM Symposium on the Theory of Computing*, pages 418–426, 1996.
8. E. Koutsoupias and C. Papadimitriou. Beyond competitive analysis. In *Proceedings of the 35th Annual IEEE Symposium on the Foundations of Computer Science*, pages 394–400, 1994.

9. S. O. Krumke, W. E. de Paepe, D. Poensgen, and L. Stougie. News from the online traveling repairman. In *Proceedings of the 26th International Symposium on Mathematical Foundations of Computer Science*, volume 2136 of *Lecture Notes in Computer Science*, pages 487–499, 2001.

10. R. Motwani and P. Raghavan. *Randomized Algorithms*. Cambridge University Press, 1995.

11. K. Pruhs and B. Kalyanasundaram. Speed is as powerful as clairvoyance. In *Proceedings of the 36th Annual IEEE Symposium on the Foundations of Computer Science*, pages 214–221, 1995.

# Routing and Admission Control in Networks with Advance Reservations

Liane Lewin-Eytan[1], Joseph (Seffi) Naor[2], and Ariel Orda[1]

[1] Electrical Engineering Dept., Technion, Haifa 32000, Israel.
liane@tx.technion.ac.il, ariel@ee.technion.ac.il,
[2] Computer Science Dept., Technion, Haifa 32000, Israel
naor@cs.technion.ac.il

**Abstract.** The provisioning of quality-of-service (QoS) for real-time network applications may require the network to reserve resources. A natural way to do this is to allow advance reservations of network resources prior to the time they are needed. We consider several two-dimensional admission control problems in simple topologies such as a line, a ring and a tree. The input is a set of connection requests, each specifying its spatial characteristics, that is, its source and destination; its temporal characteristics, that is, its start time and duration time; and, potentially, also a bandwidth requirement. In addition, each request has a profit gained by acommodating it. We address the related admission control problem, where the goal is to maximize the total profit gained by the accommodated requests. We provide approximation algorithms for several problem variations. Our results imply a $4c$-approximation algorithm for finding a maximum weight independent set of axis-parallel rectangles in the plane, where $c$ is the size of a maximum set of overlapping requests.

## 1 Introduction

### 1.1 Problem Statement and Motivation

As network capabilities increase, their usage is also expanding. At the same time, the wide range of requirements of the many applications using them calls for new mechanisms to control the allocation of network resources. However, while much attention has been devoted to resource reservation and allocation, the same does not apply to the timing of such requests. In particular, the prevailing assumption has been that requests are "immediate", i.e., made at the same time as when the network resources are needed. This is a useful base model, but it ignores the possibility, present in many other resource allocation situations, that resources might be requested in *advance* of when they are needed. This can be a useful service, not only for applications, which can then be sure that the resources they need will be available, but also for the network, as it enables better planning and more flexible management of resources. Accordingly, advance reservation of network resources has been the subject of several recent studies and proposals, e.g. [14,18,19,22]. It has also been recognized that some of the related algorithmic problems are hard [14].

K. Jansen et al. (Eds.): APPROX 2002, LNCS 2462, pp. 215–228, 2002.
© Springer-Verlag Berlin Heidelberg 2002

We investigate some fundamental admission control problems in networks with advance reservations. We concentrate on networks having special topologies such as lines, rings and trees. Yet, even for these topologies, the problems we consider remain NP-hard. These problems are in essence two-dimensional: we are presented with commodities (connection requests), each having a spatial dimension determined by its route in the specific topology, and a temporal dimension determined by its future duration. Each request specifies its source and destination, start time and duration time, and potentially also a bandwidth requirement. In addition, each request has a profit gained by accommodating it. In the line and tree topologies, each request from a source $s$ to a destination $d$ has only one possible path, while in an undirected ring topology, each request has two possible routes, namely clockwise and counter-clockwise. Thus, in all cases, the routing issue is either simple or nonexistent. We address the following admission control problem: which of the requests will be accommodated? The goal is to maximize the total profit gained by the accommodated requests. The optimal solution consists of a feasible set of requests with maximum total profit.

We consider several variants of the problems described above, all of which are NP-hard, and we provide approximation algorithms for all of them.

## 1.2   Previous Work and Our Contribution

Some of the earliest work on advance reservation in communication networks was done in the context of video-conferencing and satellite systems. Early video-conferencing systems involved high bandwidth signals between (fixed) video-conferencing sites (studios), and advance reservations were needed to ensure that adequate bandwidth was available. Similarly, early satellite systems offered the option to rent the use of transponders for specific amounts of time, which also required support for advance reservation. In those early systems, the bulk of the work (e.g., [17,21]) focused on traffic modeling and related call admission procedures, to properly size such facilities. Some more recent studies (e.g., [18,22]) have extended these early works from the circuit-switched environment they assumed to that of modern integrated packet switching networks. Most other works dealing with advance reservation in networks have focused on extensions to signalling protocols, or formulated frameworks (including signalling and resource management capabilities) to support advance reservations, e.g, [16,19]. The routing perspective of networks with advance reservations has been investigated in [14]. That study considered possible extensions to path selection algorithms in order to make them advance-reservation aware; as connections were assumed to be handled one at a time, the admission control problem was trivial. Competitive analysis of on-line admission control in general networks was studied in [4].

In this study we focus on advance reservations of *multiple connections* in specific network topologies, namely lines, rings and trees. The case of advance reservations in a line topology can be modeled as a set of axis-parallel rectangles in the plane: the $x$-axis represents the links of the network, and the $y$-axis represents the time line which is assumed to be slotted. Each rectangle corresponds to a request: its projection on the $x$-axis represents its path from the source to the destination, and its projection on the $y$-axis represents its time interval.

For a set $R$ of $n$ axis-parallel rectangles in the plane, the associated *intersection graph* is the undirected graph with vertex set equal to $R$ and an edge between two vertices if the corresponding rectangles intersect. Assuming each rectangle (corresponding to some vertex in the intersection graph) has a profit associated with it, the goal is to find the *maximum weight independent set* (MWIS) in the intersection graph. That is, to find a set of non-overlapping rectangles with maximum total profit. This problem has already been considered in the context of label placement in digital cartography. The task is to place labels on a map. The labels can be modeled as rectangles. They are assigned profit values that represent the importance of including them in the map. Often, a label can be placed in more than one position on the map. The goal is thus to compute the maximum weight independent set of label objects.

An approximation algorithm for the MWIS problem on axis-parallel rectangles achieving a factor of $O(\log n)$ was given by [1]. This factor was improved to $\log n/\alpha$ for any constant $\alpha$ by [9]. In the case where all rectangles have the same height (in our model, the same duration of time), [1] presents a 2-approximation algorithm that incurs $O(n \log n)$ time. Extending this result, using dynamic programming, [1] obtains a PTAS, i.e., a $(1 + 1/k)$-approximation algorithm whose running time is $O(n \log n + n^{2k-1})$ time, for any $k \geq 1$.

**Our Results.** For line topologies, i.e., for the MWIS problem in the intersection graph of axis-parallel rectangles, we obtain a $4c$-approximation algorithm, where $c$ denotes the maximum number of rectangles that can cover simultaneoulsy a point in the plane. This improves on the approximation factor of [1,9] for the case where $c$ is small ($c$ is $o(\log n)$). Our technique for deriving the $4c$-approximate solution implies the following result. Suppose that two rectangles are defined to be intersecting if and only if one of them covers a corner of the other. A 4-approximation for the MWIS problem is obtained in this case by using the local structure of the linear programming relaxation of the problem. It remains an intriguing open problem whether a constant approximation factor can be found for the MWIS problem in an intersection graph of arbitrary axis-parallel rectangles in the plane.

For an undirected ring topology, we can generalize the same approximation factors obtained for a line topology. We also consider the case where the ratio between the duration of different requests does not differ by more than $k$ a parameter. We present a $\log k$-approximation algorithm for this case.

For a tree topology, we consider the case where each request has a bandwidth demand which is defined to be the *width* of the request. We first present a 5-approximation algorithm for the MWIS problem in the case of one-dimensional requests in a tree (that is, their durations are ignored). Then, we provide an $O(\log n)$-approximation for the two-dimensional case. (These factors also hold for the cases of line and ring topologies.)

## 1.3    Model

The time domain over which reservations are made is composed of *time slots* $\{0, 1, 2, \ldots\}$ of equal size. The duration of each reservation is an integer number of slots. In all three topologies (lines, rings and trees) the available bandwidth of each link $l \in E$ is fixed over time (before any requests are being accommodated); we normalize it to unit size for convenience. We are presented with a set $R$ of $n$ connection requests (=commodities), each specifying its source and destination nodes, and a specific amount of bandwidth $B$ from some time slot $t_1$ up to some time slot $t_2 > t_1$. We consider two cases: the case where all demands $B$ are equal to one (that is, only a single request can be routed through link $l$ during time slot $t$), and the case where the demands are positive arbitrary numbers bounded by one. Each request $I$ has a profit $p(I)$ gained by routing it. The goal is to select a *feasible* set of requests with maximum total profit. A set is *feasible* if for all time slots $t$ and for all links $l$, the total bandwidth of requests whose time interval contains $t$ and whose route contains $l$ does not exceed 1.

# 2    Preliminaries: The Local Ratio Technique

We shall use the *local ratio* technique [6] extensively, hence we briefly present some related preliminaries.

Let $\mathbf{p} \in \mathbb{R}^n$ be a profit (or penalty) vector, and let $F$ be a set of feasibility constraints on vectors $\mathbf{x} \in \mathbb{R}^n$. A vector $\mathbf{x} \in \mathbb{R}^n$ is a *feasible solution* to a given problem $(F, \mathbf{p})$ if it satisfies all of the constraints in $F$. The *value* of a feasible solution $\mathbf{x}$ is the inner product $\mathbf{p} \cdot \mathbf{x}$. A feasible solution is *optimal* for a maximization (or minimization) problem if its value is maximal (or minimal) among all feasible solutions. A feasible solution $\mathbf{x}$ is an *r-approximate* solution, or simply an *r-approximation*, if $\mathbf{p} \cdot \mathbf{x} \geq$ (or $\leq$) $r \cdot \mathbf{p} \cdot \mathbf{x}^*$, where $\mathbf{x}^*$ is an optimal solution. An algorithm is said to have a *performance guarantee* of $r$ if it always computes $r$-approximate solutions.

Our algorithms use the local ratio technique. This technique was first developed by Bar-Yehuda and Even [7] and later extended by [5,6,8]. The local ratio theorem is as follows.

**Theorem 1 (local ratio).** *Let $F$ be a set of constraints, and let $\mathbf{p}, \mathbf{p_1}, \mathbf{p_2}$ be profit (or penalty) vectors where $\mathbf{p} = \mathbf{p_1} + \mathbf{p_2}$. If $\mathbf{x}$ is an r-approximate solution with respect to $(F, \mathbf{p_1})$ and with respect to $(F, \mathbf{p_2})$, then $\mathbf{x}$ is an r-approximate solution with respect to $(F, \mathbf{p})$.*

An algorithm that uses the local ratio technique typically proceeds as follows. Initially, the solution is empty. The idea is to find a decomposition of $\mathbf{p}$ into $\mathbf{p_1}$ and $\mathbf{p_2}$ such that $\mathbf{p_1}$ is an "easy" weight function in some respect, e.g., any solution which is maximal with respect to containment would be a good approximation to the optimal solution of $(F, \mathbf{p_1})$. The local ratio algorithm continues recursively on the instance $(F, \mathbf{p_2})$. We assume inductively that the solution returned recursively for the instance $(F, \mathbf{p_2})$ is a good approximation and need to prove that it is also a good approximation for $(F, \mathbf{p})$. This requires proving that the solution returned recursively for the instance $(F, \mathbf{p_2})$ is also a good approximation for the instance $(F, \mathbf{p_1})$. This step is usually the "heart" of the proof of the approximation factor.

## 3    Line Topology

The case of advance reservations in a line topology can be modeled as a set of axis-parallel rectangles in the plane: the $x$-axis represents the network links, and the $y$-axis represents the time line. Each rectangle corresponds to a request: its projection on the $x$-axis represents its path from the source to the destination, and its projection on the $y$-axis represents its time interval. We consider the case where all bandwidth requirements are equal to the capacity of the links.

For a set $R$ of $n$ rectangles in the plane, the associated *intersection graph* $G = (R, E)$ is the undirected graph with vertex set equal to $R$ and an edge between two vertices if and only if the corresponding rectangles intersect. We assume that each rectangle (corresponding to some vertex in the intersection graph) has a *profit* (or *weight*) associated with it. The goal is to find a *maximum weight independent set* (MWIS) in $G$, i.e., a set of non-overlapping rectangles with maximum total profit.

The problem of finding a maximum independent set in the intersection graph of unit squares is NP-complete [2]. Since unit squares are a special case of rectangles, the intractability of this problem implies the intractability of the general case.

A maximal (with respect to containment) clique $Q$ in $G$ corresponds to a point in the plane $a$ such that the vertices in $Q$ correspond to the rectangles covering $a$. We assume that the maximum clique in $G$ is of size $c$, i.e., no point is covered by more than $c$ rectangles.

A rectangle $r_1$ is said to be *vertex-intruding* into another rectangle $r_2$, if $r_2$ contains at least one of the corners of $r_1$. Two rectangles are *vertex-incident* if at least one of them is vertex-intruding into the other. Two identical rectangles are also said to be vertex-intruding into each other. Figure 1 contains several examples of vertex-incident rectangles. A set of rectangles is said to be *vertex-incident-free* if the rectangles are not pairwise vertex incident. For a rectangle $r$, we denote by $N^i[r]$ the set of rectangles that are vertex-incident to $r$. (Note that $r \in N^i[r]$).

We present a $4c$-approximation algorithm for the MWIS problem. The algorithm is structured as follows.

1. Compute a set $S$ of rectangles that are vertex-incident-free, such that the weight of $S$ is at least $1/4$ of the weight of a MWIS in $G$.
2. Find an independent set $I \subseteq S$ of rectangles, such that its weight is at least $1/c$ of the weight of $S$.
3. The output is the independent set $I$.

Clearly, the independent set $I$ is a $4c$-approximate solution. We now elaborate on the steps of the algorithm.

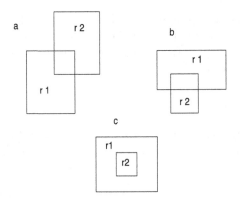

**Fig. 1.** In (a) both rectangles are vertex-intruding into each other, while in (b) and (c) $r_2$ is vertex-intruding into $r_1$.

*Step (1).* We formulate a linear program for the MWIS in $G$. We define an indicator variable $x(v)$ for each rectangle $v \in R$. If $x(v) = 1$, then rectangle $v$ belongs to the independent set. The linear relaxation of the indicator variables assigns fractions to the rectangles (requests) with the constraint that for each clique $Q$, the sum of the fractions assigned to all rectangles in $Q$ does not exceed 1. Let $\mathbf{x}$ denote the vector of indicator variables.

$$(L) \quad \text{maximize} \quad \sum_{v \in R} w(v) \cdot x(v)$$

*subject to:*

$$\text{For each clique } Q: \sum_{v \in Q} x(v) \leq 1 \tag{1}$$

$$\text{For all } v \in R: \quad x(v) \geq 0. \tag{2}$$

Note that the number of cliques in $G$ is $O(n^2)$. It is easy to see that an independent set in $G$ provides a feasible integral solution to the linear program. Thus, the value of an optimal (fractional) solution to the linear program is an upper bound on the value of an optimal integral solution.

We compute an optimal solution to (L). We now show how to round the solution obtained to get the set $S$. The core of our rounding algorithm is the following lemma.

**Lemma 1.** *Let* $\mathbf{x}$ *be a feasible solution to* (L). *Then, there exists a rectangle* $v \in R$ *satisfying:*

$$\sum_{u \in N^i[v]} x(u) \leq 4.$$

*Proof.* For two vertex incident rectangles vertices $u$ and $v$, define $y(u, v) = x(v) \cdot x(u)$. Also, define $y(u, u) = x(u)^2$. For a point $a$ in the plane, let $C(a)$ denote the set of rectangles that contain $a$. For a rectangle $r$, let $S(r)$ denote the set of four corners of $r$. We prove the lemma using a *weighted* average argument, where the weights are the values $y(u,v)$ for all pairs of vertex-incident rectangles, $u$ and $v$. We claim that

$$\sum_{v \in R} \sum_{u \in N^i[v]} y(u,v) \leq \sum_{v \in R} \sum_{a \in S(v)} \sum_{u \in C(a)} y(u,v). \tag{3}$$

We shall prove (3) by considering the different cases of vertex-incidence between two rectangles $u$ and $v$ (see Fig. 1). If $u$ and $v$ are vertex-incident, then they contribute together $2y(u,v)$ to the LHS of Equation (3). The contribution to the RHS of Equation (3) is as follows.

1. Rectangle $u$ is vertex-intruding into $v$ and vice versa (see Fig. 1(a)). In this case, $u$ and $v$ contribute $2y(u,v)$ to the RHS of Equation (3), since $u$ ($v$) contains a corner of $v$ ($u$).
2. Rectangle $v$ is vertex-intruding into $u$, but $u$ is not vertex-intruding into $v$ (see Fig. 1(b)). In this case, $v$ contributes $2y(u,v)$ to the RHS of Equation (3), since $v$ has two corners intruding into $u$.
3. Rectangle $v$ is either contained inside rectangle $u$, or $v = u$ (see Fig. 1(c)). In this case, $v$ contributes $4y(u,v)$ to the RHS of Equation (3), since $u$ contains all four corners of $v$.

Hence, there exists a rectangle $v$ satisfying

$$\sum_{u \in N^i[v]} y(u,v) \leq \sum_{a \in S(v)} \sum_{u \in C(a)} y(u,v).$$

If we factor out $x(v)$ from both sides, we obtain

$$\sum_{u \in N^i[v]} x(u) \leq \sum_{a \in S(v)} \sum_{u \in C(a)} x(u).$$

From Constraint (1) in (L) it follows that for each rectangle $v$, for all $a \in S(v)$, $\sum_{u \in C(a)} x(u) \leq 1$. Therefore,

$$\sum_{u \in N^i[v]} x(u) \leq 4.$$

Completing the proof. □

We now use a fractional version of the Local Ratio technique developed by [8]. The proof of the next lemma is immediate.

**Lemma 2 (fractional local ratio).** *Let $\mathbf{x}$ be a feasible solution to (L). Let $\mathbf{w_1}$ and $\mathbf{w_2}$ be a decomposition of the weight vector $\mathbf{w}$ such that $\mathbf{w} = \mathbf{w_1} + \mathbf{w_2}$. Suppose that $\mathbf{z}$ is a feasible integral solution vector to (L) satisfying: $\mathbf{w_1} \cdot \mathbf{z} \geq r(\mathbf{w_1} \cdot \mathbf{x})$ and $\mathbf{w_2} \cdot \mathbf{z} \geq r(\mathbf{w_2} \cdot \mathbf{x})$. Then,*

$$\mathbf{w} \cdot \mathbf{z} \geq r(\mathbf{w} \cdot \mathbf{x}).$$

The rounding algorithm will apply a local ratio decomposition of the weight vector $\mathbf{w}$ with respect to an optimal solution $\mathbf{x}$ to linear program (L). The algorithm for computing $S$ proceeds as follows.

1. Delete all rectangles with non-positive weight. If no rectangles remain, return the empty set.
2. Let $v' \in R$ be a rectangle satisfying $\sum_{u \in N^i[v']} x(u) \leq 4$. Decompose $\mathbf{w}$ by $\mathbf{w} = \mathbf{w_1} + \mathbf{w_2}$ as follows:

$$w_1(u) = \begin{cases} w(v') & \text{if } u \in N^i[v'], \\ 0 & \text{otherwise.} \end{cases}$$

(In the decomposition, the component $\mathbf{w_2}$ may be non-positive.)
3. Solve the problem recursively using $\mathbf{w_2}$ as the weight vector. Let $S'$ be the independent set returned.
4. If $v'$ is not vertex incident to any rectangle in $S'$, return $S = S' \cup \{v'\}$. Otherwise, return $S = S'$.

Clearly, the set $S$ is vertex-incident-free. We now analyze the quality of the solution produced by the algorithm.

**Theorem 2.** *Let $\mathbf{x}$ be an optimal solution to linear program (L). Then, it holds for the set $S$ computed by the algorithm that $w(S) \geq \frac{1}{4} \cdot \mathbf{w} \cdot \mathbf{x}$.*

*Proof.* The proof is by induction on the number of recursive calls. At the basis of the recursion, the set returned satisfies the theorem, since no rectangles remain. Clearly, the first step in which rectangles of non-positive weight are deleted cannot decrease the above RHS. We now prove the inductive step. Let $\mathbf{z}$ and $\mathbf{z'}$ be the indicator vectors of the sets $S$ and $S'$, respectively. Assume that $\mathbf{w_2} \cdot \mathbf{z'} \geq (1/4) \cdot \mathbf{w_2} \cdot \mathbf{x}$. Since $w_2(v') = 0$, it also holds that $\mathbf{w_2} \cdot \mathbf{z} \geq (1/4) \cdot \mathbf{w_2} \cdot \mathbf{x}$. From Step (4) of the algorithm it follows that at least one vertex from $N^i[v']$ belongs to $S$. Hence, $\mathbf{w_1} \cdot \mathbf{z} \geq (1/4) \cdot \mathbf{w_1} \cdot \mathbf{x}$. Thus, by Lemma 2, it follows that $\mathbf{w} \cdot \mathbf{z} \geq (1/4) \cdot \mathbf{w} \cdot \mathbf{x}$, i.e., $w(S) \geq \frac{1}{4} \cdot \mathbf{w} \cdot \mathbf{x}$. □

*Step (2).* The input to this step is a set of rectangles $S$ that are vertex-incident-free. We show that the intersection graph of such a family of rectangles is a perfect graph. Perfect graphs are graphs for which the chromatic number is equal to the maximum clique size and this property holds for all subgraphs.

Let the maximum clique size in $S$ be $c'$. Clearly, $c' \leq c$.

**Theorem 3.** *There exists a legal coloring of the rectangles in $S$ that uses precisely $c'$ colors.*

*Proof.* We use here ideas from [3]. Rectangles that are not vertex-incident can only intersect in the pattern described in Figure 2. Let us look at the cliques of size $c'$ in $S$, and choose in each clique, the tallest and narrowest rectangle. We observe that these rectangles are disjoint, otherwise one of them is vertex-intruding into the other. We color the rectangles chosen by the same color. Since the maximum clique among the remaining rectangles in $S$ is of size $c' - 1$, we can continue this process recursively, and get a $c'$-coloring of $S$. □

The proof of the above theorem is constructive, namely, we color the rectangles of $S$ by $c'$ colors and choose $I$ to be the color class of maximum weight. Thus,

$$w(I) \geq \frac{w(S)}{c'} \geq \frac{w(S)}{c}.$$

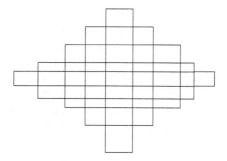

**Fig. 2.** Intersection of rectangles that are not vertex-incident

### 3.1 Ring Topology

We consider an undirected ring, that is, each request has two possible routes from its source to its destination, namely clockwise and counter-clockwise. Each request has a profit associated with it, and the goal is to find a MWIS of requests, that is, a set of non-overlapping requests with maximum total profit. We have to choose a unique path for each request such that the profit of the solution is maximized. We consider the case where all bandwidth requirements are equal to the capacity of the links.

We first remark that the approximation factors obtained for the case of a line topology can be generalized for this case too. We now consider the case where the ratio between the duration of different requests is not more than $k$, a parameter. We sketch how to obtain a $\log k$-approximation algorithm in this case. We start with a line. Partition the set of rectangles into groups such that the heights are within a factor of two in each group. There are at most $\log k$ groups. For each group, a constant-factor approximate solution can be computed using ideas from [1] (and also [12]). The output is the best of the $\log k$ solutions, yielding an approximation factor of $\log k$. This approach can be generalized to rings by cutting one link and then reducing the problem to the line case.

## 4   Tree Topology

In previous sections we assumed that each connection requested the full link capacity. We now consider the case where each request $r_i$ has a bandwidth demand $0 < w_i \leq 1$, which is defined to be the *width* of the request. We present a

($5 \log n$)-approximation algorithm. We first present a 5-approximation algorithm for the MWIS problem in the case of one-dimensional requests in a tree (that is, their durations are ignored). Then, we provide a $5 \log n$-approximation for the two-dimensional case by extending the algorithm presented in [1], which is a simple divide-and-conquer algorithm for computing a maximum independent (non-overlapping) set of $n$ axis-parallel rectangles in the plane. We remark that these results also hold for the cases of line and ring topologies. The one dimensional case in tree topologies has received much attention, e.g., [10,11]. However, to the best of our knowledge, there are no known approximation factors for the one-dimensional problem we are considering here (where requests have widths).

We now present a 5-approximation algorithm for the one-dimensional MWIS problem in a tree. We divide our instances (requests) into two sets: a set consisting of all *narrow* instances, i.e., that have width of at most $1/2$, and a set consisting of all *wide* instances, i.e., that have width greater than $1/2$. We solve our problem separately for the two sets, and return the solution with greater profit.

The problem where all instances are wide reduces to the uncapacitated case, since no pair of intersecting instances can be simultaneously in the solution. It thus reduces to the problem of finding a MWIS of one-dimensional requests (paths) in a tree which can be solved optimally (see [20]).

We describe a 4-approximation algorithm for the case all one-dimensional requests are narrow. We begin with some definitions. Let $G = (V, E)$ be a graph with vertex set $V$ and edge set $E$. If $X$ is a subset of the vertices, then $G(X)$ is the subgraph of $G$ induced by $X$. A clique $C$ is called a *separator* if $G(V \setminus C)$ is not connected. An *atom* is a connected graph with no clique separator. Let $C$ be a clique separator of $G$, and let $A, B, C$ be a vertex partition such that no vertex in $A$ is adjacent to a vertex in $B$. Then, $G$ can be decomposed into two components, namely $G' = G(A \cup C)$ and $G'' = G(B \cup C)$. By decomposing $G'$ and $G''$ in the same way, and repeating this process as long as possible, we get a *clique decomposition* of $G$. Figure 4 contains an example of a clique decomposition of a graph. $G$ is thus decomposed into atoms, each being a subgraph of $G$ containing no separator. This decomposition can be represented by a binary tree: each external node represents an atom, and each internal node represents a clique separator. The algorithm described in [20] produces a decomposition tree where the internal nodes lie on one path.

The intersection graph defined by the requests on a tree is the intersection graph of paths on a tree, known as an Edge-Path-Tree (EPT) graph. In order to consider the clique decomposition of an EPT graph [13], we define another class of graphs. A graph is a *line graph* if its vertices correspond to the edges of a multigraph (i.e., a graph with multiple edges) such that two vertices in the original graph are adjacent if and only if the corresponding edges in the multigraph share a common vertex. The following are known results [13]: a graph is a line graph if and only if it is the EPT graph of a star; the atoms of any EPT graph are line graphs. Also, recognizing a line graph and finding its star graph can be done in polynomial time [15].

**Interval Graph:**

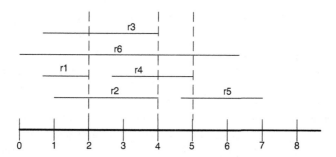

**Clique Decomposition:**

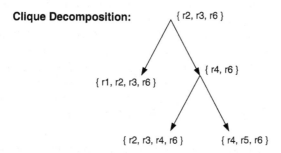

We now describe a 4-approximation algorithm for the one-dimensional narrow case, as follows.

1. Given an undirected tree and a set $n$ of paths (instances), define $G$ as the EPT graph of these paths (each vertex in $G$ is a path in the undirected tree).
2. Delete all instances with non-positive profit.
3. If no instances remain, return the empty set. Otherwise, proceed to the next step.
4. Perform the **first** decomposition step of $G$. Let $A, B, C$ be the vertex partition such that $C$ is the clique separator, no edge in $G$ joins a vertex in $A$ and a vertex in $B$, and $G(A \cup C)$ is an atom. Represent the atom $G(A \cup C)$ as a star graph (each vertex in $A \cup C$ is a path in the star), and arbitrarily choose one of the paths in $A$ as $\tilde{I}$.
5. Decompose the profit vector $p$ by $p = p_1 + p_2$.
6. Solve the problem recursively using $p_2$ as the profit function. Let $S'$ be the returned set.
7. If $S' \cup \{\tilde{I}\}$ is a feasible set, return $S = S' \cup \{\tilde{I}\}$. Otherwise, return $S = S'$.

Define $\mathcal{I}(I)$ to be the set of instances intersecting $I$. The profit decomposition is the following:

$$p_1(I) = p(\tilde{I}) \cdot \begin{cases} 1 & I = \tilde{I}, \\ \alpha \cdot w(I) & I \in \mathcal{I}(\tilde{I}), \\ 0 & \text{otherwise.} \end{cases} \tag{4}$$

**Proposition 1.** *The extended unified algorithm presented above, with* $\alpha = \frac{1}{1-w(\tilde{I})}$, *yields a 4-approximate solution.*

*Proof.* Define $b_{opt}$ to be an upper bound on the optimum $p_1$-profit and $b_{max}$ to be a lower bound on the $p_1$-profit of every $\tilde{I}$-maximal set, both are normalized by $p(\tilde{I})$. We consider an optimal solution. By the definition of $p_1$, only instances in $\tilde{I} \cup \mathcal{I}(\tilde{I})$ contribute to its $p_1$-profit. As $\tilde{I}$ is a path in a star graph, its path consists of two links. For each of these links $l$, the total width of instances in $\mathcal{I}(\tilde{I})$ that use $l$ in their path is at most 1. In case $\tilde{I}$ is not in the optimal solution, the contribution of the instances in $\mathcal{I}(\tilde{I})$ is at most $2\alpha \cdot p(\tilde{I})$. As $\tilde{I}$ is a path in $A$, it does not intersect with any other path from $B$. However, $\tilde{I}$ intersects with at most two independent groups of instances from $A \cup C$: the instances in one group are paths in the star graph intersecting the first link of $\tilde{I}$, and in the other group the instances are paths in the star graph intersecting the second link of $\tilde{I}$. Otherwise, if $\tilde{I}$ is in the optimal solution, the contribution of the instances in $\mathcal{I}(\tilde{I})$ is at most $2\alpha(1-w(\tilde{I})) \cdot p(\tilde{I})$. Thus, we get $b_{opt} = \max\{2\alpha, 1+2\alpha(1-w(\tilde{I}))\}$.

Turning to $\tilde{I}$-maximal solutions, either such a solution contains $\tilde{I}$, or else it contains a set $\mathcal{X} \neq \emptyset$ of instances intersecting $\tilde{I}$ that prevent $\tilde{I}$ from being added to the solution. The total width of instances in $\mathcal{X}$ is at least $1-w(\tilde{I})$, for otherwise $\tilde{I}$ can be added to the solution. We thus have $b_{max} = \min\{1, \alpha \cdot (1 - w(\tilde{I}))\}$.

The approximation factor of the algorithm in the case where all instances are narrow is at least

$$\frac{\min\{1, \alpha \cdot (1 - w(\tilde{I}))\}}{\max\{2\alpha, 1 + 2\alpha(1 - w(\tilde{I}))\}}. \tag{5}$$

For $\alpha = \frac{1}{1-w(\tilde{I})}$ we get an approximation factor of $1/4$. $\qquad\square$

**Corollary 1.** *By choosing the solution with greater profit out of the two sets (wide and narrow), we get an approximation factor of $1/5$ for the general one-dimensional case.*

We now present a $(5 \log n)$-approximation for the two-dimensional case. Let $R$ be the set of $n$ requests. We sort the requests by their start and end time-coordinates. Let $t_{med}$ be the median time-coordinate. That is, the number of requests whose time-coordinates are below or above $t_{med}$ is not more than $n/2$. We partition the requests of $R$ into three groups: $R_1$, $R_2$ and $R_{12}$. $R_1$ and $R_2$ contain the requests whose time-coordinates are respectively below and above $t_{med}$. $R_{12}$ contains the requests with time interval containing time slot $t_{med}$, and thus defines a one-dimensional problem: as all these requests intersect in the time line, their time intervals can be ignored. Now, we compute the approximate

MWIS $M_{12}$ of $R_{12}$ (as explained before). We recursively compute $M_1$ and $M_2$, the approximate MWIS in $R_1$ and $R_2$, respectively: if $p(M_{12}) \geq p(M_1) + p(M_2)$, then we return $M_{12}$, otherwise we return $M_1 \cup M_2$ ($p(M)$ denotes the sum of the profits of the requests in $M$). The proof of the following proposition is omitted.

**Proposition 2.** *The algorithm computes a* $5 \log n$-*approximate solution.*

## Acknowledgements

We would like to thank an anonymous referee for many insightful comments.

# References

1. P.K. Agarwal, M.van Kreveld, ans S. Suri, "Label placement by maximum independent set in rectangles", *Computational Geometry: Theory and applications*, 11(3-4): 209-218 (1998).
2. T. Asano, "Difficulty of the maximum independent set problem on intersection graphs of geometric objects", Proceedings of the Sixth Internat. Conf. on the Theory and Applications of Graphs, Western Michigan University 1988.
3. E. Asplund, B. Grunbaum, "On a coloring problem", Math. Scand. 8 (1960), 181-188.
4. B. Awerbuch, Y. Azar, and S. Plotkin, "Throughput-competitive online routing", Proceedings of the 34th Annual Symposium on Foundations of Computer Science, 1993, pp. 32-40.
5. V. Bafna, P. Berman, and T. Fujito, "A 2-approximation algorithm for the undirected feedback vertex set problem", SIAM J. on Disc. Mathematics, Vol. 12, pp. 289–297, 1999.
6. A. Bar-Noy, R. Bar-Yehuda, A. Freund, J. Naor and B. Schieber, "A unified approach to approximating resource allocation and scheduling", Journal of the ACM, Vol. 48 (2001), pp. 1069-1090.
7. R. Bar-Yehuda and S. Even, "A local-ratio theorem for approximating the weighted vertex cover problem", Annals of Discrete Mathematics, Vol. 25, pp. 27–46, 1985.
8. R. Bar-Yehuda, M. Halldorsson, J. Naor, H. Shachnai, and I. Shapira, "Scheduling split intervals", Proceedings of the 13th Annual ACM-SIAM Symposium on Discrete Algorithms, San Francisco, CA, (2002), pp. 732-741.
9. P. Berman, B. DasGupta, S. Muthukrishana, and S. Ramaswami, "Improved approximation algorithms for rectangle tiling and packing", Proceedings of the 12th Annual ACM-SIAM Symposium on Discrete Algorithms, 2001, pp. 427-436.
10. T. Erlebach and K. Jansen, "Off-line and on-line call-scheduling in stars and trees", Proceedings of the 23rd International Workshop on Graph-Theoretic Concepts in Computer Science (WG'97) LNCS 1335, Springer Verlag, 1997, pp. 199-213.
11. T. Erlebach and K. Jansen, "Maximizing the number of connections in optical tree networks", Proceedings of the Ninth Annual International Symposium on Algorithms and Computation (ISAAC'98) LNCS 1533, Springer Verlag, 1998, pp. 179-188.
12. T. Erlebach, K. Jansen, and E. Seidel, "Polynomial-time approximation schemes for geometric graphs", Proceedings of the 12th Annual ACM-SIAM Symposium on Discrete Algorithms, 2001, pp. 671-679.

13. M.C. Golumbic and R.E Jamison, "Edge and vertex intersection of paths in a tree", Discrete Math. 55 (1985), 151-159.
14. R. Guerin and A. Orda, "Networks with advance reservations: the routing perspective", Proceedings of INFOCOM 2000, Tel-Aviv, Israel, April 2000.
15. P.G.H Lehot, "An optimal algorithm to detect a line graph and output its root graph", J. ACM 21 (1974) 569-575.
16. W. Reinhardt. Advance resource reservation and its impact on reservation protocols. In *Proceedings of Broadband Islands'95*, Dublin, Ireland, September 1995.
17. J. W. Roberts and K.-Q. Liao. Traffic models for the telecommunication services with advanced reservation. In *Proceedings of the 11th Intl. Teletraffic Congress*, Kyoto, Japan, June 1985. Paper 1.3-2.
18. O. Schelén and S. Pink. Sharing resources through advance reservations agents. In *Proceedings of IWQoS'97*, New York, NY, May 1997.
19. A. Schill, F. Breiter, and S. Kühn. Design and evaluation of an advance resource reservation protocol on top of RSVP. In *Proceedings of IFIP Broadband'98*, Stuttgart, Germany, April 1998.
20. R. Tarjan, "Decomposition by clique separators", Discrete Math. 55 (1985), 221-231.
21. J. Virtamo. A model of reservation systems. *IEEE. Trans. Commun.*, 40(1):109-118, January 1992.
22. D. Wischik and A. Greenberg. Admission control for booking ahead shared resources. In *Proceedings of INFOCOM'98*, San Francisco, CA, April 1998.

# Improved Approximation Algorithms for Metric Facility Location Problems

Mohammad Mahdian[1], Yinyu Ye[2,*], and Jiawei Zhang[3,**]

[1] Department of Mathematics,
MIT, Cambridge, MA 02139, USA
mahdian@math.mit.edu
[2] Department of Management Science and Engineering,
Stanford University, Stanford, CA 94305-4026, USA
yinyu-ye@stanford.edu
[3] Department of Management Sciences,
University of Iowa, Iowa City, IA, 52242, USA
jiazhang@stanford.edu

**Abstract.** In this paper we present a 1.52-approximation algorithm for the uncapacitated metric facility location problem. This algorithm uses an idea of cost scaling, a greedy algorithm of Jain, Mahdian and Saberi, and a greedy augmentation procedure of Charikar, Guha and Khuller. We also present a 2.89-approximation for the capacitated metric facility location problem with soft capacities.

## 1  Introduction

In the uncapacitated facility location problem (UFLP), we have a set $\mathcal{F}$ of $n_f$ *facilities* and a set $\mathcal{C}$ of $n_c$ *cities*. For every facility $i \in \mathcal{F}$, a nonnegative number $f_i$ is given as the *opening cost* of facility $i$. Furthermore, for every city $j \in \mathcal{C}$ and facility $i \in \mathcal{F}$, we have a *connection cost* (a.k.a. service cost) $c_{ij}$ between city $j$ and facility $i$. The objective is to open a subset of the facilities in $\mathcal{F}$, and connect each city to an open facility so that the total cost is minimized. We will consider the *metric* version of this problem, i.e., the connection costs satisfy the triangle inequality.

The facility location problem is a central problem in operations research, and a large number of approximation algorithms using a variety of techniques have been proposed for this problem. Table 1 shows a summary of the results. The running times in this table are in terms of $n = n_f + n_c$ (the running times of linear programming rounding algorithms are dominated by the time complexity of solving associated linear programming relaxations, which is substantially higher). Regarding hardness results, Guha and Khuller [4] proved that it is impossible to get an approximation guarantee of 1.463 for the uncapacitated metric facility location problem, unless $\mathbf{NP} \subseteq \text{DTIME}[n^{O(\log \log n)}]$. For a more detailed survey on this problem, see Shmoys [10].

---

\* Research supported in part by NSF grants DMI-9908077.
\*\* Research supported in part by NSF grants DMI-9908077 through Yinyu Ye.

K. Jansen et al. (Eds.): APPROX 2002, LNCS 2462, pp. 229–242, 2002.
© Springer-Verlag Berlin Heidelberg 2002

**Table 1.** Approximation Algorithms for UFLP

| approx. factor | reference | technique/running time |
|---|---|---|
| $O(\ln n_c)$ | [5] | greedy algorithm/$O(n^3)$ |
| 3.16 | [11] | LP rounding |
| 2.41 | [4] | LP rounding + greedy augmentation |
| 1.736 | [2] | LP rounding |
| $5 + \epsilon$ | [8] | local search/$O(n^6 \log(n/\epsilon))$ |
| 3 | [7] | primal-dual method/$O(n^2 \log n)$ |
| 1.853 | [1] | primal-dual method + greedy augmentation/$O(n^3)$ |
| 1.728 | [1] | LP + primal-dual method + greedy augmentation |
| 1.861 | [9] | greedy algorithm/$O(n^2 \log n)$ |
| 1.61 | [6] | greedy algorithm/$O(n^3)$ |
| 1.582 | [12] | LP rounding |
| 1.52 | This paper | greedy algorithm + greedy augmentation/$O(n^3)$ |

In this paper, we combine the greedy algorithm of Jain, Mahdian, and Saberi [6] and the greedy augmentation of Charikar, Guha, and Khuller [1,4] to show that UFLP can be approximated within a factor of 1.52, whereas the best previously known factor was 1.582 [12]. Note that our approximation factor is very close to the lower bound of 1.463 proved by Guha and Khuller [4].

As a by-product of our analysis, we obtain an improved 2.89-approximation algorithm for the capacitated facility location problem (CFLP) where opening multiple copies of the same facility is allowed. The best previously known factor was 3 [6]. Also, Chudak and Shmoys [3] gave a 3-approximation algorithm for CFLP with uniform capacities.

The improved algorithm for UFLP and its underlying intuition are presented in Section 2. In Section 3, we prove the upper bound of 1.52 on the approximation factor of the algorithm. In Section 4, we present the 2.89-approximation for CFLP. Some possible directions for future research on these problems are discussed in Section 5.

## 2    Algorithm

Jain, Mahdian, and Saberi [6] proposed a greedy algorithm for UFLP that achieves a factor of 1.61. Here is a sketch of their algorithm:

### The JMS Algorithm

1. At the beginning, all cities are *unconnected*, all facilities are *unopened*, and the *budget* of every city $j$, denoted by $B_j$, is initialized to 0. At every moment, each city $j$ offers some money from its budget to each *unopened* facility $i$. The amount of this offer is equal to $\max(B_j - c_{ij}, 0)$ if $j$ is unconnected, or $\max(c_{i'j} - c_{ij}, 0)$ if it is already connected to some other facility $i'$.

2. While there is an unconnected city, increase the budget of each *unconnected* city at the same rate, until one of the following two events occurs:

   (a) For some unopened facility $i$, the total offer that it receives from cities is equal to the cost of opening $i$. In this case, we open facility $i$, and connect $j$ to $i$ for every city $j$ (connected or unconnected) which has a non-zero offer to $i$.

   (b) For some unconnected city $j$, and some facility $i$ that is already open, the budget of $j$ is equal to the connection cost $c_{ij}$. In this event, we connect $j$ to $i$.

One important property of the above algorithm is that it finds a solution in which there is no unopened facility that one can open to decrease the cost (without closing any other facility). This is because for a connected city $j$ and unopened facility $i$, $j$ continuously makes offer to $i$ the amount that it would save by switching service site to $i$. This is, in fact, the main advantage of the JMS algorithm over a previous algorithm of Mahdian et al. [9].

Here we use the JMS algorithm to solve UFLP with an improved approximation factor. Our algorithm has two phases. In the *first* phase, we scale up the opening costs of all facilities by a factor of $\delta(\geq 1)$ (which is a constant that will be fixed later) and then run the JMS algorithm to find a solution $SOL_1$. The technique of cost scaling has been previously used by Charikar and Guha [1] for the facility location problem, in order to take advantage of the asymmetry between the performance of the algorithm with respect to facility and connection costs. Here we use this idea for a different reason: Intuitively, facilities that are opened by the JMS algorithm with scaled-up facility costs are those that are very economical, because we weight the facility cost more than the connection cost in the objective function. Therefore, we open these facilities in the first phase of the algorithm.

In the *second* phase of the algorithm, we scale down the opening costs of facilities back to their original values all at the same rate. If at any point during this process, a facility could be opened without increasing the total cost (i.e., if the opening cost of the facility equals the total amount that cities can save by switching their "service provider" to that facility), then we open the facility and connect each city to its closest open facility. It is not difficult to see that this is equivalent to a greedy procedure introduced by Guha and Khuller [4] and Charikar and Guha [1]. In this procedure, in each iteration, we pick a facility $u$ with opening cost $f_u$ such that if by opening $u$, the total connection cost decreases from $C$ to $C'$, the ratio $(C - C' - f_u)/f_u$ is maximized. If this ratio is positive, then we open the facility $u$, and iterate; otherwise we stop.

Let $SOL_2$ denote the solution after the above greedy augmentation procedure. We will prove in the next section that the cost of $SOL_2$ is at most 1.52 times the cost of the optimal solution.

## 3    The Approximation Factor

In order to analyze the approximation factor of our algorithm, we use results of [6] and [1] that bound the cost of the solution found by the JMS algorithm and the greedy augmentation procedure.

Before we present the result of the [6], we give the following definition.

**Definition 1.** *An algorithm is called a $(\gamma_f, \gamma_c)$-approximation algorithm for UFLP, if for every instance $\mathcal{I}$ of UFLP, and for every solution SOL for $\mathcal{I}$ with facility cost $F_{SOL}$ and connection cost $C_{SOL}$, the cost of the solution found by the algorithm is at most $\gamma_f F_{SOL} + \gamma_c C_{SOL}$.*

The following theorem gives tight bounds on the tradeoff between $\gamma_f, \gamma_c$ for the JMS algorithm.

**Lemma 1.** *([6]) Let $\gamma_f \geq 1$ and $\gamma_c := \sup_k\{z_k\}$, where $z_k$ is the solution of the following optimization program (which we call the factor-revealing LP).*

$$Maximize \ \frac{\sum_{i=1}^{k} \alpha_i - \gamma_f f}{\sum_{i=1}^{k} d_i} \tag{1}$$

$$s.t. \quad \forall 1 \leq i < k : \ \alpha_i \leq \alpha_{i+1}$$
$$\forall 1 \leq j < i < k : \ r_{j,i} \geq r_{j,i+1}$$
$$\forall 1 \leq j < i \leq k : \ \alpha_i \leq r_{j,i} + d_i + d_j$$
$$\forall 1 \leq i \leq k : \ \sum_{j=1}^{i-1} \max(r_{j,i} - d_j, 0) + \sum_{j=i}^{k} \max(\alpha_i - d_j, 0) \leq f$$
$$\forall 1 \leq j \leq i \leq k : \ \alpha_j, d_j, f, r_{j,i} \geq 0$$

*Then the JMS algorithm is a $(\gamma_f, \gamma_c)$-approximation algorithm for UFLP.*

In particular, it is proved [6] that $\gamma_c \leq 1.61$ when $\gamma_f = 1.61$, and therefore the JMS algorithm is a $(1.61, 1.61)$-approximation algorithm. Here we use the above theorem with $\gamma_f = 1.11$. The following lemma shows that JMS algorithm is a $(1.11, 1.78)$-approximation algorithm for UFLP. The proof is long and technical and it can be found in the journal paper version. Here, we include a sketch of proof in Appendix A.

**Lemma 2.** *For every $k$, the solution of the maximization program (1) with $\gamma_f = 1.11$ is at most $1.78$. In other words, $\gamma_c \leq 1.78$ when $\gamma_f = 1.11$.*

We also use the following result of Charikar and Guha [1] that bounds the cost of the solution after running the greedy augmentation procedure in terms of the cost of the initial solution and an arbitrary solution.

**Lemma 3.** *([1])  For every instance $\mathcal{I}$ of UFLP and for every solution SOL of $\mathcal{I}$ with facility cost $F_{SOL}$ and connection cost $C_{SOL}$, if an initial solution has* *facility cost F and connection cost C, then after greedy augmentation the cost of the solution is at most*

$$F + F_{SOL} \max\left\{0, \ln\left(\frac{C - C_{SOL}}{F_{SOL}}\right)\right\} + F_{SOL} + C_{SOL}.$$

Using the above lemmas, we can prove the following.

**Theorem 1.** *The uncapacitated facility location problem can be approximated within a factor of 1.52 in time $O(n^3)$.*

*Proof.* Let $OPT$ be an optimal solution with facility and connection costs $F^*$ and $C^*$, respectively, and consider a pair $(\gamma_f, \gamma_c)$ given in Lemma 1. Let $SOL_1$ denote the solution found by the JMS algorithm for an instance in which facility costs are scaled by a factor of $\delta$ ($\delta \geq 1$). By Lemma 1, the cost of this solution, evaluated using scaled-up facility costs, is at most $\gamma_f \delta F^* + \gamma_c C^*$. Therefore, if $F_{SOL_1}$ and $C_{SOL_1}$ denote the facility and connection costs of $SOL_1$, evaluated with the original costs, then we have

$$\delta F_{SOL_1} + C_{SOL_1} \leq \gamma_f \delta F^* + \gamma_c C^*. \tag{2}$$

Also, by Lemma 3 the cost of the solution returned by the greedy augmentation procedure is at most

$$\text{cost}(SOL_2) \leq F_{SOL_1} + F^* \max\left\{0, \ln\left(\frac{C_{SOL_1} - C^*}{F^*}\right)\right\} + F^* + C^* \tag{3}$$

Now, we consider two cases based on whether $C_{SOL_1} < F^* + C^*$ or $C_{SOL_1} \geq F^* + C^*$. In the first case, using inequality (2) we have

$$
\begin{aligned}
F_{SOL_1} + C_{SOL_1} &= \frac{\delta F_{SOL_1} + C_{SOL_1}}{\delta} + \left(1 - \frac{1}{\delta}\right) C_{SOL_1} \\
&\leq \frac{\gamma_f \delta F^* + \gamma_c C^*}{\delta} + \left(1 - \frac{1}{\delta}\right)(F^* + C^*) \\
&= \left(\gamma_f + 1 - \frac{1}{\delta}\right) F^* + \left(1 + \frac{\gamma_c - 1}{\delta}\right) C^*.
\end{aligned}
$$

$$\tag{4}$$

Therefore, since the greedy augmentation procedure never increases the cost, the cost of the final solution $SOL_2$ of our algorithm is at most

$$\text{cost}(SOL_2) \leq \max\left(\gamma_f + 1 - \frac{1}{\delta}, 1 + \frac{\gamma_c - 1}{\delta}\right) \text{cost}(OPT). \tag{5}$$

In the second case ($C_{SOL_1} \geq F^* + C^*$), by inequality (2) we have

$$C_{SOL_1} \leq \gamma_f \delta F^* + \gamma_c C^* - \delta F_{SOL_1}. \tag{6}$$

Also, since $C_{SOL_1} \geq F^* + C^*$, we have $\ln\left(\frac{C_{SOL_1} - C^*}{F^*}\right) \geq 0$. Therefore, by inequalities (3) and (6) we have

$$
\begin{aligned}
\text{cost}(SOL_2) &\leq F_{SOL_1} + F^* \ln\left(\frac{C_{SOL_1} - C^*}{F^*}\right) + F^* + C^* \\
&\leq F_{SOL_1} + F^* \ln\left(\frac{\gamma_f \delta F^* + (\gamma_c - 1)C^* - \delta F_{SOL_1}}{F^*}\right) + F^* + C^*
\end{aligned}
$$

$$(7)$$

Considering $F_{SOL_1}$ as a variable while all others are fixed, we have the above term maximized at $F_{SOL_1} = (\gamma_f - 1)F^* + \frac{\gamma_c - 1}{\delta}C^*$. Therefore,

$$
\begin{aligned}
\text{cost}(SOL_2) &\leq (\gamma_f + \ln \delta)F^* + \left(1 + \frac{\gamma_c - 1}{\delta}\right)C^* \\
&\leq \max\left(\gamma_f + \ln \delta, 1 + \frac{\gamma_c - 1}{\delta}\right)\text{cost}(OPT)
\end{aligned}
$$

$$(8)$$

Inequalities (5) and (8) show that in either case, our algorithm finds a solution whose cost is at most a factor of $\alpha := \max\left(\gamma_f + \ln \delta, 1 + \frac{\gamma_c - 1}{\delta}, \gamma_f + 1 - \frac{1}{\delta}\right)$ more than the optimal solution. By Lemma 2, we can pick $(\gamma_f, \gamma_c) = (1.11, 1.78)$. By minimizing $\alpha$ over the choice of $\delta$, we obtain $\delta = 1.504$ and $\alpha \approx 1.519 < 1.52$. Therefore, our algorithm is a 1.52-approximation algorithm for the uncapacitated facility location problem. It is easy to see that this algorithm can be implemented in $O(n^3)$ time ([6] and [1]).

## 4    Capacitated Facility Location with Soft Capacities

In this section, we present a 2.89-approximation algorithm for the metric capacitated facility location problem (CFLP) with soft capacities. CFLP is similar to UFLP except that there is a *capacity* $u_i$ associated with each facility $i$, that means that facility $i$ can only serve at most $u_i$ cities. This problem has two variants: CFLP with soft capacities, and CFLP with hard capacities. In CFLP with soft capacities, it is allowed to open multiple copies of each facility, while in CFLP with hard capacities each facility can be opened at most once. In this section, we will only talk about the variant with soft capacities.

More precisely, CFLP can be formulated by the following integer program.

$$
\begin{aligned}
\text{Minimize} \quad & \sum_{i \in \mathcal{F}} f_i y_i + \sum_{i \in \mathcal{F}} \sum_{j \in \mathcal{C}} c_{ij} x_{ij} & (9) \\
\text{s.t.} \quad & \forall i \in \mathcal{F}, j \in \mathcal{C}: \ x_{ij} \leq y_i \\
& \forall i \in \mathcal{F}: \ \sum_{j \in \mathcal{C}} x_{ij} \leq u_i y_i \\
& \forall j \in \mathcal{C}: \ \sum_{i \in \mathcal{F}} x_{ij} = 1
\end{aligned}
$$

$$\forall i \in \mathcal{F}, j \in \mathcal{C}: \ x_{ij} \in \{0,1\}$$
$$\forall i \in \mathcal{F}: \ y_i \text{ is a nonnegative integer}$$

It is known that using the Lagrangian relaxation technique of Jain and Vazirani [7] one can get a $2\gamma$-approximation algorithm for CFLP from an $(\gamma, \gamma)$-approximation algorithm for the UFLP; and an $(1+\gamma)$-approximation algorithm for CFLP from a $(1, \gamma)$-approximation algorithm for UFLP. Our next theorem generalizes the above results.

**Theorem 2.** *Any $(\gamma_f, \gamma_c)$-approximation algorithm for UFLP implies a $(\gamma_f + \gamma_c)$-approximation algorithm for CFLP with soft capacities.*

*Proof.* Let $\lambda = \frac{\gamma_c}{\gamma_f + \gamma_c} \in [0,1]$. For any solution $(x_{ij}, y_i)$ of integer program (9), the inequality $\sum_{j \in \mathcal{C}} x_{ij} \leq u_i y_i$ implies that $\sum_{i \in \mathcal{F}}(1-\lambda)f_i y_i + \sum_{i \in \mathcal{F}} \sum_{j \in \mathcal{C}}(c_{ij} + \lambda \cdot \frac{f_i}{u_i})x_{ij} \leq \sum_{i \in \mathcal{F}} f_i y_i + \sum_{i \in \mathcal{F}} \sum_{j \in \mathcal{C}} c_{ij} x_{ij}$. Therefore, the following integer program is a relaxation of the integer program (9), i.e., its solution is a lower bound on the solution of (9).

$$\text{Minimize} \sum_{i \in \mathcal{F}}(1 - \lambda)f_i y_i + \sum_{i \in \mathcal{F}} \sum_{j \in \mathcal{C}}(c_{ij} + \lambda \cdot \frac{f_i}{u_i})x_{ij} \qquad (10)$$
$$\text{s.t.} \quad \forall i \in \mathcal{F}, j \in \mathcal{C}: \ x_{ij} \leq y_i$$
$$\forall j \in \mathcal{C}: \ \sum_{i \in \mathcal{F}} x_{ij} = 1$$
$$\forall i \in \mathcal{F}, j \in \mathcal{C}: \ x_{ij} \in \{0,1\}$$
$$\forall i \in \mathcal{F}: \ y_i \text{ is a nonnegative integer}$$

It is easy to see that this relaxation is an uncapacitated facility location problem where the connection cost between facility $i$ and city $j$ is $c_{ij} + \lambda_i \cdot \frac{f_i}{u_i}$ and the opening cost of facility $i$ is $(1 - \lambda_i)f_i$.

Our algorithm for CFLP is as follows:

1. Construct the UFLP instance (10).
2. Scale the facility costs by a factor of $\gamma_c/\gamma_f$ in this instance.
3. Solve the resulting instance using the $(\gamma_f, \gamma_c)$-approximation algorithm.
4. Interpret the output as a solution to CFLP, by opening $\lceil (\sum_{j \in \mathcal{C}} x_{ij})/u_i \rceil$ copies of facility $i$, for every $i \in \mathcal{F}$.

We prove that the approximation ratio of this algorithm is at most $\gamma_f + \gamma_c$. Consider an optimal solution of the program (10) and denote its facility and connection costs by $F_{UFLP}$ and $C_{UFLP}$, respectively. By the above argument, $F_{UFLP} + C_{UFLP}$ is a lower bound on the optimal solution of the program (9), which we denote by $OPT_{CFLP}$. Also, let $SOL = (x_{ij}, y_i)$ denote the solution found in the third step of the above algorithm. If we consider $SOL$ as a solution of the UFLP instance constructed in step 1 of the above algorithm, its facility and connection costs will be

$$F_{SOL} = \sum_{i \in \mathcal{F}}(1 - \lambda)f_i y_i \qquad C_{SOL} = \sum_{i \in \mathcal{F}} \sum_{j \in \mathcal{C}}(c_{ij} + \lambda \cdot \frac{f_i}{u_i})x_{ij}. \qquad (11)$$

Also, the facility and connection costs of the solution found by the above algorithm is give by

$$F = \sum_{i\in\mathcal{F}}\lceil(\sum_{j\in C}x_{ij})/u_i\rceil f_i \leq \sum_{i\in\mathcal{F}}\left(y_i + (\sum_{j\in C}x_{ij})/u_i\right)f_i \qquad C = \sum_{i\in\mathcal{F}}\sum_{j\in C}c_{ij}x_{ij},$$

$$(12)$$

where the inequality follows from the fact that $\sum_{j\in C}x_{ij} > 0$ implies than $y_i = 1$.

Since the algorithm that we use in step 3 of the above algorithm is a $(\gamma_f, \gamma_c)$-approximation, and the facility costs are scaled by a factor of $\gamma_c/\gamma_f$ in step 2, we have

$$\frac{\gamma_c}{\gamma_f}F_{SOL} + C_{SOL} \leq \gamma_f \cdot \frac{\gamma_c}{\gamma_f}F_{UFLP} + \gamma_c C_{UFLP}$$

$$= \gamma_c \cdot (F_{UFLP} + C_{UFLP})$$

$$\leq \gamma_c \cdot OPT_{CFLP}. \qquad (13)$$

On the other hand,

$$\frac{\gamma_c}{\gamma_f}F_{SOL} + C_{SOL} = \frac{\gamma_c}{\gamma_f}\sum_{i\in F}\frac{\gamma_f}{\gamma_f + \gamma_c}f_i y_i + \sum_{i\in F}\sum_{j\in C}(c_{ij} + \frac{\gamma_c f_i}{(\gamma_f + \gamma_c)u_i})x_{ij}$$

$$= \gamma_c\gamma_f + \gamma_c\sum_{i\in F}(\frac{\sum_{j\in C}x_{ij}}{u_i} + y_i)f_i + \sum_{i\in F}\sum_{j\in C}c_{ij}x_{ij}$$

$$\geq \frac{\gamma_c}{\gamma_f + \gamma_c}F + C$$

$$\geq \frac{\gamma_c}{\gamma_f + \gamma_c} \cdot (F + C) \qquad (14)$$

Inequalities (13) and (14) implies the desired inequality

$$F + C \leq (\gamma_f + \gamma_c) \cdot OPT_{CFLP}.$$

Theorem 2, together with Lemma 2, implies the main result of this section:

**Theorem 3.** *The metric CFLP with soft capacities can be approximated within a factor of* 2.89 *in time* $O(n^3)$.

## 5    Concluding Remarks

In Section 3, we have proved an upper bound of 1.52 on the approximation factor of our algorithm for UFLP. The reader may ask why we choose the pair $(\gamma_f, \gamma_c) = (1.11, 1.78)$. In fact, we have numerically computed many pairs of $(\gamma_f, \gamma_c)$ for $k = 100$ using CPLEX. Then, we have selected the pair to minimize the approximation bound. For example, the pair $(\gamma_f, \gamma_c) = (1.00, 2.00)$ would be easy to prove, but it only gives us a bound of 1.57 on the approximation factor of the algorithm. Similarly, the same pair $(\gamma_f, \gamma_c) = (1.11, 1.78)$ minimizes the approximation ratio $\gamma_f + \gamma_c$ for our CFLP algorithm.

We do not know whether the bound of 1.52 that we proved on the approximation factor of our algorithm is tight or not. The important open question is whether or not our algorithm can close the gap with the approximability lower bound of 1.463 [4]. The main ingredients of our analysis are Lemmas 1 (for the analysis of the first phase of our algorithm) and 3 (for the second phase). Lemma 1 is tight, and the estimate proved in Lemma 2 for the value of $\gamma_c$ is also very close to the correct value of $\gamma_c$. We do not know whether the bound proved in Lemma 3 is tight. It might be possible to apply a method similar to the one used in [6] (i.e., deriving a factor-revealing LP and analyzing it) to analyze both phases of our algorithm in one shot. This might give us a tighter bound on the approximation factor of the algorithm.

Jain et al. [6] show that the existence of a $(\gamma_f, \gamma_c)$-approximation algorithm for UFLP with $\gamma_c < 1 + 2e^{-\gamma_f}$, would imply that $\mathbf{NP} \subseteq \mathrm{DTIME}[n^{O(\log \log n)}]$. Thus, since $\gamma_f + 1 + 2e^{-\gamma_f} \geq 2 + \ln 2 > 2.693$, one can not hope to get an approximation ratio better than 2.693 for CFLP using Theorem 2.

### Acknowledgments

The authors would like to thank Chaitanya Swamy for pointing out a typo in the earlier draft of the paper.

## References

1. M. Charikar and S. Guha. Improved Combinatorial algorithms for facility location and $k$-median problems. In *Proceeding of the 40th Annual IEEE Symposium on Foundations of Computer Science*, pages 378-388, 1999.
2. F.A. Chudak, Improved approximation algorithms for uncapacitated facility location. In R.E. Bixby, E.A. Boyd, and R.Z. Ríos-Mercado, editors, *Integer Programming and Combinatorial Optimization*, volume 1412 of *Lecture Notes in Computer Science*, pages 180-194. Springer, Berlin, 1998.
3. F.A. Chudak and D.B. Shmoys. Improved approximation algorithms for the capacitated facility location problem. In *Proc. 10th Annual ACM-SIAM Symposium on Discrete Algorithms*, pages 875-876, 1999.
4. S. Guha and S. Khuller. Greedy strikes back: Improved facility location algorithms. *Journal of Algorithms*, 31:228-248, 1999.
5. D.S. Hochbaum. Heuristics for the fixed cost median problem. *Mathematical Programming*, 22(2):148-162, 1982.
6. K. Jain, M. Mahdian, and A. Saberi. A new greedy approach for facility location problems, In *Proceedings of STOC 2002*.
7. K. Jain and V.V. Vazirani. Approximation algorithms for metric facility location and k-median problems using the primal-dual schema and lagrangian relaxation *Journal of the ACM*, 48:274-296, 2001.
8. M.R. Korupolu, C.G. Plaxton, and R. Rajaraman. Analysis of a local search heuristic for facility location problems. In *Proceedings of the 9th Annual ACM-SIAM Symposium on Discrete Algorithms*, pages 1-10, 1998.

9. M. Mahdian, E. Markakis, A. Saberi, and V.V. Vazirani. A Greedy Facility Location Algorithm Analyzed using Dual Fitting. In *Proceedings of 5th International Workshop on Randomization and Approximation Techniques in Computer Science*, Volume 2129 of *Lecture Notes in Compuer Science*, page 127-137. Springer-Verlag, 2001.

10. D.B. Shmoys. Approximation algorithms for facility location problems. In K. Jansen and S. Khuller, editors, *Approximation Algorithms for Combinatorial Optimization*, volume 1913 of *Lecture Notes in Computer Science*, pages 27-33. Springer, Berlin, 2000.

11. D.B. Shmoys, E. Tardos, and K.I. Aardal. Approximation algorithms for facility location problems. In *Proceedings of the 29th Annual ACM Symposium on Theory of Computing*, pages 265-274, 1997.

12. M. Sviridenko. An improved approximation algorithm for the metric uncapacitated facility location problem. In W.J. Cook and A.S. Schulz, editors, *Integer Programming and Combinatorial Optimization*, volume 2337 of *Lecture Notes in Computer Science*, pages 240-257. Springer, Berlin, 2002.

## A     Proof of Lemma 2

In this appendix, we prove Lemma 2. We first state the following lemma which allows us to restrict our attention to large $k$'s. The proof of this Lemma is exactly the same as the proof of Lemma 12 in [6].

**Lemma 4.** *If $z_k$ denotes the solution to the factor-revealing LP, then for every $k$, $z_k \leq z_{2k}$.*

Now, it is enough to prove the following.

**Lemma 5.** *Let $\gamma_f = 1.11$. Then for every sufficiently large $k$, the solution of the maximization program (1) is at most 1.78.*

*Proof.* This is a sketch of proof. Consider a feasible solution of the factor-revealing LP. Let $x_{j,i} := max(r_{j,i} - d_j, 0)$. The fourth inequality of the factor-revealing LP implies that for every $i \leq i'$,

$$(i' - i + 1)\alpha_i - f \leq \sum_{j=i}^{i'} d_j - \sum_{j=1}^{i-1} x_{j,i}. \tag{15}$$

Now, we define $l_i$ as follows:

$$l_i = \begin{cases} p_2 k & \text{if } i \leq p_1 k \\ k & \text{if } i > p_1 k \end{cases}$$

where $p_1$ and $p_2$ are two constants (with $p_1 < p_2$) that will be fixed later. Consider Inequality (15) for every $i \leq p_2 k$ and $i' = l_i$:

$$(l_i - i + 1)\alpha_i - f \leq \sum_{j=i}^{l_i} d_j - \sum_{j=1}^{i-1} x_{j,i}. \tag{16}$$

For every $i = 1, \ldots, k$, we define $\theta_i$ as follows. Here $p_3$ and $p_4$ are two constants (with $p_1 < p_3 < 1 - p_3 < p_2$ and $p_4 \leq 1 - p_2$) that will be fixed later.

$$
\theta_i = \begin{cases}
\frac{1}{l_i - i + 1} & \text{if } i \leq p_3 k \\
\frac{1}{(1-p_3)k} & \text{if } p_3 k < i \leq (1-p_3)k \\
\frac{p_4 k}{(k-i)(k-i+1)} & \text{if } (1-p_3)k < i \leq p_2 k \\
0 & \text{if } i > p_2 k
\end{cases} \tag{17}
$$

By multiplying both sides of inequality (16) by $\theta_i$ and adding up this inequality for $i = 1, \ldots, p_1 k$, $i = p_1 k + 1, \ldots, p_3 k$, $i = p_3 k + 1 \ldots, (1-p_3)k$, and $i = (1-p_3)k + 1, \ldots, p_2 k$, we get the following inequalities.

$$
\sum_{i=1}^{p_1 k} \alpha_i - \left(\sum_{i=1}^{p_1 k} \theta_i\right) f \leq \sum_{i=1}^{p_1 k} \sum_{j=i}^{p_2 k} \frac{d_j}{p_2 k - i + 1} - \sum_{i=1}^{p_1 k} \sum_{j=1}^{i-1} \frac{\max(r_{j,i} - d_j, 0)}{p_2 k - i + 1} \tag{18}
$$

$$
\begin{aligned}
&\sum_{i=p_1 k+1}^{p_3 k} \alpha_i - \left(\sum_{i=p_1 k+1}^{p_3 k} \theta_i\right) f \leq \\
&\sum_{i=p_1 k+1}^{p_3 k} \sum_{j=i}^{k} \frac{d_j}{k-i+1} - \sum_{i=p_1 k+1}^{p_3 k} \sum_{j=1}^{i-1} \frac{\max(r_{j,i} - d_j, 0)}{k-i+1}
\end{aligned} \tag{19}
$$

$$
\begin{aligned}
&\sum_{i=p_3 k+1}^{(1-p_3)k} \frac{k-i+1}{(1-p_3)k} \alpha_i - \left(\sum_{i=p_3 k+1}^{(1-p_3)k} \theta_i\right) f \leq \\
&\sum_{i=p_3 k+1}^{(1-p_3)k} \sum_{j=i}^{k} \frac{d_j}{(1-p_3)k} - \sum_{i=p_3 k+1}^{(1-p_3)k} \sum_{j=1}^{i-1} \frac{\max(r_{j,i} - d_j, 0)}{(1-p_3)k}
\end{aligned} \tag{20}
$$

$$
\begin{aligned}
&\sum_{i=(1-p_3)k+1}^{p_2 k} \frac{p_4 k}{k-i} \alpha_i - \left(\sum_{i=(1-p_3)k+1}^{p_2 k} \theta_i\right) f \leq \\
&\sum_{i=(1-p_3)k+1}^{p_2 k} \sum_{j=i}^{k} \frac{p_4 k d_j}{(k-i)(k-i+1)} - \sum_{i=(1-p_3)k+1}^{p_2 k} \sum_{j=1}^{i-1} \frac{p_4 k \max(r_{j,i} - d_j, 0)}{(k-i)(k-i+1)}
\end{aligned} \tag{21}
$$

We define $s_i := \max_{l \geq i}(\alpha_l - d_l)$. Using this definition and the second and third inequalities of the maximization program (1) we obtain

$$
\forall i : r_{j,i} \geq s_i - d_j, \text{ which further implies } \max(r_{j,i} - d_j, 0) \geq \max(s_i - 2d_j, 0) \tag{22}
$$

$$
\forall i : \alpha_i \leq s_i + d_i \tag{23}
$$

$$
s_1 \geq s_2 \geq \ldots \geq s_k \ (\geq 0) \tag{24}
$$

We assume $s_k \geq 0$ here because that, if on contrary $\alpha_k < d_k$, we can always set $\alpha_k$ equal $d_k$ without violating any constraint in the factor-revealing LP (1) and increase $z_k$.

Inequality (23) and $p_4 \leq 1 - p_2$ imply

$$
\begin{aligned}
&\sum_{i=p_3 k+1}^{(1-p_3)k} \left(1 - \frac{k-i+1}{(1-p_3)k}\right) \alpha_i + \sum_{i=(1-p_3)k+1}^{p_2 k} \left(1 - \frac{p_4 k}{k-i}\right) \alpha_i + \sum_{i=p_2 k+1}^{k} \alpha_i \\
&\leq \sum_{i=p_3 k+1}^{(1-p_3)k} \frac{i - p_3 k - 1}{(1-p_3)k}(s_i + d_i) + \sum_{i=(1-p_3)k+1}^{p_2 k} \left(1 - \frac{p_4 k}{k-i}\right)(s_i + d_i) \\
&\quad + \sum_{i=p_2 k+1}^{k} (s_i + d_i)
\end{aligned} \tag{25}
$$

Let $\zeta := \sum_{i=1}^{k} \theta_i$. Thus,

$$\zeta = \sum_{i=1}^{p_1 k} \frac{1}{p_2 k - i + 1} + \sum_{i=p_1 k+1}^{p_3 k} \frac{1}{k - i + 1} + \sum_{i=p_3 k+1}^{(1-p_3)k} \frac{1}{(1 - p_3)k}$$

$$+ \sum_{i=(1-p_3)k+1}^{p_2 k} \left( \frac{p_4 k}{k - i} - \frac{p_4 k}{k - i + 1} \right)$$

$$= \ln \left( \frac{p_2}{p_2 - p_1} \right) + \ln \left( \frac{1 - p_1}{1 - p_3} \right) + \frac{1 - 2p_3}{1 - p_3} + \frac{p_4}{1 - p_2} - \frac{p_4}{p_3} + o(1). \quad (26)$$

By adding the inequalities (18), (19), (20), (21), (25) and using (22), (24), and the fact that $\max(x, 0) \geq \delta x$ for every $0 \leq \delta \leq 1$, we eventually obtain

$$\sum_{i=1}^{k} \alpha_i - \zeta f$$

$$\leq \sum_{j=1}^{k} \lambda_j d_j - \sum_{j=1}^{p_1 k} \frac{j - 1}{2(p_2 k - j + 1)} s_j - \sum_{j=p_1 k+1}^{p_3 k} \frac{j - 1}{k - j + 1} s_j$$

$$- \sum_{j=p_3 k+1}^{(1-p_3)k} \frac{p_3 k}{(1 - p_3)k} s_j + \sum_{j=(1-p_3)k+1}^{p_2 k} \left( 1 - \frac{p_4 k}{k - j} \right) s_j$$

$$+ \left( 1 - p_2 - (1 - 2p_3) \left( \frac{p_4}{1 - p_2} - \frac{p_4}{p_3} \right) \right) k s_{p_2 k + 1}$$

$$- \left( \frac{p_4}{1 - p_2} - \frac{p_4}{p_3} \right) \sum_{j=1}^{p_3 k} \max(s_{p_2 k + 1} - 2d_j, 0), \quad (27)$$

where $\lambda_j$ equals

$$\begin{cases} \ln(\frac{p_2}{p_2 - p_1}) + 2\ln(\frac{1 - p_1}{1 - p_3}) + \frac{2(1 - 2p_3)}{1 - p_3} + o(1) & \text{if } 1 \leq j \leq p_1 k \\ \ln(\frac{p_2}{p_2 - p_1}) + \ln(\frac{1 - p_1}{1 - p_3}) + \frac{2(1 - 2p_3)}{1 - p_3} + H_{k-j} - H_{(1-p_3)k} + o(1) & \text{if } p_1 k < j \leq p_3 k \\ \ln(\frac{p_2}{p_2 - p_1}) + \ln(\frac{1 - p_1}{1 - p_3}) + \frac{2(1 - 2p_3)}{1 - p_3} + \frac{2p_4}{1 - p_2} - \frac{2p_4}{p_3} + o(1) & \text{if } p_3 k < j \leq (1 - p_3)k \\ \ln(\frac{p_2}{p_2 - p_1}) + \ln(\frac{1 - p_1}{1 - p_3}) + \frac{1 - 2p_3}{1 - p_3} + 1 - \frac{p_4}{p_3} + o(1) & \text{if } (1 - p_3)k < j \leq p_2 k \\ \ln(\frac{1 - p_1}{1 - p_3}) + \frac{1 - 2p_3}{1 - p_3} + 1 + \frac{p_4}{1 - p_2} - \frac{p_4}{p_3} + o(1) & \text{if } p_2 k < j \leq k. \end{cases}$$

For every $j \leq p_3 k$, we have

$$\lambda_{(1-p_3)k} - \lambda_j \leq \frac{2p_4}{1 - p_2} - \frac{2p_4}{p_3} \Rightarrow \delta_j := (\lambda_{(1-p_3)k} - \lambda_j) \bigg/ \left( \frac{2p_4}{1 - p_2} - \frac{2p_4}{p_3} \right) \leq 1. \quad (28)$$

Also, if we choose $p_1, p_2, p_3, p_4$ in a way that

$$\ln(\frac{1 - p_1}{1 - p_3}) < \frac{2p_4}{1 - p_2} - \frac{2p_4}{p_3}, \quad (29)$$

then for every $j \leq p_3 k$, $\lambda_j \leq \lambda_{(1-p_3)k}$ and therefore $\delta_j \geq 0$. Then, since $0 \leq \delta_j \leq 1$, we can replace $\max(s_{p_2 k+1} - 2d_j, 0)$ by $\delta_j(s_{p_2 k+1} - 2d_j)$ in (27). This gives us

$$\sum_{i=1}^{k} \alpha_i - \zeta f \leq \lambda_{(1-p_3)k} \sum_{j=1}^{(1-p_3)k} d_j + \sum_{j=(1-p_3)k+1}^{k} \lambda_j d_j + \sum_{j=1}^{p_2 k+1} \mu_j s_j, \tag{30}$$

where

$$\mu_j = \begin{cases} -\frac{j-1}{2(p_2 k - j + 1)} & \text{if } 1 \leq j \leq p_1 k \\ -\frac{j-1}{k-j+1} & \text{if } p_1 k < j \leq p_3 k \\ -\frac{p_3}{1-p_3} & \text{if } p_3 k < j \leq (1-p_3)k \\ 1 - \frac{p_4 k}{k-j} & \text{if } (1-p_3)k < j \leq p_2 k \end{cases} \tag{31}$$

and

$$\begin{aligned} &\mu_{p_2 k+1} \\ &= \left( \ln(\frac{1-p_1}{1-p_3}) + 2 - 2p_2 - p_3 + p_1 - 2(1-p_3)\left(\frac{p_4}{1-p_2} - \frac{p_4}{p_3}\right) + o(1) \right) \frac{k}{2} \end{aligned} \tag{32}$$

Now, if we pick $p_1, p_2, p_3, p_4$ in such a way that $\lambda_j \leq \gamma$ for every $j \geq (1-p_3)k$, i.e.,

$$\ln(\frac{p_2}{p_2 - p_1}) + \ln(\frac{1-p_1}{1-p_3}) + \frac{2(1-2p_3)}{1-p_3} + \frac{2p_4}{1-p_2} - \frac{2p_4}{p_3} < \gamma \tag{33}$$

$$\ln(\frac{p_2}{p_2 - p_1}) + \ln(\frac{1-p_1}{1-p_3}) + \frac{1-2p_3}{1-p_3} + 1 - \frac{p_4}{p_3} < \gamma \tag{34}$$

and

$$\ln(\frac{1-p_1}{1-p_3}) + \frac{1-2p_3}{1-p_3} + 1 + \frac{p_4}{1-p_2} - \frac{p_4}{p_3} < \gamma. \tag{35}$$

then the term $\lambda_{(1-p_3)k} \sum_{j=1}^{(1-p_3)k} d_j + \sum_{j=(1-p_3)k+1}^{k} \lambda_j d_j$ on the right-hand side of (30) is at most $\gamma \sum_{j=1}^{k} d_j$. Also, if for every $i \leq p_2 k + 1$, we have

$$\mu_1 + \mu_2 + \cdots + \mu_i \leq 0, \tag{36}$$

then by inequality (24), we have $\sum_{j=1}^{p_2 k+1} \mu_j s_j \leq 0$. Therefore, if $p_1, p_2, p_3, p_4$ are chosen in such a way that in addition to the above inequalities, we have

$$\ln\left(\frac{p_2}{p_2 - p_1}\right) + \ln\left(\frac{1-p_1}{1-p_3}\right) + \frac{1-2p_3}{1-p_3} + \frac{p_4}{1-p_2} - \frac{p_4}{p_3} < 1.11, \tag{37}$$

then inequality (30) can be written as

$$\sum_{i=1}^{k} \alpha_i - 1.11 f \leq \gamma \sum_{j=1}^{k} d_j, \tag{38}$$

which shows that the solution of the maximization program (1) is at most $\gamma$. From (31), it is clear that $\mu_j \leq 0$ for every $j \leq (1-p_3)k$ and $\mu_j \geq 0$ for every

$(1 - p_3)k \le j \le p_2 k$. Therefore, it is enough to check inequality (36) for $i = p_2 k$ and $i = p_2 k + 1$. We have

$$\sum_{j=1}^{p_2 k} \mu_j = -\sum_{j=1}^{p_1 k} \frac{p_2 k - p_2 k + j - 1}{2(p_2 k - j + 1)} - \sum_{j=p_1 k+1}^{p_3 k} \frac{k - k + j - 1}{k - j + 1} - \frac{p_3(1 - 2p_3)k}{1 - p_3}$$

$$+ (p_2 - 1 + p_3)k - \sum_{j=(1-p_3)k+1}^{p_2 k} \frac{p_4 k}{k - j}$$

$$= -\frac{p_2 k}{2}(H_{p_2 k} - H_{(p_2 - p_1)k}) + \frac{p_1 k}{2} - k(H_{(1-p_1)k} - H_{(1-p_3)k}) + (p_3 - p_1)k$$

$$- \frac{p_3(1 - 2p_3)k}{1 - p_3} + (p_2 - 1 + p_3)k - p_4 k(H_{p_3 k} - H_{(1-p_2)k})$$

$$= \left( -\frac{p_1}{2} + p_2 + 2p_3 - 1 - \frac{p_2}{2}\ln(\frac{p_2}{p_2 - p_1}) - \ln(\frac{1 - p_1}{1 - p_3}) - \frac{p_3(1 - 2p_3)}{1 - p_3} \right.$$

$$\left. - p_4 \ln(\frac{p_3}{1 - p_2}) + o(1) \right) k \tag{39}$$

Therefore, inequality (36) is equivalent to the following two inequalities.

$$-\frac{p_1}{2} + p_2 + 2p_3 - 1 - \frac{p_2}{2}\ln(\frac{p_2}{p_2 - p_1}) - \ln(\frac{1 - p_1}{1 - p_3}) - \frac{p_3(1 - 2p_3)}{1 - p_3} - p_4 \ln(\frac{p_3}{1 - p_2}) < 0 \tag{40}$$

$$-\frac{p_1}{2} + p_2 + 2p_3 - 1 - \frac{p_2}{2}\ln(\frac{p_2}{p_2 - p_1}) - \ln(\frac{1 - p_1}{1 - p_3}) - \frac{p_3(1 - 2p_3)}{1 - p_3}$$

$$- p_4 \ln(\frac{p_3}{1 - p_2}) + \frac{1}{2}\ln(\frac{1 - p_1}{1 - p_3}) + 1 - p_2 - \frac{p_3}{2} + \frac{p_1}{2}$$

$$- (1 - p_3)\left( \frac{p_4}{1 - p_2} - \frac{p_4}{p_3} \right) < 0 \tag{41}$$

Now, it is enough to observe that if we let $p_1 = 0.225, p_2 = 0.791, p_3 = 0.305$, $p_4 = 0.06984$, and $\gamma = 1.7764$, then inequalities (29), (33), (34), (35), (37), (40), and (41) are all satisfied. Therefore, the solution of the optimization program (1) is at most $1.7764 < 1.78$.

*Remark 1.* Numerical computations using CPLEX show that $z_{500} \approx 1.7743$ and therefore $\gamma_c > 1.774$ for $\gamma_f = 1.11$. Thus, the estimate provided by Lemma 2 for the value of $\gamma_c$ is close to its actual value.

# Complexity of Makespan Minimization for Pipeline Transportation of Petroleum Products[*]

Ruy Luiz Milidiú, Artur Alves Pessoa, and Eduardo Sany Laber

PUC-Rio, Informatics Department, Rua Marquês de São Vicente 225, RDC,
4° andar, CEP 22453-900, Rio de Janeiro – RJ, Brazil,
{milidiu,artur,laber}@inf.puc-rio.br

**Abstract.** SPTP is a model for the pipeline transportation of petroleum products. It uses a directed graph $G$, where arcs represent pipes and nodes represent locations. In this paper, we analyze the complexity of finding a minimum makespan solution to SPTP. This problem is called SPTMP. We prove that, for any $\epsilon > 0$, there is no $\eta^{1-\epsilon}$-approximate algorithm for the SPTMP unless $\mathcal{P} = \mathcal{NP}$, where $\eta$ is the input size. This result also holds if $G$ is both planar and acyclic. If $G$ is acyclic, then we give a $m$-approximate algorithm to SPTMP, where $m$ is the number of arcs in $G$.

## 1 Introduction

P etroleum products are typically transported through pipelines. Pipelines are different from all other transportation methods since they use stationary carriers whose cargo moves rather than moving carriers of stationary cargo. An important characteristic of pipelines is that they must be always full. Hence, assuming incompressible fluids, an elementary pipeline operation is the following: pump an amount of product into the pipeline and remove the same amount of product from the opposite side. T ypically each oil pipeline is a few inches wide and several miles long. As a result, reasonable amounts of distinct products can be transported through the same pipeline with a very small loss due to mixing at liquid boundaries.

Optimizing the transportation through oil pipelines is a problem of high relevance, since a non negligible component of a petroleum product's price depends on its transportation cost. Nevertheless, as far as we know, just a few authors have specifically addressed this problem [HR95,Cam95,MLPR99,MPL02]. Let us define an *order* as a requirement to transport a given amount of some product from one location to another. In [HR95], Hane and Ratliff present a model that assumes cyclic orders. In this case, the same orders always repeat after the completion of a given time period. In [MPL00,MPL02], the PTP (Pipeline T ransportation Problem) model is proposed for the pipeline transportation of petroleum products with non-cyclic orders. PTP models a pipeline

---

[*] Sponsored by CTPetro/FINEP, CENPES/Petrobras, and FAPERJ

K. Jansen et al. (Eds.): APPROX 2002, LNCS 2462, pp. 243–255, 2002.

system through a directed graph $G$, where each of the $n$ nodes represents a location and each of the $m$ directed arcs represents a pipeline, with a corresponding flow direction. In this sense, PTP is more general than Hane's model, where the pipeline system must be represented by a directed tree. As in Hane's model, the flow inside each pipeline is assumed to be unidirectional.

Throughout this paper, we use the term *batch* to denote the amount of product that corresponds to a given order. Each batch is defined by both its initial position and its associated destination node. The initial position of a batch may be either a node or a pipeline. Moreover, PTP assumes that all batches have unitary volumes and that no batch can be split during its transportation. In general, PTP allows multiple batches corresponding to the same order. In this paper, we assume an one-to-one correspondence between batches and orders. Observe that this assumption makes our lower bounds stronger since they apply to a more restricted model.

Let $L$ be the set of $r$ batches. Since pipelines must always be full, some batches must be used to fill the pipelines at the end of the schedule. Observe that these batches are not delivered. Due to this fact, PTP defines a subset $F \subset L$ of *further* batches that are not necessarily delivered at the end of a feasible pumping sequence. As result, a feasible solution is a pumping sequence that delivers all *non-further* batches in $L - F$.

In [MPL02], the problem of finding a feasible solution to PTP is proved to be $\mathcal{NP}$-hard, even if $G$ is acyclic. Moreover, the authors introduce the Synchronous PTP (SPTP), a special case of PTP where all batches in $F$ are initially stored at nodes. The problem of finding a minimum pumping cost solution to SPTP is called SPTOP. In this work, the authors also introduce the BPA algorithm, that finds feasible solutions to SPTP in polynomial time. If $G$ is acyclic, then these solutions are also optimal for the SPTOP.

In this paper, we analyze the complexity of finding minimum makespan solution to SPTP. This problem is called the SPTMP (Synchronous Pipeline Transportation Makespan Problem). We prove that, for any fixed $\epsilon > 0$, there is no $\eta^{1-\epsilon}$-approximate algorithm for SPTMP unless $\mathcal{P} = \mathcal{NP}$, where $\eta$ is the input size. This result also holds if the graph $G$ is both planar and acyclic.

Next, we give an overview of our proof for this lower bound. Let us assume that $\mathcal{P} \neq \mathcal{NP}$. First, we consider two pipeline operations $\pi_1$ and $\pi_2$ that must be executed in a special class of instances of the SPTMP. Then, we prove that it is $\mathcal{NP}$-complete to find feasible solutions to this class of instances where $\pi_1$ is not executed before $\pi_2$. Let $I$ be an instance in this class. We propose a construction of another instance $I^\alpha$ by *enchaining* $\alpha$ copies of $I$. This construction is such that the operation $\pi_2$ for the $i$-th copy of $I$ is the same as the operation $\pi_1$ for the $(i+1)$-th copy, for $i = 1, 2, \ldots, \alpha - 1$. Hence, any solution for $I^\alpha$ with makespan smaller than $\alpha$, does not execute $\pi_1$ before $\pi_2$ in at least one copy of $I$. On the other hand, the construction of $I^\alpha$ assures that it has a feasible solution with makespan $O(|I|)$ whenever $\pi_1$ does not need to be executed before $\pi_2$, where $|I|$ is the number of bits required to represent $I$. Our approximation lower bound is obtained by assigning an appropriate value to $\alpha$ as a function of $|I|$.

For completeness, we also show that the BPA algorithm can be modified to find a $m$-approximate solution to SPTMP, for acyclic graphs. Although this approximation factor is very high, the lower bound proved in this paper prevents one to do much better unless $\mathcal{P} = \mathcal{NP}$.

This paper is organized as follows. In Section 2, we describe the SPTMP. In Section 3, we prove our approximability bounds. In the last section, we present our final remarks.

## 2    The SPTMP Model

In this section, we describe the SPTMP model. Our description includes its pipeline system, orders, pipeline contents, allowed operations, and objective function.

### Pipeline System

Let $G = (N, A)$ be a directed graph, where $N$ is the set of $n$ nodes and $A$ is the set of $m$ arcs. Given an arc $a = (i, j) \in A$, we say that $i$ is the start node of $a$ and $j$ is the end node of $a$. Arcs represent pipes and nodes represent locations. Each arc $a \in A$ has an associated integer capacity $v(a)$. Moreover, we divide each arc $a$ into $v(a)$ pipeline positions. We also define the set of all pipeline positions $A' = \{(a, l) | a \in A \text{ and } l \in \{1, \ldots, v(a)\}\}$.

### Orders

Let $L$ be a set of $r$ unitary volume batches, where $F \subset L$ is the subset of *further* batches and $L - F$ is the subset of *non-further* batches. Each $b \in L$ corresponds to a transportation order which is a commitment to deliver $b$ at $d(b) \in N$.

### Pipeline Contents

Pumping a batch into a pipeline requires a non negligible amount of time. However, we only consider the instants where each arc $a \in A$ contains exactly $v(a)$ integral batches. As a result, any solution to this model generates a discrete sequence of states, where the positions of all batches are well-defined.

Let us use $p_t(b)$ to denote the position of batch $b$ at state $t$. If $p_t(b) = (a, l) \in A'$, then batch $b$ is located at the $l$th position of arc $a$ at state $t$. Otherwise, if $p_t(b) = i \in N$, then batch $b$ is stored at node $i$. Furthermore, the content of a given arc $a$ at a given state $t$ is represented by a list of batches $[b_1, b_2, \ldots, b_{v(a)}]$. In this case, $b_l$ is a batch such that $p_t(b_l) = (a, l)$, for $l = 1, 2, \ldots, v(a)$.

As an example, Figure 1.(a) represents the pipeline contents corresponding to the graph of Figure 1.(b). Observe that the system has two pipelines $a_1 = (1, 2)$ and $a_2 = (1, 3)$, whose flow direction is indicated by the corresponding arcs. The capacities of $a_1$ and $a_2$ are $v(a_1) = 3$ and $v(a_2) = 1$, respectively. Let us assume that Figure 1.(a) corresponds to state $t$. In this case, we have $p_t(b_1) = (a_1, 1)$, $p_t(b_2) = (a_1, 2)$, $p_t(b_3) = (a_1, 3)$, and $p_t(b_4) = (a_2, 1)$, since the contents of $a_1$ and $a_2$ are respectively represented by the lists $[b_1, b_2, b_3]$ and $[b_4]$. Furthermore, we have $p_t(b_5) = p_t(b_6) = 1$ since both $b_5$ and $b_6$ are stored at node 1.

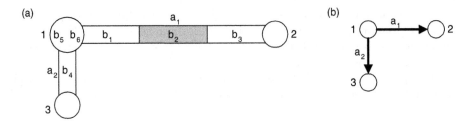

**Fig. 1.** (a) The contents of a pipeline system; (b) the corresponding graph.

At the initial state (state 0), the position $p_0(b)$ of each batch $b$ is given. As in the SPTP, we assume that every further batch $b$ has $p_0(b) \in N$.

### Operations

A solution for the model is a set $Q$ of elementary pipeline operations (EPO), defined as follows. Let $a = (i, j)$ be an arc of $G$, whose contents at a given state $t$ are given by the list $[b_1, b_2, \ldots, b_{v(a)}]$. Moreover, let $b$ be a batch stored at node $i$ at this moment. An EPO $(b, a, t)$ is to pump $b$ into $a$ during the time interval $[t, t+1)$. As a result of this operation, the contents of $a$ at state $t+1$ are given by the list $[b, b_1, b_2, \ldots, b_{v(a)-1}]$ and $b_{v(a)}$ is stored at the node $j$. We point out that some EPO's may be simultaneously executed. Formally, given two EPO's $(b_1, a_1, t_1)$ and $(b_2, a_2, t_2)$, if we have $t_1 = t_2$, then we must have $b_1 \neq b_2$ and $a_1 \neq a_2$.

Let $q = \max\{t + 1 | (b, a, t) \in Q\}$. $Q$ is feasible when the following two conditions hold:

1. every batch $b \in L - F$ is stored in node $d(b)$, when the state is $q$;
2. for every batch $b \in F$ there is a path in $G$ containing $p_q(b)$ and terminating at node $d(b)$.

### Objective Function

The SPTMP is to find a minimum makespan set $Q$ of EPO's. Hence, the value of $q$ shall be minimum.

## 3   Complexity of SPTMP

In this section, we analyze the complexity of SPTMP. Here, we also assume that the graph $G$ is both acyclic and planar, which makes our lower bounds stronger.

Now, let us introduce some terminology. Let us use the term *source (tail) node* of $p_t(b)$ to denote:

1. the start (end) node of $a$ if $p_t(b) = (a, l) \in A'$;
2. the node $i$, if $p_t(b) = i \in N$.

We say that $p_q(b)$ is a *valid final position* for $b \in F$ when there is a path in $G$ connecting the tail node of $p_0(b)$ to the source node of $p_q(b)$ and another path connecting the tail node of $p_q(b)$ to $d(b)$. If $b \in L - F$ then $d(b)$ is the only *valid final position* for $b$. A *valid final state* for a pipeline system is a state where each batch $b \in L$ is located at a valid final position. Now, we are ready to present a theorem proved in [MPL02].

**Theorem 1.** *An instance $I$ of the SPTMP is feasible if and only if there exists an assignment from $F$ to $A'$ with the following two properties:*

*1. to each pipeline position $(a, l) \in A'$, it is assigned exactly one batch of $F$;*
*2. for every batch $b$ assigned to $(a, l) \in A'$, $(a, l)$ is a valid final position for $b$.*

By the previous theorem, one can construct an instance $B$ of the Assignment Problem [AMO93] that is feasible if and only if $I$ is feasible. Observe that any feasible assignment $X$ of $B$ gives a valid final position of each batch $b \in F$ as follows: if $b$ is assigned to $(a, l) \in A'$, then $p_q(b) = (a, l)$. Moreover, any valid final state for $I$ corresponds to an assignment of $B$. This fact is used in some of our proofs.

Finally, we say that an arc $a$ is *allowed* to a batch $b$ when there are both a path from $p_0(b)$ to the start node of $a$ and another path from the end node of $a$ to $d(b)$. Observe that a batch $b$ can be pumped only into allowed arcs.

### 3.1 Precedence Pipeline Problem

Here, we prove that, for a given instance $I$ of the SPTMP and two given EPO's $\pi_1$ and $\pi_2$, finding a feasible solution to $I$ where $\pi_1$ is not executed before $\pi_2$ is a $\mathcal{NP}$-complete problem. Formally, given an instance $I$ of the SPTMP, two batches $\bar{b}_1, \bar{b}_2 \in L$, and two arcs $\bar{a}_1, \bar{a}_2 \in A$, the *Precedence Pipeline Problem* (PPP) is to find a feasible solution $Q$ to $I$ containing both the EPO's $\pi_1 = (\bar{a}_1, \bar{b}_1, t_1)$ and $\pi_2 = (\bar{a}_2, \bar{b}_2, t_2)$, for $t_1 \geq t_2$. In the next theorem, we prove that PPP is a $\mathcal{NP}$-complete problem by showing a polynomial reduction from the Vertex Cover Problem (VCP) to PPP.

Given an undirected graph $H = (V, E)$ and a positive integer $k < |V|$, the VCP is to find a subset $S \subset V$ of vertices with $|S| \leq k$ such that, for all $e = (i, j) \in E$, either $i \in S$ or $j \in S$ (or both). Here, we consider a special case of VCP (say 3-VCP) where every vertex degree in $H$ is at most 3. We point out that 3-VCP is also $\mathcal{NP}$-complete [GJ79].

**Theorem 2.** *PPP is $\mathcal{NP}$-complete.*

*Proof:* First, we prove that PPP belongs to $\mathcal{NP}$. Let $I$ be an instance of the SPTMP. Since $G$ is acyclic[1], for any feasible solution $Q$ to $I$, each batch can be pumped into at most $m$ arcs. Hence, $Q$ has no more than $rm$ EPO's. Let $I'$ be

---

[1] If $G$ has one or more cycles, then whether PPP belongs to $\mathcal{NP}$ or not is an open question.

an instance of PPP given by $I$, $\bar{b}_1, \bar{b}_2 \in L$, and $\bar{a}_1, \bar{a}_2 \in A$. Since any certificate to $I'$ is also a feasible solution to $I$, PPP belongs to $\mathcal{NP}$.

Next, we show a polynomial reduction from an instance of 3-VCP represented by both $H$ and $k$ to an instance $I'$ of PPP. For the sake of simplicity, we number the vertices of $H$ from 3 to $|V|+2$ and the edges of $H$ from $|V|+4$ to $|E|+|V|+3$.

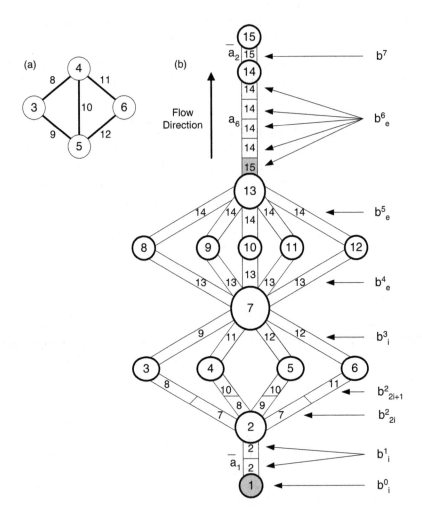

**Fig. 2.** (a) an example of a graph $H$; (b) the corresponding instance $I'$ of PPP, for $k = 2$.

Figure 2.(a) shows an example of a graph $H$. For $k = 2$, Figure 2.(b) shows the corresponding instance $I'$ of PPP. Later, we explain the construction of $I'$. This figure represents nodes, arcs and arc contents as in Figure 1.(a). The number of each node is inside the corresponding ellipse. Each batch $b$ contained in an arc is labeled inside by $d(b)$. The initial positions of $\bar{b}_1$ and $\bar{b}_2$ are represented in gray. The arcs $\bar{a}_1 = (1, 2)$, $\bar{a}_2 = (14, 15)$ and $a_6 = (13, 14)$ are also indicated. The flow directions of all pipelines are defined by a single arrow. Finally, this figure shows the notation used for each group of batches on the right side of the pipeline system. Clearly, the graph $G$ that corresponds to the pipeline system of Figure 2.(b) is both acyclic and planar.

Now, let us consider a general instance of 3-VCP represented by both $H$ and $k$. We construct a corresponding instance $I'$ of PPP as follows:

1. create $|E| + |V| + 6$ nodes in $N$, numbered from 1 to $|E| + |V| + 6$;
2. create the following arcs in $A$:
   (a) $\bar{a}_1 = (1, 2)$ with initial content $[b_1^1, b_2^1, \ldots, b_{|V|-k}^1]$, where $d(b_j^1) = 2$, for $j = 1, \ldots, |V| - k$;
   (b) $(2, i)$ and $(i, |V| + 3)$, for all $i \in V$, where:
      i. $(2, i)$ has initial content $[b_{2i}^2, b_{2i+1}^2]$;
      ii. $(i, |V| + 3)$ has initial content $[b_i^3]$;
      iii. $d(b_{2i}^2)$, $d(b_{2i+1}^2)$ and $d(b_i^3)$ are respectively given by the numbers of the edges adjacent to $i$ in $H$;
      iv. if $i$ has degree $\delta < 3$, then the remaining $3 - \delta$ batches are destined to the node $|V| + 3$;
   (c) $(|V| + 3, e)$, for all $e \in E$, with initial content $[b_e^4]$, where $d(b_e^4) = |E| + |V| + 4$;
   (d) $(e, |E| + |V| + 4)$, for all $e \in E$, with initial content $[b_e^5]$, where $d(b_e^5) = |E| + |V| + 5$;
   (e) $a_6 = (|E| + |V| + 4, |E| + |V| + 5)$ with initial content $[b_1^6, b_2^6, \ldots, b_{|E|}^6]$, where $d(b_2^6) = d(b_3^6) = \cdots = d(b_{|E|}^6) = |E| + |V| + 5$ and $d(b_1^6) = |E| + |V| + 6$;
   (f) $\bar{a}_2 = (|E| + |V| + 5, |E| + |V| + 6)$, with initial content $[b^7]$, where $d(b^7) = |E| + |V| + 6$;
3. for each arc $a \in A$, create $v(a)$ further batches $b_1^a, b_2^a, \ldots, b_{v(a)}^a$ initially stored at the start node of $a$ and designated to its end node;
4. create $|V|$ batches $b_3^0, b_4^0, \ldots, b_{|V|+2}^0$ in $L - F$, initially stored at node 1, such that $d(b_i^0) = i$.
5. create the batch $\bar{b}_1$ in $L - F$, initially stored at node 1, with $d(\bar{b}_1) = 2$.
6. let $\bar{b}_2 = b_1^6$;

Observe that no batch initially contained in an arc is a further batch.

Now, let us consider the feasibility condition of Theorem 1. In every assignment from $F$ to $A'$ with the two properties of this theorem, the batch $b_i^a$ is assigned to some position of arc $a$, for all $a \in A$ and $i = 1, 2, \ldots, v(a)$. This is true because, for each arc $a \in A$, $b_1^a, b_2^a, \ldots, b_{v(a)}^a$ are the only further batches that can be assigned to the pipeline positions of $a$. Hence, we obtain that $p_q(b_i^a) \in \{(a, 1), (a, 2), \ldots, (a, v(a))\}$ in any certificate to $I'$, for all $a \in A$ and $i = 1, 2, \ldots, v(a)$. As a result, since $G$ is acyclic, $b_1^a, b_2^a, \ldots, b_{v(a)}^a$ are necessarily the last $v(a)$ batches pumped into $a$.

Next, we use the previous fact to show that any certificate to $I'$ gives a vertex cover to $H$ with no more than $k$ vertices. Let $Q$ be a certificate to $I'$. For that, we consider, from the batches $b_3^0, b_4^0, \ldots, b_{|V|+2}^0$, which ones leave the arc $\bar{a}_1$ before $\bar{b}_2$ is pumped into $\bar{a}_2$, according to $Q$. Let us refer to these batches as the *selected batches*. Observe that exactly $|V| - k$ batches must stay at $\bar{a}_1$ to keep it filled. Moreover, $\bar{b}_1$ cannot be pumped into $\bar{a}_1$ before $\bar{b}_2$ is pumped into $\bar{a}_2$. Furthermore, as a consequence of theorem 1, the batches $b_1^{\bar{a}_1}, b_2^{\bar{a}_1}, \ldots, b_{|V|-k}^{\bar{a}_1}$ must be pumped into $\bar{a}_1$ after $\bar{b}_1$. Since $\bar{a}_1$ is not allowed to any other batch, we obtain that at most $k$ batches can leave $\bar{a}_1$ before $\bar{b}_2$ is pumped into $\bar{a}_2$. As a result, we have no more than $k$ selected batches.

Let us refer to the vertices of $H$ that correspond to the selected batches as the *selected vertices*. We claim that the set of selected vertices is a vertex cover of $H$.

First, we give an overview of our proof for this claim. We show that $\bar{b}_2$ reaches the start node of $\bar{a}_2$ only after pumping the batches $b_e^5$, for all $e \in E$, into the arc $a_6$. Then, we prove that this can be done only after pumping at least one batch into the arc $(|V| + 3, e)$, for each $e \in E$. On the other hand, we show that each selected batch $b_i^0$ allows the batches $b_{2i}^2, b_{2i+1}^2, b_i^3$ to reach the node $|V| + 3$. By construction of $I'$, the destination nodes of these batches correspond to edges that are adjacent to the selected vertices in $H$, which leads to the desired result.

Next, we present a detailed proof of our claim. First, observe that $\bar{b}_2$ reaches the start node of $\bar{a}_2$ only after pumping $|E|$ batches into $a_6$. Moreover, as a consequence of Theorem 1, $b_{|V|+4}^5, b_{|V|+5}^5, \ldots, b_{|E|+|V|+3}^5$ are necessarily pumped into $a_6$ before $b_1^{a_6}, b_2^{a_6}, \ldots, b_{|E|}^{a_6}$. In addition, $a_6$ is not allowed to any other batch. On the other hand, $b_e^5$ reaches the start node of $a_6$ only after pumping at least one batch into $(e, |E| + |V| + 4)$, for each $e \in E$. As a result, $\bar{b}_2$ reaches the start node of $\bar{a}_2$ only after pumping at least one batch into $(e, |E| + |V| + 4)$, for all $e \in E$. As a consequence of Theorem 1, $b_e^4$ must be pumped into $(e, |E| + |V| + 4)$ before $b_1^{(e, |E|+|V|+4)}$, for each $e \in E$. Moreover, $(e, |E| + |V| + 4)$ is not allowed to any other batch. Since $b_e^4$ is initially contained at the arc $(|V| + 3, e)$, we obtain that $\bar{b}_2$ reaches the start node of $\bar{a}_2$ only after pumping at least one batch into $(|V| + 3, e)$, for all $e \in E$. As a consequence of Theorem 1, $b_1^{(|V|+3,e)}$ cannot be pumped into $(|V| + 3, e)$ before any non-further batch $b$ with $d(b) = e$. By construction, for each $e = (i, j) \in E$, we have exactly two non-further batches destined to the node $e$. These two batches must pass respectively by the arc $(i, |V|+3)$ and the arc $(j, |V|+3)$ before reaching node $|V|+3$. Since $(|V|+3, e)$ is not allowed to any other batch, we have that, for all $e = (i, j) \in E$, either the content of $(i, |V| + 3)$ or the content of $(j, |V| + 3)$ must move before $\bar{b}_2$ reaches the start node of $\bar{a}_2$. Let $S \subset V$ be a set containing every index $i$ such that the content of $(i, |V| + 3)$ moves before $\bar{b}_2$ reaches the start node of $\bar{a}_2$, according to $Q$. By the previous discussion, we have that $S$ is a vertex cover of $H$. Moreover, for each $i \in S$, $b_1^{(i,|V|+3)}$ is pumped into $(i, |V|+3)$ only after both $b_{2i}^2$ and $b_{2i+1}^2$. Again, this is a consequence of Theorem 1. Since $(i, |V|+3)$ is not allowed to any other batch, we obtain that the contents of $(2, i)$ are moved before $\bar{b}_2$ reaches the start node of $\bar{a}_2$, for all $i \in S$. Furthermore, as a consequence of Theorem 1, $b_1^{(2,i)}$ is pumped into $(2,i)$ only after pumping $b_i^0$ into $(2,i)$, for each $i \in S$. Since no other batch is allowed to $(2,i)$, we obtain that all indexes in $S$ correspond to selected batches. As a result, the set of selected vertices is a vertex cover of $H$.

Now, let $S = \{i_1, i_2, \ldots, i_k\}$ be a vertex cover for $H$ and let $V - S = \{i_{k+1}, i_{k+2}, \ldots, i_{|V|}\}$. If there is a vertex cover with less than $k$ vertices to $H$, than we arbitrarily insert other vertices in it to obtain $S$. In this case, we construct a corresponding certificate $Q$ to $I'$ as follows:

1. For $t = 1, 2, \ldots, |V|$, create the EPO $(b_{i_t}^0, (1,2), t-1)$. Since $v((1,2)) = |V| - k$, $b_{i_t}^0$ is stored at the node 2 at the state $t + |V| - k$, for $t = 1, 2, \ldots, k$.

2. For $t = 1, 2, \ldots, k$, create the EPO's $(b_{i_t}^0, (2, i_t), t + |V| - k)$, $(b_1^{(2,i_t)}, (2, i_t), t + |V| - k + 1)$, and $(b_2^{(2,i_t)}, (2, i_t), t + |V| - k + 2)$.

3. For $t = 1, 2, \ldots, k$, create the EPO's $(b_{2i_t+1}^2, (i_t, |V| + 3), t + |V| - k + 1)$, $(b_{2i_t}^2, (i_t, |V| + 3), t + |V| - k + 2)$, and $(b_1^{(i_t, |V|+3)}, (i_t, |V| + 3), t + |V| - k + 3)$.

4. For $t = 1, 2, \ldots, k$, create the EPO's $(b_{i_t}^3, (|V| + 3, d(b_{i_t}^3)), t + |V| - k + 2)$, $(b_{2i_t+1}^2, (|V| + 3, d(b_{2i_t+1}^2)), t + |V| + 2)$, and $(b_{2i_t}^2, (|V| + 3, d(b_{2i_t}^2)), t + |V| + k + 2)$. Since $S$ is a vertex cover for $H$, at least one batch is pumped into $(|V|+3, e)$, for all $e \in E$. Hence, $b_e^4$ is stored at node $e$, at the state $|V|+2k+3$.

5. For all $e \in E$, create the EPO's $(b_e^4, (e, |E| + |V| + 4), |V| + 2k + 3)$, and $(b_1^{(e, |E|+|V|+4)}, (e, |E| + |V| + 4), |V| + 2k + 4)$.

6. For all $e \in E$, create the EPO's $(b_e^5, a_6, e+2k)$, and $(b_{e-|V|-3}^{a_6}, a_6, e+|E|+2k)$.

7. Create the EPO's $\pi_1 = (\bar{b}_1, \bar{a}_1, |E| + |V| + 2k + 4)$, $\pi_2 = (\bar{b}_2, \bar{a}_2, |E| + |V| + 2k + 4)$, and $(b_1^{\bar{a}_2}, \bar{a}_2, |E| + |V| + 2k + 5)$.

8. For $t = 1, 2, \ldots, |V| - k$, create the EPO's $(b_t^{\bar{a}_1}, \bar{a}_1, t + |E| + |V| + 2k + 4)$. As a result, $b_{i_t}^0$ is stored at the node 2 at the state $t + |E| + |V| + k + 5$, for $t = k+1, k+2, \ldots, |V|$.

9. For $t = k+1, k+2, \ldots, |V|$, create the EPO's $(b_{i_t}^0, (2, i_t), t + |E| + |V| + k + 5)$, $(b_1^{(2,i_t)}, (2, i_t), t + |E| + |V| + k + 6)$, and $(b_2^{(2,i_t)}, (2, i_t), t + |E| + |V| + k + 7)$.

10. For $t = k+1, k+2, \ldots, |V|$, create the EPO's $(b_{2i_t+1}^2, (i_t, |V| + 3), t + |E| + |V| + k + 6)$, $(b_{2i_t}^2, (i_t, |V| + 3), t + |E| + |V| + k + 7)$, and $(b_1^{(i_t, |V|+3)}, (i_t, |V| + 3), t + |E| + |V| + k + 8)$.

11. For $t = k+1, k+2, \ldots, |V|$, create the EPO's $(b_{i_t}^3, (|V| + 3, d(b_{i_t}^3)), t + |E| + |V| + k + 7)$, $(b_{2i_t+1}^2, (|V| + 3, d(b_{2i_t+1}^2)), t + |E| + 2|V| + 7)$, and $(b_{2i_t}^2, (|V| + 3, d(b_{2i_t}^2)), t + |E| + 3|V| - k + 7)$.

12. For all $e \in E$, create the EPO $(b_1^{(|V|+3,e)}, (|V| + 3, e), |E| + 4|V| - k + 8)$.

It can be verified that $Q$ is a certificate to $I'$, and we are done. ∎

## 3.2 Approximability Lower Bound

In this section, we prove our lower bound on the approximability of SPTMP. For that, we use the following approach. For any instance $J$ of SPTMP, let us use $|J|$ to denote the number of bits required to represent $J$. Given an instance of

3-VCP represented by both $H$ and $k$, and a corresponding instance $I$ of SPTMP constructed as in the proof of theorem 2, we construct an instance $I^\alpha$ of SPTMP by *enchaining* $\alpha$ copies of $I$. Later, we explain this construction. After that, we prove that, if $H$ has a vertex cover with no more than $k$ vertices, then $I^\alpha$ has a feasible solution with a makespan equal to $t(|I|) = O(|I|)$. Otherwise, $I^\alpha$ has no feasible solution with makespan smaller than $\alpha$. We also show that $|I^\alpha|$ is $O(\alpha|I|)$. Now, let us consider an $|J|^{1-\epsilon}$-approximation algorithm $\mathcal{A}$ that runs in $O(|J|^c)$ time, for any instance $J$ of SPTMP and a given constant $c$. For $\alpha = |I|^{(3/\epsilon)-1}$, we have that $\mathcal{A}$ finds an $O(|I|^{(3/\epsilon)-3})$-approximate solution to $I^\alpha$ in $O(|I|^{3c/\epsilon})$ time. Since $\alpha/t(|I|) = \Omega(|I|^{(3/\epsilon)-2})$, if $|I|$ is sufficiently large, then $\mathcal{A}$ can be used to decide whether $H$ has a vertex cover with no more than $k$ vertices.

Hence, we have the following theorem.

**Theorem 3.** *For any fixed $\epsilon > 0$, there is no $\eta^{1-\epsilon}$-approximate algorithm for SPTMP unless $\mathcal{P} = \mathcal{NP}$, where $\eta$ is the input size. This result also holds if the graph $G$ is both planar and acyclic.*

*Proof:* By the previous discussion, it is enough to construct an instance $I^\alpha$ with the following three properties:

1. if $H$ has a vertex cover with no more than $k$ vertices, then $I^\alpha$ has a feasible solution with an $O(|I|)$ makespan;
2. if every vertex cover to $H$ has more than $k$ vertices, then $I^\alpha$ has no feasible solution with makespan smaller than $\alpha$;
3. $|I^\alpha|$ is $O(\alpha|I|)$.

Now, let $I^{(1)}, I^{(2)}, \ldots, I^{(\alpha)}$ be $\alpha$ copies of $I$. In order to construct $I^\alpha$, for each $j = 1, 2, \ldots, \alpha - 1$, do the following five steps:

1. remove from $I^{(j)}$ the following:
   (a) the node $|E| + |V| + 6$;
   (b) the arc $\bar{a}_2$;
   (c) the two batches $b^7$ and $b_1^{\bar{a}_2}$;
2. connect the two pipeline networks of $I^{(j)}$ and $I^{(j+1)}$ by replacing both the node $|E| + |V| + 5$ of $I^{(j)}$ and the node 1 of $I^{(j+1)}$ by a single node;
3. let the arc $\bar{a}_1$ of $I^{(j+1)}$ be also the new arc $\bar{a}_2$ for $I^{(j)}$;
4. remove the batch $\bar{b}_1$ of $I^{(j+1)}$;
5. let the batch $\bar{b}_2$ of $I^{(j)}$ be also the new batch $\bar{b}_1$ for $I^{(j+1)}$;
6. destinate this batch to the node 2 of $I^{(j+1)}$.

Clearly, $I^\alpha$ satisfies Property 3. Let us use $b^{(j)}$ and $a^{(j)}$ to denote the batch $\bar{b}_1$ and the arc $\bar{a}_1$ for $I^{(j)}$, respectively, for $j = 1, 2, \ldots, \alpha$. Let also $b^{(\alpha+1)}$ and $a^{(\alpha+1)}$ be respectively the batch $\bar{b}_2$ and the arc $\bar{a}_2$ for $I^{(\alpha)}$. Figure 3 represents the connections between the instances $I^{(j-1)}$, $I^{(j)}$, and $I^{(j+1)}$, in $I^\alpha$. In this figure, circles represent nodes, rectangles represent pipelines, and each cloud represents the remaining of the pipeline network corresponding to each copy of $I$. In addition, the three batches $b^{(j)}$, $b^{(j+1)}$, and $b^{(j+2)}$ are gray colored.

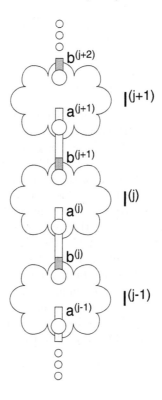

**Fig. 3.** Connections between the instances $I^{(j-1)}$, $I^{(j)}$, and $I^{(j+1)}$, in $I^\alpha$.

Observe that $b^{(j)}$ represents both the batch $\bar{b}_2$ for $I^{(j-1)}$ and the batch $\bar{b}_1$ for $I^{(j)}$, for $j = 2, 3, \ldots, \alpha$. Moreover, $a^{(j)}$ represents both the arc $\bar{a}_2$ for $I^{(j-1)}$ and the arc $\bar{a}_1$ for $I^{(j)}$. Furthermore, pumping $b^{(j)}$ into $a^{(j)}$ before pumping $b^{(j+1)}$ into $a^{(j+1)}$ is essentially the same as pumping $\bar{b}_1$ into $\bar{a}_1$ before pumping $\bar{b}_2$ into $\bar{a}_2$, in $I$. Hence, by Theorem 2, if every vertex cover to $H$ has more than $k$ vertices, then any feasible solution to $I^\alpha$ pumps $b^{(j)}$ into $a^{(j)}$ before pumping $b^{(j+1)}$ into $a^{(j+1)}$, for $j = 1, 2, \ldots, \alpha - 1$. Observe that Property 2 of $I^\alpha$ immediately follows from this claim.

Now, let $Q^{(j)}$ be a feasible solution to $I^{(j)}$ constructed as in the proof of theorem 2. If $H$ has a vertex cover with no more than $k$ vertices, then a feasible solution $Q$ to $I^\alpha$ with an $O(|I|)$ makespan is constructed as follows:

1. for $j = 1, 2, \ldots, \alpha$, remove from $Q^{(j)}$ the EPO $(b_1^{\bar{a}_2}, \bar{a}_2, |E| + |V| + 2k + 5)$;
2. for $j = 1, 2, \ldots, \alpha - 1$, replace both the EPO $(\bar{b}_2, \bar{a}_2, |E| + |V| + 2k + 4)$ of $Q^{(j)}$ and the EPO $(\bar{b}_1, \bar{a}_1, |E| + |V| + 2k + 4)$ of $Q^{(j+1)}$ by a single EPO $(b^{(j)}, a^{(j)}, |E| + |V| + 2k + 4)$;
3. $Q = Q^{(1)} \cup Q^{(2)} \cup \cdots \cup Q^{(\alpha)}$.

It follows from this construction that $I^\alpha$ satisfies Property 1, which completes our proof. ∎

### 3.3  Approximability Upper Bound

Here, we show that the BPA algorithm [MPL02] can be modified to find a $m$-approximate solution to SPTMP, for acyclic graphs.

BPA assumes that every order has a corresponding batch weight. If $G$ is acyclic, then BPA finds a minimum cost solution to SPTMP, where the cost of an EPO $(b, a, t)$ is equal to the sum of the weights of all batches contained in $a$ during the execution of this EPO. In this case, both the batch that enters $a$ and the batch that leaves $a$ have only one half of their weights added to this cost. Moreover, BPA generates EPO's that are sequentially executed. Hence, the obtained solution has a makespan equal to the number of generated EPO's.

Now, let us consider that cost of any EPO is exactly one. In this case, we point out that BPA can still be used to find a minimum cost solution to SPTMP, with minor modifications. Let $\hat{q}$ be the minimum number of EPO's for an instance $I$ of SPTMP. Since each arc can execute at most one EPO per time unit, we obtain that any feasible solution to $I$ has a makespan not smaller then $\hat{q}/m$. Moreover, since BPA gives a solution to $I$ with makespan $\hat{q}$, this solution is $m$-approximate.

## 4  Final Remarks

In this paper, we prove that, for any fixed $\epsilon > 0$, there is no $\eta^{1-\epsilon}$-approximate algorithm for SPTMP unless $\mathcal{P} = \mathcal{NP}$, where $\eta$ is the input size. In [Kan94], V. Kann investigates the class of *polynomially bounded minimization problems* (Min PB). The author shows that Min PB-complete problems cannot be approximated within $\eta^\epsilon$, for some $\epsilon > 0$. Moreover, some of these problems are proved to have the same approximability bound as SPTMP. Hence, whether SPTMP is Min PB-complete is an interesting open question.

For completeness, we also give a $m$-approximate algorithm for the SPTMP, for acyclic graphs. An interesting open problem is to design an $O(\delta)$-approximate algorithm for the SPTMP, where $\delta$ is the maximum number of arcs in a simple path of $G$. Observe that such algorithm does not conflict with the previous lower bound.

## References

[AMO93]  R. Ahuja, T. Magnanti, and J. Orlin. *Network Flows: Theory, Algorithms and Applications*. Prentice Hall, 1993.

[Cam95]  Eduardo Camponogara. A-teams para um problema de transporte de derivados de petróleo. Master's thesis, Departamento de Ciência da Computação, IMECC - UNICAMP, December 1995.

[GJ79]      M. R. Garey and D. S. Johnson. *Computers and Intractability: A Guide to the Theory of NP-Completeness*. W. H. Freeman and Company, 1979.

[HR95]      Christopher A. Hane and H. Donald Ratliff. Sequencing inputs to multi-commodity pipelines. *Annals of Operations Research*, 57, 1995. Mathematics of Industrial Systems I.

[Kan94]     Viggo Kann. Polynomially bounded minimization problems that are hard to approximate. *Nordic Journal of Computing*, 1(3):317–331, 1994.

[MLPR99]    Ruy L. Milidiú, Eduardo S. Laber, Artur A. Pessoa, and Pablo A. Rey. Petroleum products scheduling in pipelines. In *The International Workshop on Harbour, Maritime & Industrial Logistics Modeling and Simulation*, september 1999.

[MPL00]     Ruy L. Milidiú, Artur A. Pessoa, and Eduardo S. Laber. Transporting petroleum products in pipelines (abstract). In *ISMP 2000 – 17th International Symposium on Mathematical Programming*, pages 134–135, Atlanta, Georgia, USA, August 2000.

[MPL02]     Ruy L. Milidiú, Artur A. Pessoa, and Eduardo S. Laber. Pipeline transportation of petroleum products with no due dates. In *Proceedings of the LATIN'2002*, pages 248–262, Canún, Mexico, april 2002.

# Primal-Dual Algorithms
# for Connected Facility Location Problems

Chaitanya Swamy* and Amit Kumar**

Department of Computer Science, Cornell University, Ithaca, NY 14853, USA
{swamy, amitk}@cs.cornell.edu

**Abstract.** We consider the *Connected Facility Location* problem. We are given a graph $G = (V, E)$ with cost $c_e$ on edge $e$, a set of facilities $\mathcal{F} \subseteq V$, and a set of demands $\mathcal{D} \subseteq V$. We are also given a parameter $M \geq 1$. A solution opens some facilities, say $F$, assigns each demand $j$ to an open facility $i(j)$, and connects the open facilities by a Steiner tree $T$. The cost incurred is $\sum_{i \in F} f_i + \sum_{j \in \mathcal{D}} d_j c_{i(j)j} + M \sum_{e \in T} c_e$. We want a solution of minimum cost. A special case is when all opening costs are 0 and facilities may be opened anywhere, i.e., $\mathcal{F} = V$. If we *know* a facility $v$ that is open, then this problem reduces to the *rent-or-buy* problem.

We give the first primal-dual algorithms for these problems and achieve the best known approximation guarantees. We give a 9-approximation algorithm for connected facility location and a 5-approximation for the rent-or-buy problem. Our algorithm integrates the primal-dual approaches for facility location [7] and Steiner trees [1,2]. We also consider the connected $k$-median problem and give a constant-factor approximation by using our primal-dual algorithm for connected facility location. We generalize our results to an edge capacitated version of these problems.

## 1 Introduction

Facility location problems have been widely studied in the Operations Research community(see for e.g. [14]). These problems can be described as follows: we are given a graph $G = (V, E)$, a set of facilities $\mathcal{F} \subseteq V$, and a set of demands $\mathcal{D} \subseteq V$. Facilities may have *opening costs*. We want to *open* some facilities from the set $\mathcal{F}$ and assign each demand to one of these open facilities. We consider a setting where besides opening the facilities, we also want to connect them by a Steiner tree. This will allow the facilities to communicate easily with each other. For example, the facilities could be caches or file servers which need to communicate with each other to maintain consistent data, and the clients could be users or processes requesting data items. Another example is telecommunication network design. Designing the network involves selecting a subset of core nodes, connecting the selected core nodes, and routing traffic from the endnodes to the

---

* Research partially supported by NSF grant CCR-9912422.
** Research supported by NSF grant CCR-9820951 and NSF ITR/IM grant IIS-0081334.

K. Jansen et al. (Eds.): APPROX 2002, LNCS 2462, pp. 256–270, 2002.

selected core nodes. Here the clients are the endnodes of the network, and the facilities are the core nodes. The opening cost of a facility corresponds to the switch cost of the corresponding core node.

The problems mentioned above are instances of the *Connected Facility Location* problem (ConFL). We are given a graph $G = (V, E)$ with cost $c_e$ on edge $e$, a set of facilities $\mathcal{F} \subseteq V$ and a set of demand nodes or clients $\mathcal{D} \subseteq V$. Let $c_{ij}$ be the shortest path distance between $i$ and $j$ (with respect to the costs $c_e$). We are also given a parameter $M \geq 1$. Client $j$ has $d_j$ units of demand and facility $i$ has an opening cost of $f_i$. A solution has to *open* a set of facilities $F$, assign each demand $j$ to an open facility $i(j)$, and further has to connect the open facilities by a Steiner tree $T$. The cost of connecting facilities is simply the cost of the Steiner tree $T$ scaled by a factor of $M$. The total cost of this solution is $\sum_{i \in F} f_i + \sum_{j \in \mathcal{D}} d_j c_{i(j)j} + M \sum_{e \in T} c_e$. We want to find a solution of minimum cost. This problem has attracted the interest of both the operations research community [10,12,13] and the computer science community [5,8,9].

**The Rent-or-Buy Problem.** A special case of this problem is when all opening costs are 0 and facilities may be opened anywhere, i.e., $\mathcal{F} = V$. Suppose we *know* that facility $v$ is opened by an optimal solution. Then the problem becomes a special case of the *single-sink buy-at-bulk* problem with two cable types, also known as the *rent-or-buy* problem. Here we want to route traffic in a minimum-cost way from the clients to the sink $v$ by installing capacity on edges. We can either *rent* capacity on an edge paying a cost proportional to the capacity rented, or *buy* unlimited capacity on an edge by paying a large fixed cost of $M$.

This problem arises in various scenarios. Karger & Minkoff [8] reduced the *maybecast* problem to this special case of ConFL. Gupta et. al. [5] arrived at this problem by considering the problem of provisioning a virtual private network.

**Our Results.** We give a primal-dual 9-approximation algorithm for the connected facility location problem and a 5-approximation algorithm for the rent-or-buy problem. Previously the best known approximation guarantees for these problems were 10.66 and 9.001 respectively [5]. But these results were obtained by solving an exponential size linear program using the ellipsoid method, making the algorithm very inefficient in practice. Karger & Minkoff [8] gave a combinatorial algorithm, but the constant guarantee was much larger.

In many settings there is an additional requirement that at most $k$ facilities can be opened. We call this variant of ConFL the Connected $k$-Median problem. We use our primal-dual algorithm to get a 20-approximation algorithm for this problem. To the best of our knowledge, this is the first time anyone has considered this problem, though the connected $k$-center problem has been considered earlier [4]. at before [4]. We generalize our results to an edge capacitated version of these problems. These differ from the uncapacitated versions in the facility location aspect. We now require clients to be connected to facilities via cables which have a fixed cost of $\sigma$ per unit length and a capacity of $u$. Multiple cables may be laid along an edge. The cost of connecting facilities is still $M$ times the cost of the tree $T$. We give a constant-factor approximation for these capacitated variants.

**Our Techniques.** Connected Facility Location has elements of both facility location and the Steiner tree problem. Without the connectivity requirement, the problem is simply the uncapacitated facility location problem. If we know which facilities are open we only need to connect them by a Steiner tree. However, simply running a facility location algorithm and then a Steiner tree algorithm does not work, since we are ignoring the connectivity requirement. In the rent-or-buy problem, this would just open a facility at each demand point, but connecting all the open facilities might incur a huge cost. The connectivity requirement implicitly imposes a facility opening cost so that it is only profitable to open a facility if it serves a significant demand. Previously [8] the clustering of demands around facilities was achieved by solving a Load Balanced Facility Location (LBFL) instance, where we want each open facility to serve at least $M$ clients. The disadvantage with this approach is that (1) we only know a bicriteria approximation for LBFL, so the demand lower bound on a facility is only approximately satisfied, and (2) the LBFL instance is solved using a black box, so we do not use anything specific to the ConFL instance. In particular we make no use of the fact that the need to cluster demands is imposed by the connectivity requirement of ConFL.

Our algorithm is based on a novel application of the primal-dual schema. The algorithm is in two phases. First, we decide which facilities to open, connect demands to facilities, and cluster demands at each open facility. At the end of this phase, we obtain a feasible dual solution and a primal facility location solution where each open facility serves *at least $M$* demand points, satisfying the demand lower bound. We do this by *charging some of the cost incurred to the Steiner tree portion of the dual solution*, thereby exploiting the fact that any ConFL solution also needs to connect the open facilities. Despite the added clustering requirement, our algorithm has a fairly simple description. Each demand $j$ keeps raising its dual variable, $\alpha_j$, till it gets connected to a facility and is 'near' a point at which $M$ demands are clustered. All other variables simply respond to this change trying to maintain feasibility or complementary slackness. Phase 2 is a *Steiner* phase where we connect the open facilities by a Steiner tree. The dual solution constructed in this phase is not feasible, but the infeasibility is bounded by a small *additive* factor.

Roughgarden [11] gave a primal-dual algorithm Very recently, Kumar et al. [11] have subsequently obtained a constant-factor approximation for a multi-commodity rent-or-buy problem. Their algorithm is however much more involved and they get a much worse approximation factor. A very interesting open problem is to see whether the techniques used here and in [11] can be extended to solve the multiple source-sink buy-at-bulk problem with multiple cable types.

**Previous Work on Primal-Dual Algorithms.** Our work reinforces the belief that the primal-dual schema is extremely versatile. The first truly primal-dual approximation algorithm was given by Bar-Yehuda & Even(see [3]) for the vertex cover problem. Subsequently, primal-dual algorithms have especially flourished in the area of network-design problems. One of the first such algorithms was by Agrawal, Klein & Ravi [1] for the generalized Steiner problem on networks.

Goemans & Williamson [2] further refined the primal-dual method and extended it to a large class of network-design problems; see [3,17] for a survey of this and earlier work. The basic mechanism involves raising the dual variables and setting primal variables till an integral primal solution is found satisfying the primal complementary slackness conditions. Next a *reverse delete* step is used to remove any redundancies in the primal solution. This relaxes the dual slackness conditions. The approximation ratio of the algorithm is this relaxation factor.

Jain & Vazirani [7] gave an elegant primal-dual algorithm for various facility location problems which could not be solved by the earlier schema. They remove redundancies while relaxing the primal slackness conditions. They also show that their algorithm can be used to solve other facility location variants, most notably the $k$-median problem using a Lagrangian relaxation.

## 2   A Linear Programming Relaxation

In what follows, $i$ will be used to index facilities, $j$ to index the clients and $e$ to index the edges in $G$. We will use the terms client and demand point, and connection cost and assignment cost interchangeably.

ConFL can be formulated naturally as an Integer Program. Suppose we *know* that a facility $v$ is opened and hence belongs to the Steiner tree constructed by the optimal solution. We can make this assumption because we can try all $|\mathcal{F}|$ different possibilities for $v$.

We can now write an integer program (IP) for ConFL as follows:

$$\min \ \sum_i f_i y_i + \sum_j d_j \sum_i c_{ij} x_{ij} + M \sum_e c_e z_e \tag{IP}$$

$$\text{s.t.} \ \sum_i x_{ij} \geq 1 \qquad\qquad \text{for all } j \tag{1}$$

$$x_{ij} \leq y_i \qquad\qquad \text{for all } i,j \tag{2}$$

$$y_v = 1 \tag{3}$$

$$\sum_{i \in S} x_{ij} \leq \sum_{e \in \delta(S)} z_e \qquad\qquad \text{for all } S \subseteq V, v \notin S, \ j \tag{4}$$

$$x_{ij}, y_i, z_e \in \{0,1\} \tag{5}$$

Here $y_i$ indicates if facility $i$ is open, $x_{ij}$ indicates if client $j$ is connected to facility $i$ and $z_e$ indicates if edge $e$ is included in the Steiner tree. Relaxing the integrality constraints (5) to $x_{ij}, y_i, z_e \geq 0$ gives us a linear program (LP).

## 3   A Primal-Dual Approximation Algorithm

We now show that the integrality gap of (LP) is at most 9 by giving a primal-dual algorithm for this problem. For simplicity, we assume that all $d_j$ are equal to 1. We show how to get rid of this assumption in section 5.

### 3.1 The Rent-or-Buy Problem

We first consider the case where all opening costs are 0 and $\mathcal{F} = V$, i.e., a facility can be opened at any vertex of $V$. The linear program (LP) now simplifies to:

$$\min \quad \sum_j \sum_i c_{ij} x_{ij} + M \sum_e c_e z_e \tag{P1}$$

$$\text{s.t.} \quad \sum_i x_{ij} \geq 1 \qquad\qquad \text{for all } j$$

$$\sum_{i \in S} x_{ij} \leq \sum_{e \in \delta(S)} z_e \qquad \text{for all } S \subseteq V, v \notin S, \ j$$

$$x_{ij}, z_e \geq 0$$

The dual of this linear program is:

$$\max \quad \sum_j \alpha_j \tag{D1}$$

$$\text{s.t.} \quad \alpha_j \leq c_{ij} + \sum_{S \subseteq V : i \in S, v \notin S} \theta_{S,j} \qquad \text{for all } i \neq v, j \tag{6}$$

$$\alpha_j \leq c_{vj} \qquad\qquad\qquad \text{for all } j \tag{7}$$

$$\sum_j \sum_{S \subseteq V : e \in \delta(S), v \notin S} \theta_{S,j} \leq M c_e \qquad \text{for all } e \tag{8}$$

$$\alpha_j, \theta_{S,j} \geq 0$$

Intuitively, $\alpha_j$ is the *payment* that demand $j$ is willing to make towards constructing a feasible primal solution. Constraint (6) says that a part of the payment $\alpha_j$ goes towards assigning $j$ to a facility $i$. The remaining part goes towards constructing the part of the Steiner tree which joins $i$ to $v$.

### 3.2 Algorithm Description

We begin with a simplifying assumption. We assume that a facility can be opened *anywhere along an edge*. We collectively refer to vertices in $V$ and internal points on an edge as *locations*. We reserve the term facility for vertices in $\mathcal{F}$. Clearly the metric $c$ can be extended to a metric on locations.

The intuition behind our algorithm is as follows. Suppose all demands were of size at least $M$. Then, the optimal solution would locate a facility at each of these demands and connect them by a Steiner tree. So, our algorithm first *clusters* the demands in groups of $M$ and then builds a Steiner tree joining these clusters.

Initially, all the dual variables are 0. The algorithm runs in two phases. In the first phase, we *cluster* the demands in groups of $M$. Once we have this, we run the second phase where we build the Steiner tree.

*Phase 1.* We raise the dual variables $\alpha_j$ for all demands in this phase. We have a notion of time, $t$. Initially $t = 0$. At some point of time, we say that demand $j$ is *tight* with a location $i$ if $\alpha_j \geq c_{ij}$. Let $S_j$ be the set of vertices which $j$ is tight with at some point of time. When we raise $\alpha_j$, we also raise $\theta_{S_j,j}$ at the same rate. This will ensure feasibility of constraints (6). So, it is enough to describe how to raise the dual variables $\alpha_j$.

Initially, all locations are closed. We shall *tentatively open* some locations. Initially $v$ is tentatively open. Demands can be in two states: *frozen* or *unfrozen.* When a demand $j$ gets frozen, we stop raising its dual variable $\alpha_j$. After $j$ is frozen, it does not become tight with any new location, i.e., a location not in $S_j$. Initially, all demands are unfrozen.

We start raising the $\alpha_j$ of all demands at the same rate until one of the following events happen (if several events happen, consider them in any order):

1. $j$ becomes tight with a tentatively open location $i$: $j$ becomes frozen.
2. There is a closed location $i$ with which at least $M$ demands are tight: tentatively open $i$. All of the demand points tight with $i$ become frozen.

We now raise the $\alpha_j$ of unfrozen demands only. We continue this process till all demands become frozen. Note that although there is a continuum of points along an edge, to implement the above process we only need to know the time when the next event will take place. This can be obtained by keeping track of, for every edge and every $j$, the portion of the edge that is tight with $j$.

Now we decide which locations to open. Let $F'$ be the set of tentatively open locations. We say that $i, i' \in F'$ are dependent if there is demand $j$ which is tight with both these locations. We say that a set of locations is *independent* if no two locations in this set are dependent. We find a maximal independent set $F$ of locations in $F'$ as follows: arrange the locations in $F'$ in the order they were tentatively opened. Consider the locations in this order and add a location to $F$ if no dependent location is already present in $F$. We open the locations in $F$. Observe that $v \in F$.

We assign a demand $j$ to an open location as follows. If $j$ is tight with some $i \in F$, assign $j$ to $i$. Otherwise let $i$ be the location in $F'$ that caused $j$ to become frozen. So $j$ is tight with $i$. There must be some previously opened location $i' \in F$ such that $i$ and $i'$ are dependent. We assign $j$ to $i'$.

We still have to build a Steiner tree on $F$. First we augment the graph $G$ to include edges incident on open non-vertex locations. Let $\{i_1, \ldots, i_k\}$ be the open locations on edge $e = (u, w)$ ordered by increasing distance from $u$, with $i_1 \neq u, i_k \neq w$. We add edges $(u, i_1), (i_1, i_2), \ldots, (i_{k-1}, i_k), (i_k, w)$ to $G$.

*Phase 2.* For a location $i \in F$, let $D_i$ be the set of demands tight with $i$. Let $D' = \bigcup_{i \in F - \{v\}} D_i$. Initially, we set $\alpha_j = 0$ for all $j$. We raise the $\alpha_j$ value of demands in $D'$ only, and simulate the primal-dual algorithm for the (rooted) Steiner tree problem.

Initially, the minimal violated sets (MVS) are the singleton sets $\{i\}$ for $i \in F - \{v\}$. For a set $S$, define $D_S = \bigcup_{i \in S \cap F} D_i$. The tree $T$ that we shall construct is empty to begin with. For each MVS $S$, $j \in D_S$, we raise $\alpha_j$ at rate $1/|D_S|$.

We also raise $\theta_{S,j}$, at the same rate. This ensures that $\sum_j \theta_{S,j}$ grows at rate 1 for any MVS $S$. Note that we are not ensuring feasibility of constraints (6), (7).

We say that an edge goes tight if (8) holds with equality for that edge. We raise the dual variables till an edge $e$ goes tight. We add $e$ to $T$ and update the minimal violated sets. This process continues till there is no violated set, i.e., we have only one component (so $v$ is in this component). Now we perform a reverse delete step to remove any redundant edges from $T$. This is our final solution.

**Analysis.** Let $(\alpha^1, \theta^1)$, $(\alpha^2, \theta^2)$ be the value of the dual variables at the end of Phases 1 and 2 respectively.

**Lemma 1.** *The dual solution* $(\alpha^1, \theta^1)$ *is feasible.*

*Proof.* It is easy to see that (6) is satisfied. Indeed, once $j$ gets tight with $i$, $\alpha_j$ and $\sum_{S:i \in S, v \notin S} \theta_{S,j}$ are raised at the same rate. Similarly, (7) is satisfied.

Now consider an edge $e = (u, w)$. Let $l(j)$ be the contribution of $j$ to the left hand side of (8) for this edge, i.e., $l(j) = \sum_{S:e \in \delta(S), v \notin S} \theta_{S,j}$. Suppose $c_{ju} \leq c_{jw}$. So, $j$ becomes tight with $u$ before it gets tight with $w$. Consider a point $p$ on the edge $(u, w)$ at distance $x$ from $u$. If $p$ were the last point on this edge with which $j$ became tight with (before it became frozen), then $l(j) \leq x$. Define $f(j, x)$ as 1 if $j$ is tight with $p$ and $j$ was not frozen at the time at which it became tight with $p$, otherwise $f(j, x)$ is 0. So, we can write $l(j) \leq \int_0^{c_e} f(j, x) dx$. Interchanging the summation and the integral in (8), we get

$$\sum_j \sum_{S \subseteq V: e \in \delta(S), v \notin S} \theta_{S,j} \leq \sum_j \int_0^{c_e} f(j, x) dx = \int_0^{c_e} \sum_j f(j, x) dx$$

Now for any $x$, $\left( \sum_j f(j, x) \right) \leq M$. Otherwise, we have more than $M$ demands tight with a point such that none of these demands are frozen — a contradiction. So the integral above is at most $M c_e$ which proves the lemma.     □

**Lemma 2.** *At the end of Phase 1, demand $j$ is assigned to an open location $i$ such that* $c_{ij} \leq 3\alpha_j^1$.

*Proof.* This clearly holds if $j$ is tight with a location in $F$. Otherwise let $j$ be assigned to $i$. Let $i'$ be the tentatively open facility that caused $j$ to become frozen. It must be the case that $i$ and $i'$ are dependent. So there is a demand $k$ which is tight with both $i$ and $i'$. Let $t_{i'}$ be the time at which $i'$ was tentatively opened. Define $t_i$ similarly. It is clear that $\alpha_j \geq t_{i'}$.

Now, $c_{ij} \leq c_{ik} + c_{ki'} + c_{i'j} \leq 2\alpha_k^1 + \alpha_j^1$. Also, $\alpha_k^1 \leq t_{i'}$. Otherwise, at time $t = \alpha_k^1$, $k$ is tight with both $i$ and $i'$. Suppose it becomes tight with $i$ first (the other case is similar). If $i$ is tentatively open at this time, then $k$ will freeze and so it will never become tight with $i'$. Therefore, $i$ can not be tentatively open at this time. But then, $k$ must freeze by the time $i$ becomes tentatively open, i.e., $\alpha_k^1 \leq t_i \leq t_{i'}$. So, $\alpha_k^1 \leq t_{i'} \leq \alpha_j$. This implies that $c_{ij} \leq 3\alpha_j^1$.     □

**Lemma 3.** *Let $i$ be an open location. If $j$ is tight with $i$, then the assignment cost of $j$ is at most $\alpha_j^1$.*

We now bound the cost of the tree $T$. Define $D_V$ as $\cup_{i \in F} D_i$.

**Lemma 4.** $cost(T) \le 2 \cdot \sum_{j \in D_V} \alpha_j^2$.

*Proof.* Consider Phase 2. At any point in time, define the variable $\theta_S$, where $S$ is a minimal violated set, as $\sum_j \theta_{S,j}$. We observed that $\theta_S$ grows at rate 1. Thus, Phase 2 simulates the primal dual algorithm for the rooted Steiner tree problem with $v$ as the root. So, the cost of the tree is bounded by $2 \cdot \sum_S \theta_S^2$ [3,1,17], where the sum is over all subsets of vertices $S$. But $\sum_S \theta_S^2 = \sum_{j \in D_V} \alpha_j^2$.  □

**Lemma 5.** *Consider a demand $j$. If $i \ne v$, then $\alpha_j^2 \le \alpha_j^1 + c_{ij} + \sum_{S \subseteq V : i \in S, v \notin S} \theta_{S,j}^2$. Further, $\alpha_j^2 \le \alpha_j^1 + c_{vj}$.*

*Proof.* Fix a demand $j$ and facility $i$, $i \ne v$. During the execution of Phase 2, let $S_t$ be the component to which $j$ contributes at time $t$. Consider the earliest time $t'$ for which $i \in S_{t'}$. After this time, both the left hand side and right hand side of (6) increase at the same rate, so we only need to bound the increase in $\alpha_j$ by time $t'$. Let $l$ be the location that $j$ is assigned to in Phase 1. Since we are raising $\alpha_j$, it must be the case that $j \in D_l$ and so, $c_{lj} \le \alpha_j^1$. We claim that $t' \le Mc_{li}$. This is true since $S_t$ always contains $l$, and by time $t = Mc_{li}$ all of the edges along the shortest path between $l$ and $i$ would have grown tight.

Note that $\alpha_j$ rises at a rate of at most $1/M$. Indeed, initially, $|D_{\{i\}}| \ge M$ for any open location $i$, and as new components $S$ form, $|D_S|$ can only increase. So, the increase in $\alpha_j$ by time $t'$ can then be bounded by $\frac{Mc_{li}}{M} \le c_{lj} + c_{ij} \le \alpha_j^1 + c_{ij}$. This proves the first inequality. The second inequality is proved similarly.  □

It is clear that the $\theta_{S,j}^2$ values satisfy (8), so we have shown that $(\alpha', \theta^2)$ is a feasible dual solution, where $\alpha_j' = \max(\alpha_j^2 - \alpha_j^1, 0)$. We can now prove the main theorem. Let $OPT$ be the cost of the optimal solution.

**Theorem 1.** *The above algorithm produces a solution of cost at most $5 \cdot OPT$.*

*Proof.* Note that $\alpha_j^2 \le \alpha_j' + \alpha_j^1$. So, Lemma 4 implies that the cost of $T$ is at most $2\sum_j \alpha_j' + 2\sum_{j \in D_V} \alpha_j^1 \le 2 \cdot OPT + 2\sum_{j \in D_V} \alpha_j^1$.

If $j \in D_V$, Lemma 3 implies that its assignment cost is at most $\alpha_j^1$. Otherwise by Lemma 2, its assignment cost is at most $3\alpha_j^1$. Adding all terms, we see that the cost of our solution is at most $5 \cdot OPT$.  □

Our solution may be infeasible since a non-vertex location may be opened as a facility. Let $e = (u, w)$ be an edge and suppose we open locations on the internal points of $e$. Let $D_u$ be the set of demands that reach their assigned location on $e$ via $u$, i.e., $c_{i(j)j} = c_{uj} + c_{i(j)u}$ for $j \in D_u$. $D_w$ is defined similarly. $T$ must contain at least one of $u$ or $w$. If both $u, w \in T$, we assign clients in $D_u$ to $u$ and clients in $D_w$ to $w$ without increasing the cost. Suppose $u \in T, w \notin T$. We assign all demands in $D_u$ to $u$. If $|D_w| < M$, we assign clients in $D_w$ to $u$ and remove edges in $T$ that lie along $e$; otherwise we reassign all clients in $D_w$ to $w$ and add all of $e$ to $T$. It is easy to see that the total cost only decreases.

### 3.3   The General Case

We now consider the case where $\mathcal{F}$, need not be $V$ and facility $i$ has an opening cost $f_i \geq 0$. For convenience we assume that $f_v = 0$. Clearly, this does not affect the approximation ratio of the algorithm. The primal and dual LPs are:

$$\min \ \sum_{i \neq v} f_i y_i + \sum_j \sum_i c_{ij} x_{ij} + M \sum_e c_e z_e \tag{P2}$$

$$\begin{aligned}
\text{s.t.} \quad & \sum_i x_{ij} \geq 1 && \text{for all } j \\
& x_{ij} \leq y_i && \text{for all } i \neq v, j \\
& x_{vj} \leq 1 \\
& \sum_{i \in S} x_{ij} \leq \sum_{e \in \delta(S)} z_e && \text{for all } S \subseteq V, v \notin S, \ j \\
& x_{ij}, y_i, z_e \geq 0
\end{aligned}$$

$$\max \ \sum_j \alpha_j - \sum_j \beta_{vj} \tag{D2}$$

$$\begin{aligned}
\text{s.t.} \quad & \alpha_j \leq c_{ij} + \beta_{ij} + \sum_{S \subseteq V : i \in S, v \notin S} \theta_{S,j} && \text{for all } i \neq v, j && (9) \\
& \alpha_j \leq c_{vj} + \beta_{vj} && \text{for all } j && (10) \\
& \sum_j \beta_{ij} \leq f_i && \text{for all } i \neq v && (11) \\
& \sum_j \sum_{S \subseteq V : e \in \delta(S), v \notin S} \theta_{S,j} \leq M c_e && \text{for all } e && (12) \\
& \alpha_j, \beta_{ij}, \theta_{S,j} \geq 0
\end{aligned}$$

*Phase 1.* Most of the changes are in this phase. We now also have to pay for opening facilities. Besides opening facilities and connecting clients to facilities, we will also form some components. These will act as the terminals for the Steiner tree constructed in Phase 2. A location still refers to a vertex in $V$ or a point along an edge. We will only open facilities at locations in $\mathcal{F} \subseteq V$.

Initially all dual variables are 0 and only facility $v$ is tentatively open. As before, a demand can be frozen or unfrozen. Further, a demand may be *connected* or *unconnected*. Initially, all demands are unfrozen and unconnected. As before, we say that a demand $j$ gets tight with a location $i$ if $\alpha_j \geq c_{ij}$. We say that a facility $i$ has been *paid* for if $\sum_j \beta_{ij} = f_i$. The *weight* of a location $l$ is defined as the number of *connected* demands $j$ which are tight with $l$.

The basic idea is similar to the algorithm in the previous section. Earlier we tentatively opened any location with which $M$ demands became tight, but we cannot do that here because of two reasons — (1) we cannot open *any* location; the set of candidate facillity locations, $\mathcal{F}$, may be a very small subset of $V$, (2) we need to pay a facility opening cost before we can open a facility.

At any point of time, define $S_j$ to be the set of facilities that a demand $j$ is tight with. When $j$ becomes tight with a facility $i$, we have two options – we can raise $\beta_{ij}$ or we can raise $\theta_{S_j,j}$. If none of the facilities in $S_j$ have been paid for, we raise $\beta_{ij}$ for all $i \in S_j$ at the same rate. If there is a facility $i \in S_j$ which has been paid for, then we raise $\theta_{S_j,j}$ and do not raise $\beta_{ij}$ for any $i \in S_j$. Thus, it is enough to describe how the $\alpha_j$s get raised.

We now describe the algorithm in more detail. We raise the $\alpha_j$ of all unfrozen demands uniformly till one of the following events happen:

1. An unconnected demand $j$ becomes tight with a tentatively open facility $i$: $j$ becomes connected to $i$. If $i = v$, freeze $j$. Otherwise, as described above, we raise $\theta_{S_j,j}$ at the same rate as $\alpha_j$. Further we do not raise any variable $\beta_{i'j}$ for any facility $i'$ from now on.
2. A facility $i$ gets paid for, i.e., $\sum_j \beta_{ij} = f_i$ : tentatively open $i$. If an unconnected demand $j$ is tight with $i$, connect $j$ to $i$. From this point on we only raise $\theta_{S_j,j}$ as described above.
3. The weight of some location $l$ becomes at least $M$: declare $l$ to be a *terminal location*. Freeze all unfrozen demands which are tight with $l$.
4. A connected demand $j$ becomes tight with a terminal location $l$: freeze $j$.

We continue this process until all $j$ become frozen. Let $(\alpha^1, \beta^1, \theta^1)$ be the dual solution obtained. Note that $\beta^1_{vj}$ is 0 for all $j$.

Let $L$ be the set of all terminal locations. As in the previous section, we greedily select an independent set of terminal locations from $L$ and assign $j$ to a terminal location $\sigma(j)$. We say that locations $l$ and $l'$ in $L$ are dependent if there is a demand tight with both these locations. We look at the locations in $L$ in the order they were declared to be a terminal location, and greedily select a maximal independent subset $L'$ of $L$. If demand $j$ is tight with a location $l' \in L'$, set $\sigma(j) = l'$. Otherwise let $l$ be a location in $L$ that caused $j$ to get frozen, and $l' \in L'$ be some location such that $l$ and $l'$ are dependent. Set $\sigma(j) = l'$. Note that if $j$ is tight with $v$, $\sigma(j) = v$.

Now, consider a location $l \in L'$. $l$ may not be in the set $\mathcal{F}$ of candidate facilities. So, we need to locate a facility $i \in \mathcal{F}$ *near* $l$ and open it. Let $j$ be the demand tight with $l$ having the smallest value of $\alpha^1_j$. $j$ is connected to a tentatively open facility $i$. Call $i$ a *terminal facility*. We say that $i$ is the terminal facility corresponding to the terminal location $l$. Let $F$ be the set of all terminal facilities. Add $v$ to $F$. Again, we have a notion of dependence among facilities in $F$. We say that two facilities $i, i'$ are dependent if there is a demand $j$ with both $\beta^1_{ij}, \beta^1_{i'j} > 0$. We select a maximal independent set from $F$ – call it $F'$. Note that $v \in F'$ because $\beta^1_{vj} = 0$ for all $j$. We open all the facilities in $F'$.

A demand $j$ is assigned to a facility in $F'$ as follows: if there is a facility $i \in F'$ such that $\beta^1_{ij} > 0$, assign $j$ to $i$. If $\sigma(j) = v$, assign $j$ to $v$. Otherwise, let $i$ be the be terminal facility corresponding to the terminal location $\sigma(j) \in L'$. If $i \in F'$, assign $j$ to $i$. Otherwise, there is a facility $i' \in F'$ such that $i$ and $i'$ are dependent. We assign $j$ to $i'$. Let $i(j)$ be the facility that $j$ is assigned to.

Let $L_1$ be the terminal locations in $L'$ such that the terminal facilities corresponding to them are in $F'$. We now add some Steiner edges. We initialize the Steiner tree $T$ to the empty set. For each terminal location $l \in L_1$ with corresponding terminal facility $i \in F'$, we add all edges along a shortest path between $l$ and $i$ to the set $T$. Break any cycles by deleting edges.

*Phase 2.* This phase is very similar to that of the previous section. For any $l \in L$, let $D_l$ be the set of demands which are tight with $l$. Define $D_{L_1} = \bigcup_{l \in L_1} D_l$. $G$ is augmented as before to include edges incident on locations $l \in L_1$. We initialize our minimal violated sets to the components of $T$. All dual variables are initially 0. We do not raise any $\beta_{ij}$ in this phase. We shall raise the $\alpha_j$ value of demands in $D_{L_1}$ only. For a set $S$, define $D_S$ to be $\bigcup_{l \in S \cap L_1} D_l$. The rest of the procedure is identical to Phase 2 of the previous section. This yields the tree $T$ connecting all the open facilities. Let $(\alpha^2, \theta^2)$ be the dual solution constructed by this phase.

**Analysis.** The proof of the following lemma is very similar to the proof of Lemma 1.

**Lemma 6.** $(\alpha^1, \beta^1, \theta^1)$ *is a feasible dual solution.*

**Lemma 7.** *Consider a demand $j$ with $\sigma(j) = l$. Let $i$ be the terminal facility corresponding to $l$. Then, $c_{ij} \leq 5\alpha_j^1$. If $j \in D_l$ then $c_{ij} \leq 3\alpha_j^1$.*

*Proof.* Let $k$ be the demand with smallest $\alpha_k$ which is tight with $l$. $k$ is connected to $i$. So, $c_{ij} \leq c_{lj} + 2\alpha_k^1$. If $j$ is tight with $l$, then $c_{lj} \leq \alpha_j^1$, otherwise by Lemma 2, $c_{lj} \leq 3\alpha_j^1$. Further, $\alpha_k^1 \leq \alpha_j^1$ (this is true if if $j \in D_l$, otherwise we can argue as in Lemma 2). So $c_{ij} \leq 3\alpha_j^1$ if $j \in D_l$ and $c_{lj} \leq 5\alpha_j^1$ otherwise.     $\square$

**Lemma 8.** *The cost of opening facilities and connecting demands to facilities is at most $3\sum_{j \in D_{L_1}} \alpha_j^1 + 7\sum_{j \notin D_{L_1}} \alpha_j^1$.*

*Proof.* For an open facility $i$, define $C_i$ as the set of demands $j$ for which $\beta_{ij}^1 > 0$. Note that the sets $C_i$ are disjoint, and all demands in $C_i$ are assigned to $i$. We charge the cost of opening a facility at $i$ to the demands in $C_i$. Each $j \in C_i$ is charged $\beta_{ij}^1$. Let $C_{F'} = \bigcup_{i \in F'} C_i$. So, the cost of opening facilities and connecting demands in $C_{F'}$ to facilities is at most $\sum_{j \in C_{F'}} \left( c_{i(j)j} + \beta_{i(j)j}^1 \right) \leq \sum_{j \in C_{F'}} \alpha_j^1$.

If $j \in D_{L_1}$, we know by the previous lemma that $c_{i(j)j} \leq 3\alpha_j^1$. So, assume $j \notin D_{L_1} \cup C_{F'}$. By the previous lemma, we know that there is a terminal location $l$ such that the terminal facility $i$ corresponding to $l$ is at most $5\alpha_j^1$ from $j$. If $i$ is open, we are done. Otherwise, there is a facility $i'$ and a demand $j'$ such that $i'$ is open and $\beta_{ij'}, \beta_{i'j'} > 0$.

Since $i$ is the terminal facility corresponding to $l$, there is a demand $k$ such that $\alpha_k$ is smallest among all the demands tight with $l$ and $k$ is connected to $i$. Let $t_k$ and $t_{j'}$ be the times at which $k$ and $j'$ get connected respectively. Let $t_i$ and $t_{i'}$ be the times at which $i$ and $i'$ become tentatively open. Since both $\beta_{ij'}, \beta_{i'j'} > 0$, we have $c_{ij'}, c_{i'j'} \leq t_{j'}$ and $t_{j'} \leq t_i, t_{i'}$. Since $k$ is connected to $i'$, $t_{i'} \leq t_k \leq \alpha_k^1$. Further, $\alpha_k^1 \leq \alpha_j^1$. So, $c_{ij} \leq c_{li} + c_{ij'} + c_{i'j'} \leq 5\alpha_j^1 + 2t_{i'} \leq 7\alpha_j^1$.     $\square$

**Lemma 9.** *The total cost of the Steiner edges added to the set $T$ in Phase 1 is at most $2\sum_{j\in D_{L_1}} \alpha_j^1$.*

*Proof.* Consider a terminal location $l \in L_1$ with terminal facility $i$. Let $k$ be the demand in $D_l$ with smallest $\alpha_k$ So, $k$ is connected to $i$ and $c_{li} \le 2\alpha_k^1$. Note that $|D_l| \ge M$. Further if $j \in D_l$, then $\alpha_j^1 \ge \alpha_k^1$. So, $2\sum_{j\in D_l} \alpha_j^1 \ge Mc_{li}$. □

**Theorem 2.** *The above algorithm produces a solution of total cost at most $9 \cdot OPT$ and is thus a 9-approximation algorithm for ConFL.*

*Proof.* The cost of opening facilities and connecting clients to facilities in Phase 1 is bounded by $3\sum_{j\in D_{L_1}} \alpha_j^1 + 7\sum_{j\notin D_{L_1}} \alpha_j^1$ (Lemma 8). Let $T'$ be the set of edges added to $T$ in Phase 2. The cost of tree $T$ is at most $cost(T') + 2\sum_{j\in D_{L_1}} \alpha_j^1$ by Lemma 9. Finally $cost(T') \le 2 \cdot OPT + 2\sum_{j\in D_{L_1}} \alpha_j^1$ since $(\alpha', 0, \theta^2)$ is a feasible dual solution where $\alpha_j' = \max(\alpha_j^2 - \alpha_j^1, 0)$ (Lemma 5). Adding, the total cost is at most $2 \cdot OPT + 7\sum_j \alpha_j^1 \le 9 \cdot OPT^1$. □

## 4 The Connected $k$-Median Problem

The Connected $k$-Median problem is the same as ConFL with the additional constraint that at most $k$ facilities can be be opened. Since $v$ is already open, this extra constraint adds the following inequality to the linear program (P2) for ConFL: $\sum_{i\ne v} y_i \le k-1$. This changes the objective function of the dual (D2) to max $\sum_j \alpha_j - \sum_j \beta_{vj} - k'\lambda$, where $k' = k-1$. Constraint (11) in the dual LP gets replaced by $\sum_j \beta_{ij} \le f_i + \lambda$.

Let $(F^*, C^*, S^*)$ be the facility, assignment and Steiner tree cost respectively of an optimal ConFL solution. Phase 1 generates a partial primal solution $(x, y)$ and a feasible dual solution $(\alpha^1, \beta^1, \theta^1)$ satisfying $\beta_{vj}^1 = 0$ for all $j$. We can bound the cost of the Steiner tree on open facilities by $2S^* + 2(C^* + \sum_{j\in D_{L_1}, i} c_{ij}x_{ij})$. We first modify the primal-dual algorithm for ConFL so that after Phase 1,

$$9\sum_i f_i y_i + 3\sum_{j\in D_{L_1}, i} c_{ij}x_{ij} + \sum_{j\notin D_{L_1}, i} \le 9\sum_j \alpha_j^1. \tag{13}$$

Now fix $\lambda$. We modify the facility opening cost to $f_i + \lambda$ for all $i \ne v$, and run Phase 1 of the algorithm to get primal and dual solutions $(x, y)$ and $(\alpha^1, \beta^1, \theta^1)$. Let $z$ be denote the Steiner tree on the open facilities. Suppose it so happens that $\sum_{i\ne v} y_i = k'$. Then, $(x, y, z)$ and $(\alpha, \beta, \theta, \lambda)$ are feasible solutions to the primal and dual programs respectively for the connected $k$-median problem. Further from (13) we get that $9\sum_i f_i y_i + \sum_{j,i} c_{ij}x_{ij} + M\sum_e c_e z_e \le 9(\sum_j \alpha_j^1 - k'\lambda) + 2(S^* + C^*) \le 11 \cdot OPT$. The trick then to *guess* the right value of $\lambda$ so that when the facility cost is updated to $f_i + \lambda$, we end up opening $k$ facilities. This idea was first used by Jain & Vazirani [7].

---

[1] In Phase 2, if we use the 1.55-approximation algorithm [16] for the Steiner tree problem we get a slightly better guarantee of 8.55 (4.55 for the rent-or-buy problem).

We can show that there is a value of $\lambda$ such that depending on how we break ties, we get two ConFL solutions after Phase 1 — one opening $k_1 < k'$ facilities and the other opening $k_2 > k'$ facilities. These two solutions can be found in polynomial time. A *convex combination* of these two solutions yields a fractional solution $(x, y, z)$ that opens $k'$ facilities and satisfies $9\sum_i f_i y_i + 3\sum_{j \in D_{L_1}, i} c_{ij} x_{ij} + \sum_{j \notin D_{L_1}, i} \leq 9 \cdot OPT$. Now, we can round this solution (as in [7]) to get a solution which opens $k$ facilities (including $v$) losing a factor of 2. Finally we build a Steiner tree on the open facilities. We can show the following.

**Theorem 3.** *There is a 20-approximation algorithm for the Connected k-Median problem.*

To satisfy (13), we do not add any edges to the set $T$ in Phase 1. Instead the Steiner tree in Phase 2 is built with the terminals being the open facilities.

## 5   Extensions and Refinements

**Arbitrary Demands.** All our results generalize to the case where instead of unit demands, client $j$ may have a demand $d_j \geq 0$. We can reduce this to the unit demand case by making $d_j$ copies of client $j$, but this makes the algorithm run in pseudo-polynomial time. But we can easily simulate this reduction by raising $\alpha_j$ at a rate proportional to $d_j$ wherever necessary. All $d_j$ units of demand at $j$ behave *identically*. The analogues of lemmas proved in section 3 are easily shown to be true and consequently we get the same approximation ratios.

**Generalization to Edge Capacities.** We can extend our algorithm to the following more general problem. We have two types of cables — the first type has a fixed cost of $\sigma$ per unit length and a *capacity* of $u$ units. The second cable has a fixed cost of $M$ per unit length but unlimited capacity. We wish to open facilities and lay a network of cables so that clients are connected to open facilities using the first kind of cable, and facilities are connected by a Steiner tree using cables of type 2. This differs from ConFL only in the specification of the first cable type. Assuming integer demands, setting $\sigma = u = 1$ reduces this to ConFL. Ravi & Sinha [15] gave an algorithm if we only have cables of type 1 and do not have the connectivity requirement.

We get a constant-factor approximation for this problem by solving a relaxed ConFL instance and a relaxed Steiner tree instance and combining the two solutions. The approximation ratios we get are 7.55, 15.55 and 31.1 for the capacitated versions of the rent-or-buy problem, ConFL, and the connected $k$-median problem respectively. We get better guarantees if all demands are 1.

**The Case $M = 1$.** We can get significantly better results for this case. In Phase 1, we run the facility location algorithm of Jain & Vazirani [7]. For each open facility $i$ we identify a client $j$ that is tight with $i$, and add edges connecting $i$ and $j$ to the set $T$. In Phase 2 a Steiner tree is built joining the components of $T$. We show that this is a 4-approximation algorithm. This gives a 8-approximation for the the $k$-median version. This also gives better guarantees for the capacitated versions of these problems.

## Acknowledgments

We thank Jon Kleinberg and David Shmoys for useful discussions, reading through drafts of this paper and pointing out suggestions.

# References

1. A. Agrawal, P. Klein, and R. Ravi. When trees collide: An approximation algorithm for the generalized Steiner problem on networks. *SIAM Journal on Computing* 24(3):440–456, 1995.                                                                    ,
2. M. X. Goemans and D. P. Williamson. A general approximation technique for constrained forest problems. *SIAM Journal on Computing*, 24:296–317, 1995.
3. M. X. Goemans and D. P. Williamson. The primal-dual method for approximation algorithms and its application to network design problems. In D. S. Hochbaum, editor, *Approximation Algorithms for NP-Hard Problems*, chapter 4, pages 144–191. PWS Publishing Company, 1997.
4. S. Guha and S. Khuller. Connected facility location problems. *DIMACS Series in Discrete Mathematics and Theoretical Computer Science*, 40:179–190, 1997.
5. A. Gupta, J. Kleinberg, A. Kumar, R. Rastogi, and B. Yener. Provisioning a virtual private network: A network design problem for multicommodity flow. In *Proceedings of the 33rd Annual ACM Symposium on Theory of Computing (STOC)*, pages 389–398, 2001.
6. R. Hassin, R. Ravi, and F. S. Selman. Approximation algorithms for a capacitated network design problem. In *Proceedings of 4th APPROX*, pages 167–176, 2000.
7. K. Jain and V. V. Vazirani. Primal-dual approximation algorithms for metric facility location and $k$-median problems. *Journal of the ACM*, 48:274–296, 2001.
8. D. R. Karger and M. Minkoff. Building Steiner trees with incomplete global knowledge. In *Proceedings of the 41st Annual IEEE Symposium on Foundations of Computer Science (FOCS)*, pages 613–623, 2000.
9. S. Khuller and A. Zhu. The general Steiner tree-star problem. *Information Processing Letters*, 2002. To appear.
10. Tae Ung Kim, Timothy J. Lowe, Arie Tamir, and James E. Ward. On the location of a tree-shaped facility. *Networks*, 28(3):167–175, 1996.
11. A. Kumar, A. Gupta, and T. Roughgarden. A constant-factor approximation algorithm for the multicommodity rent-or-buy problem. In *Proceedings of the 43rd Annual IEEE Symposium on Foundations of Computer Science (FOCS)*, 2002. To appear.
12. M. Labbé, G. Laporte, I. Rodrígues Martin, and J. J. Salazar González. The median cycle problem. Technical Report 2001/12, Department of Operations Research and Multicriteria Decision Aid at Université Libre de Bruxelles, 2001.
13. Y. Lee, S. Y. Chiu, and J. Ryan. A branch and cut algorithm for a Steiner tree-star problem. *INFORMS Journal on Computing*, 8(3):194–201, 1996.
14. P. Mirchandani and R. Francis, eds. *Discrete Location Theory*. John Wiley and Sons, Inc., New York, 1990.
15. R. Ravi and A. Sinha. Integrated logistics: Approximation algorithms combining facility location and network design. In *Proceedings of 9th IPCO*, pages 212–229, 2002.
16. G. Robins and A. Zelikovsky. Improved steiner tree approximation in graphs. In *Proceedings of the 11th Annual ACM-SIAM Symposium on Discrete Algorithms (SODA)*, pages 770–779, 2000.
17. D. P. Williamson. The primal-dual method for approximation algorithms. *Mathematical Programming, Series B*, 91(3):447–478, 2002.

# Author Index

# Lecture Notes in Computer Science

For information about Vols. 1–2380
please contact your bookseller or Springer-Verlag

Vol. 2417: M. Ishizuka, A. Sattar (Eds.), PRICAI 2002: Trends in Artificial Intelligence. Proceedings, 2002. XX, 623 pages. 2002. (Subseries LNAI).

Vol. 2418: D. Wells, L. Williams (Eds.), Extreme Programming and Agile Methods – XP/Agile Universe 2002. Proceedings, 2002. XII, 292 pages. 2002.

Vol. 2419: X. Meng, J. Su, Y. Wang (Eds.), Advances in Web-Age Information Management. Proceedings, 2002. XV, 446 pages. 2002.

Vol. 2420: K. Diks, W. Rytter (Eds.), Mathematical Foundations of Computer Science 2002. Proceedings, 2002. XII, 652 pages. 2002.

Vol. 2421: L. Brim, P. Jančar, M. Křetínský, A. Kučera (Eds.), CONCUR 2002 – Concurrency Theory. Proceedings, 2002. XII, 611 pages. 2002.

Vol. 2422: H. Kirchner, Ch. Ringeissen (Eds.), Algebraic Methodology and Software Technology. Proceedings, 2002. XI, 503 pages. 2002.

Vol. 2423: D. Lopresti, J. Hu, R. Kashi (Eds.), Document Analysis Systems V. Proceedings, 2002. XIII, 570 pages. 2002.

Vol. 2425: Z. Bellahsène, D. Patel, C. Rolland (Eds.), Object-Oriented Information Systems. Proceedings, 2002. XIII, 550 pages. 2002.

Vol. 2426: J.-M. Bruel, Z. Bellahsène (Eds.), Advances in Object-Oriented Information Systems.Prodings, 2002. IX, 314 pages. 2002.

Vol. 2430: T. Elomaa, H. Mannila, H. Toivonen (Eds.), Machine Learning: ECML 2002. Proceedings, 2002. XIII, 532 pages. 2002. (Subseries LNAI).

Vol. 2431: T. Elomaa, H. Mannila, H. Toivonen (Eds.), Principles of Data Mining and Knowledge Discovery. Proceedings, 2002. XIV, 514 pages. 2002. (Subseries LNAI).

Vol. 2432: R. Bergmann, Experience Management. XXI, 393 pages. 2002. (Subseries LNAI).

Vol. 2434: S. Anderson, S. Bologna, M. Felici (Eds.), Computer Safety, Reliability and Security. Proceedings, 2002. XX, 347 pages. 2002.

Vol. 2435: Y. Manolopoulos, P. Návrat (Eds.), Advances in Databases and Information Systems. Proceedings, 2002. XIII, 415 pages. 2002.

Vol. 2436: J. Fong, C.T. Cheung, H.V. Leong, Q. Li (Eds.), Advances in Web-Based Learning. Proceedings, 2002. XIII, 434 pages. 2002.

Vol. 2438: M. Glesner, P. Zipf, M. Renovell (Eds.), Field-Programmable Logic and Applications. Proceedings, 2002. XXII, 1187 pages. 2002.

Vol. 2439: J.J. Merelo Guervós, P. Adamidis, H.-G. Beyer, J.-L. Fernández-Villacañas, H.-P. Schwefel (Eds.), Parallel Problem Solving from Nature – PPSN VII. Proceedings, 2002. XXII, 947 pages. 2002.

Vol. 2440: J.M. Haake, J.A. Pino (Eds.), Groupware: Design, Implementation and Use. Proceedings, 2002. XII, 285 pages. 2002.

Vol. 2442: M. Yung (Ed.), Advances in Cryptology – CRYPTO 2002. Proceedings, 2002. XIV, 627 pages. 2002.

Vol. 2443: D. Scott (Ed.), Artificial Intelligence: Methodology, Systems, and Applications. Proceedings, 2002. X, 279 pages. 2002. (Subseries LNAI).

Vol. 2444: A. Buchmann, F. Casati, L. Fiege, M.-C. Hsu, M.-C. Shan (Eds.), Technologies for E-Services. Proceedings, 2002. X, 171 pages. 2002.

Vol. 2445: C. Anagnostopoulou, M. Ferrand, A. Smaill (Eds.), Music and Artificial Intelligence. Proceedings, 2002. VIII, 207 pages. 2002. (Subseries LNAI).

Vol. 2446: M. Klusch, S. Ossowski, O. Shehory (Eds.), Cooperative Information Agents VI. Proceedings, 2002. XI, 321 pages. 2002. (Subseries LNAI).

Vol. 2447: D.J. Hand, N.M. Adams, R.J. Bolton (Eds.), Pattern Detection and Discovery. Proceedings, 2002. XII, 227 pages. 2002. (Subseries LNAI).

Vol. 2448: P. Sojka, I. Kopeček, K. Pala (Eds.), Text, Speech and Dialogue. Proceedings, 2002. XII, 481 pages. 2002. (Subseries LNAI).

Vol. 2449: L. Van Gool (Ed.), Pattern Recognotion. Proceedings, 2002. XVI, 628 pages. 2002.

Vol. 2451: B. Hochet, A.J. Acosta, M.J. Bellido (Eds.), Integrated Circuit Design. Proceedings, 2002. XVI, 496 pages. 2002.

Vol. 2452: R. Guigó, D. Gusfield (Eds.), Algorithms in Bioinformatics. Proceedings, 2002. X, 554 pages. 2002.

Vol. 2453: A. Hameurlain, R. Cicchetti, R. Traunmüller (Eds.), Database and Expert Systems Applications. Proceedings, 2002. XVIII, 951 pages. 2002.

Vol. 2454: Y. Kambayashi, W. Winiwarter, M. Arikawa (Eds.), Data Warehousing and Knowledge Discovery. Proceedings, 2002. XIII, 339 pages. 2002.

Vol. 2455: K. Bauknecht, A M. Tjoa, G. Quirchmayr (Eds.), E-Commerce and Web Technologies. Proceedings, 2002. XIV, 414 pages. 2002.

Vol. 2456: R. Traunmüller, K. Lenk (Eds.), Electronic Government. Proceedings, 2002. XIII, 486 pages. 2002.

Vol. 2458: M. Agosti, C. Thanos (Eds.), Research and Advanced Technology for Digital Libraries. Proceedings, 2002. XVI, 664 pages. 2002.

Vol. 2462: K. Jansen, S. Leonardi, V. Vazirani (Eds.), Approximation Algorithms for Combinatorial Optimization. Proceedings, 2002. VIII, 271 pages. 2002.

Vol. 2463: M. Dorigo, G. Di Caro, M. Sampels (Eds.), Ant Algorithms. Proceedings, 2002. XIII, 305 pages. 2002.

Vol. 2464: M. O'Neill, R.F.E. Sutcliffe, C. Ryan, M. Eaton, N. Griffith (Eds.), Artificial Intelligence and Cognitive Science. Proceedings, 2002. XI, 247 pages. 2002. (Subseries LNAI).

Vol. 2469: W. Damm, E.-R. Olderog (Eds.), Formal Techniques in Real-Time and Fault-Tolerant Systems. Proceedings, 2002. X, 455 pages. 2002.

Vol. 2470: P. Van Hentenryck (Ed.), Principles and Practice of Constraint Programming – CP 2002. Proceedings, 2002. XVI, 794 pages. 2002.

Vol. 2479: M. Jarke, J. Koehler, G. Lakemeyer (Eds.), KI 2002: Advances in Artificial Intelligence. Proceedings, 2002. XIII, 327 pages. (Subseries LNAI).

Vol. 2483: J.D.P. Rolim, S. Vadhan (Eds.), Randomization and Approximation Techniques in Computer Science. Proceedings, 2002. VIII, 275 pages. 2002.